NITROGEN FIXATION

Volume I Ecology

NITROGEN FIXATION

Volume I Ecology

Edited by

W. J. BROUGHTON

CLARENDON PRESS · OXFORD

1981

Oxford University Press, Walton Street, Oxford OX2 6DP
London Glasgow New York Toronto
Delhi Bombay Calcutta Madras Karachi
Kuala Lumpur Singapore Hong Kong Tokyo
Nairobi Dar es Salaam Cape Town Salisbury
Melbourne Auckland

and associate companies in
Beirut Berlin Ibadan Mexico City

Published in the United States by Oxford University Press, New York

British Library Cataloguing in Publication Data

Nitrogen fixation.
Vol. 1: Ecology
1. Nitrogen - Fixation
I. Broughton, W J
574.1'33 QR89.7 80-40612

ISBN 0-19-854540-1

Typeset by Oxprint Ltd, Oxford
Printed in Great Britain
at the University Press, Oxford
by Eric Buckley
Printer to the University

Preface

Nitrogen and phosphorus availability limit photosynthesis and therefore plant and animal productivity in many terrestrial and aquatic environments. Unfortunately for all concerned it is difficult to estimate not only the levels available to plants, but the absolute amounts of both elements. Yet just as atomic absorption spectroscopy revolutionized other aspects of mineral nutrition, the introduction of the acetylene reduction method for estimating nitrogen fixation (in the 1960s) sparked renewed interest in ecological aspects of nitrogen accretion. Coinciding as it did with recognition of the need to conserve energy and husband the environment, a great burst of research activity resulted. Laboratories, whether well equipped, or possessing little more than a gas chromatograph, began to examine plant, or soil, or water, or even coral and wood samples for tell-tale ethylene peaks on their recorders. On this evidence, nitrogen fixation was reported from such widely different habitats as the tundra and thermal springs.

The seeds of this book sprang from a UNESCO commission to C. A. (Lex) Parker and myself to prepare a chapter for the fifth 'Global Impacts of Applied Microbiology' Conference (Bangkok, 1977). In accepting this task, we assumed that in the decade since the introduction of the acetylene-reduction assay there would be an abundance of ecologically significant data on nitrogen fixation. For this reason, we optimistically entitled the chapter 'Microbial contributions to world nitrogen economy'. Our 118-page manuscript (which, like the present work, did not consider grain legumes) bears testament to the explosion in interest generated by this cheap and readily available technique. Assay of acetylene reduction is not without problems, however, the most serious of which concern the relationship between the amount of acetylene reduced and nitrogen fixed. Unfortunately there is no simple way to overcome this except by comparing the rates of acetylene reduction with the classical means of estimating nitrogen fixation, and integrating these data over considerable periods of time. Our review was finished by the end of 1975, and although it was never properly published, we felt then that understanding of nitrogen cycling in various ecosystems had reached the point where it deserved broader treatment than we were able to give it. This volume, to which experts from all over the world have contributed, is an attempt to do just that.

Obviously, much thought and effort by various people have contributed to this, the first volume in the series. Foremost among those I must thank are the fifteen authors: I shall be ever grateful for their unstinting help. In the period from conception to execution (almost four years) I was attached to four institutions, each of which unquestioningly extended their facilities. Special mention should be made of Professor Satwant S. Dhaliwal and of Cik Rohani binte Muslim of the University of Malaya, Kuala Lumpur for providing the initial facilities and for typing all the early correspondence respectively; to Professor Fritz Lenz and Fräulein Antonia Maria Schäfer of the University of Bonn, who allowed the momentum to continue and for preparing the index; to Professor Ab van Kammen and to G. Wil Landeweerd of the Agricultural University, Wageningen for allowing me to devote some of the precious time I had in The Netherlands to this subject; and to Professor Jeff Schell and Fräulein Elisabeth Schölzel of the Max-Planck-Institut für Züchtungsforschung in Köln for guiding the process to completion. Sustenance for the editor over these years came from the University of Malaya, Kuala Lumpur, the Alexander von Humdoldt-Stiftung, Bonn, the International Agricultural Centre, Wageningen, and the Max-Planck-Gesellschaft, München.

W.J.B.

Cologne
October 1979

Contents

List of Contributors

A. D. L. Akkermans,
Laboratory of Microbiology,
Agricultural University,
Wageningen,
The Netherlands.

J. Baker,
School of Marine Science,
College of William and Mary,
Gloucester Point,
Virginia 23062,
USA.

S. Brotonegoro,
Laboratorium Treub,
Pusat Penelitian Nasional,
d.a. Kebun Raya Bogor,
Bogor,
Indonesia.

W. J. Broughton,
Max-Planck-Institut für Züchtungsforschung,
D-5000, Köln 30, (Vogelsang),
West Germany.

C. van Dijk,
Institute for Ecological Research,
Department of Dune Research,
Weevers' Duin,
Oostvoorne,
The Netherlands.

P. Fay,
Department of Botany and Biochemistry,
Westfield College,
University of London,
Kidderpore Avenue,
London NW3 7ST,
UK.

J. French,
Conservation and Biodegradation Section,
CSIRO Division of Building Research,
Graham Road,
Highett,
Victoria 3190,
Australia.

E. F. Henzell,
CSIRO Division of Tropical Pastures,
Mill Road,
St. Lucia,
Queensland 4067,
Australia.

V. Jensen,
Department of Microbiology and Microbial Ecology,
Royal Veterinary and Agricultural University,
21 Rolighedsvej,
1958 Copenhagen V,
Denmark.

K. Jones,
University of Lancaster,
Lancaster LA1 4YQ,
Lancashire,
UK.

T. A. Lie,
Laboratory of Microbiology,
Agricultural University,
Wageningen,
The Netherlands.

H. W. Paerl,
Institute of Marine Sciences,
University of North Carolina,
Morehead City,
NC 28557,
USA.

I. Watanabe,
International Rice Research Institute,
PO Box 933,
Manila,
The Philippines.

G. J. Waughman,
Should Shields Marine and Technical College,
Tyne and Wear,
UK.

K. L. Webb,
Virginia Institute of Marine Science,
Gloucester Point,
Virginia,
USA.

W. J. Wiebe,
Department of Microbiology,
University of Georgia,
Athens,
Georgia 30602,
USA.

1 Photosynthetic micro-organisms

P. FAY

1.1 Introduction

Representatives of the two groups of photosynthetic prokaryotes, the blue–green algae and photosynthetic bacteria, display a unique and remarkable autotrophic capacity to assimilate both carbon dioxide and elemental nitrogen, and convert them into cell material.

Blue–green algae (Cyanophyta, Cyanobacteria) and photosynthetic bacteria show a typical prokaryotic cellular organization which incorporates an elaborate cytoplasmic membrane system, the site of photosynthetic activity. The two groups differ fundamentally in their type of photosynthesis: being accompanied by oxygen evolution (oxygenic) in blue–green algae, and not associated with oxygen evolution (anoxygenic) in photosynthetic bacteria.

Both groups comprise a great variety of unicellular and simple multicellular forms, and display a whole spectrum of nutritional patterns from obligate autotrophy, through mixotrophy, to a more pronounced heterotrophic ability. On the basis of their pigment complement and of the nature of electron donors for photosynthesis, photosynthetic bacteria are classified traditionally in three main groups: the green sulphur bacteria (Chlorobacteriaceae or Chlorobiineae), the purple sulphur bacteria (Thiorhodaceae or Chromatiaceae), and the purple non-sulphur bacteria (Athiorhodaceae or Rhodospirillaceae) (Pfennig 1977). Blue–green algae are more uniform with respect of the organization and function of the photosynthetic system, and more diverse in terms of morphology, physiological characters, and natural distribution (see Fogg, Stewart, Fay, and Walsby 1973).

Information on general aspects of the biology of photosynthetic bacteria and blue–green algae may be sought in the books of Kondrat'eva (1965), Carr and Whitton (1973), and Fogg *et al.* (1973) and in the review articles of Pfennig (1967, 1975, 1977), Wolk (1973), and Stanier and Cohen-Bazire (1977). Nitrogen fixation in photosynthetic micro-organisms has lately been reviewed by Fogg (1974) and

Stewart (1973, 1977), and the ecology of nitrogen fixation by blue–green algae by Mague (1976). The proceedings of a recent symposium on the environmental role of nitrogen-fixing blue–green algae and asymbiotic bacteria include original reports as well as review articles (Granhall 1978).

1.2 The natural occurrence and importance of nitrogen-fixing photosynthetic prokaryotes

Photosynthetic bacteria and blue–green algae commonly occur in a great variety of natural habitats and are often abundant in cultivated lands. They seem to utilize preferentially combined (inorganic or organic) forms of nitrogen, and many are able to respond with the synthesis of nitrogenase to a temporary or prolonged shortage of combined nitrogen in their environment, thus enabling them to utilize the enormous store of elemental nitrogen in the atmosphere. The ability to fix nitrogen confers on these photosynthetic prokaryotes a significant competitive advantage over other (prokaryotic and eukaryotic) photosynthetic micro-organisms which are fully dependant for their growth on the availability of combined nitrogen in the environment.

Although their metabolism is based on photosynthetic carbon dioxide fixation, many photosynthetic bacteria and blue–green algae can in addition use simple organic substances as sources of carbon and energy. Depending on species, assimilation of organic substrates may take place only in the light, or both in light and dark, but is invariably enhanced by illumination. The basic requirements of nitrogen fixation (ATP, reductant, and carbon skeletons) can thus be produced equally in light and dark metabolic processes. Apart from a few environmental influences, like the partial pressure of oxygen or the concentration of combined nitrogen in the medium, which could directly and specifically affect the activity and synthesis of nitrogenase, most environmental factors act through their effects on photosynthesis and other metabolic and biosynthetic processes.

Distribution

The distribution of photosynthetic bacteria in nature is determined by their photosynthetic mode of nutrition and their sensitivity towards free oxygen. Photosynthetic bacteria occupy an ecological niche which provides reducing conditions and to which sufficient light can penetrate for photosynthesis. Such conditions are present at the mud surface in shallow waters, or in the upper parts of the hypolimnion in deeper lakes (see Pfennig 1967, 1975), apart from certain more peculiar habitats like small polluted ponds or sulphur springs. In general, reducing conditions in such an environment are maintained by the action of sulphate-reducing bacteria, while carbon dioxide required for photosynthesis is released during fermentation of organic substrates by the anaerobic microflora, and elemental nitrogen is liberated by the action of denitrifying bacteria.

Blue–green algae, on the other hand, are distinctly aerobic micro-organisms, though they are able to survive in microaerobic or anaerobic environments (Stewart and Pearson 1970; Castenholz 1976, 1977). Moreover, several species can perform anoxygenic photosynthesis with sulphide as the electron donor in a photosystem I driven reaction, in a similar way to photosynthetic bacteria (Cohen, Padan, and Shilo 1975; Garlick, Oren, and Padan 1977). More commonly, blue–green algae are typical of well aerated aquatic and terrestrial habitats, and are the principal agents of biological nitrogen fixation in fresh and sea waters (see Fogg 1971, 1978; Stewart 1971).

Freshwater habitats

There is very little information available on the contribution of photosynthetic bacteria to the nitrogen economy of fresh waters (Stewart 1968; Knowles 1976). Blue–green algae are thought to be primarily responsible for nitrogen fixation in temperate and tropical lakes (Fogg 1971). Even so, nitrogen fixed by blue–green algae may constitute only a small proportion of the total nitrogen income, as shown by Horne and Fogg (1970) in some English lakes. Nitrogen-fixing activity is associated with dense populations of heterocystous planktonic (gas vacuolate) forms, mainly of the genera *Anabaena, Gloeotrichia,* and *Aphanizomenon* (Dugdale, Dugdale, Nees, and Goering 1959; Dugdale and Dugdale 1962; Goering and Nees 1964; Stewart, Fitzgerald, and Burris 1968; Ogawa and Carr 1969; Granhall and Lundgren 1971; Horne and Viner 1971; Horne 1972).

Benthic blue–green algae associated with macrophytes and growing over littoral sediments may be responsible for the occasional high rates of nitrogen fixation measured in oligotrophic lakes (Moeller and Roskoski 1978). They may be a significant source of nitrogen but little is known of their distribution and contribution to the nitrogen budget in oligotrophic lakes.

Nitrogen fixation in thermal springs was found to be related to the occurrence and abundance of *Calothrix* and *Mastigocladus* species (Fogg 1951; Stewart 1970).

Marine habitats

It is rather puzzling that heterocyst-bearing blue–green algae are generally absent or insignificant in the plankton of the open oceans, considering that sea waters are alkaline and generally poor in available combined nitrogen (Fogg 1978). Significant nitrogen-fixing activity is mainly associated with the often spectacular blooms of the non-heterocystous marine *Oscillatoria (Trichodesmium)*. Although nitrogen-fixing activity in natural populations of *Oscillatoria* was demonstrated conclusively by several workers using both the isotopic (^{15}N) and the acetylene-reduction assays (Dugdale, Menzel, and Ryther 1961; Goering, Dugdale, and Menzel 1966; Bunt, Cooksey, Heeb, Lee, and Taylor 1970; Taylor, Lee, and Bunt 1973; Carpenter and McCarthy 1975), the evidence for *Oscillatoria* being the agent responsible remains inconclusive. Owing to the repeated failure to grow the alga in pure culture, it is yet not possible to test nitrogen-fixing activity in the absence of bacteria. To add further to the confusion, reports concerning nitrogen-fixing ability of bacteria isolated from *Oscillatoria* blooms are unfortunately conflicting (Maruyama, Taga, and Matsuda 1970; Maruyama 1975; Carpenter and McCarthy 1975).

Nitrogenase activity measured in planktonic populations of the marine diatoms *Rhizoselenia* and *Chaetoceros*, is almost certainly associated with the presence of a heterocystous blue–green algal endosymbiont, *Richelia intracellularis* (Mague, Weare, and Holm-Hansen 1974; Venrick 1974).

Heterocyst-bearing cyanophytes (*Nostoc, Calothrix, Dichothrix, Rivularia, Scytonema*) and non-heterocystous forms (*Oscillatoria, Schizothrix, Lyngbya*) are common in the intertidal region of coral reefs. They probably contribute significantly, through both carbon and nitrogen fixation, to the high productivity of reef communities (Wiebe, Johannes, and Webb 1975; Burris 1976; Capone, Taylor, and Taylor 1977).

Epiphytic blue–green algae, like *Dichothrix*, common on *Sargassum* in the Western Sargasso Sea, and *Calothrix*, found on sea grasses in the Gulf of Mexico, have been shown to fix nitrogen, but their importance is doubtful (Carpenter 1972; Carpenter and Cox 1974). Blue–green algae, especially *Calothrix scopulorum* and *Rivularia* spp, are widely distributed and often abundant in the supralittoral fringe of temperate oceans, and their contribution to fixed nitrogen is probably more significant (Allen 1963; Stewart 1965, 1967*a*, 1971; Wärmling 1973).

Henriksson (1971) recorded light-dependent nitrogen fixation in marine sand containing photosynthetic bacteria. Several strains of purple non-sulphur bacteria were isolated from inshore muds in the Irish Sea and found to display nitrogenase activity (Wynn-Williams and Rhodes 1974). Photosynthetic bacteria, which occur in an unusual situation at the interface between an anaerobic bottom layer of sea water and an aerobic top layer of fresh water in a Norwegian lake, were shown to fix nitrogen but rates were relatively low (Stewart 1968). However, in most brackish waters and salt marshes tested, nitrogen-fixing blue–green algae (*Anabaena, Calothrix, Nodularia, Nostoc, Scytonema,*

Tolypothrix) are the dominant forms (Stewart 1965, 1967*a*, 1967*b*).

Terrestrial habitats

Blue–green algae are frequently encountered in temperate soils though they are less common than green algae or diatoms probably because of the acidic reaction of most of these soils. In forest and grassland soil, nitrogenase activity is associated mainly with the presence of heterotrophic bacteria (Jurgensen and Davey 1971; Paul, Myers, and Rice 1971). Light-dependent nitrogen-fixing activity was measured in a variety of terrestrial habitats in Scotland but the total nitrogen input which results from this activity is probably of minor importance (Stewart, Sampaio, Isichei, and Sylvester-Bradley 1978). Nitrogen fixation by blue–green algae may be more important to soil fertility in the neutral and alkaline soils of Sweden (Granhall and Henriksson 1969).

Blue–green algae are almost ubiquitous in tropical soils and often display lavish growths in waterlogged fields. The most abundant species belong to the genera *Anabaena, Anabaenopsis, Aulosira, Calothrix, Cylindrospermum, Gloeotrichia, Hapalosiphon, Nostoc, Scytonema, Stigonema* and *Tolypothrix* (Singh 1961; Watanabe and Yamamoto 1971; Venkataraman 1975; Stewart, Sampaio, Isichei, and Sylvester-Bradley 1978). Durrel (1964) found that blue–green algae represent a large proportion of the soil microflora in the Caribbean islands and in Central and South America. Soil crust samples taken from a great variety of terrestrial habitats in Nigeria showed without exception significant nitrogenase activities, when moistened and tested under simulated field conditions (Stewart *et al.* 1978).

In paddy fields, nitrogen fixation is significant only under flooded conditions, and decreases rapidly after drainage (Ishizawa, Suzuki, and Araragi 1975). While heterocystous blue–green algae are mainly active in surface waters, photosynthetic bacteria appear to be primarily responsible for nitrogen fixation at the anaerobic mud surface (Materassi and Balloni 1965; MacRae and Castro 1967; Kobayashi, Takahashi, and Kawaguchi 1967; Kobayashi and Haque 1971; Kobayashi 1975; Ishizawa *et al.* 1975).

Blue–green algae are common inhabitants also of soils in the polar region (Jurgensen and Davey 1968). They were found to make up a large proportion of the terrestrial algal flora in the Antarctic (Hirano 1965). Fourteen out of 42 sites tested in Signy Islands showed [^{15}N]-nitrogen-fixing activity (Fogg and Stewart 1968). Among the potential (heterocyst-bearing) nitrogen-fixing species, *Nostoc commune* was found to be the most abundant. It was estimated that nitrogen fixation by blue–green algae may contribute an annual increase in total nitrogen of about 7 per cent to the Antarctic habitat (Horne 1972).

Blue–green algae, especially those epiphytic on mosses, were found to be the dominant nitrogen-fixing (acetylene-reducing) micro-organisms in the Arctic meadows and peats (Jordan, McNicol, and Marshall 1978).

Blue–green algae are known to be amongst the first colonizers of arid ground (Fogg *et al.* 1973). They were observed to be actively fixing nitrogen on lava soil in Heimaey, Iceland, only eighteen months after a volcanic eruption (Englund 1976). Tests for nitrogen fixation performed 3½ years after the outbreak indicate that the abundant blue–green algal community, with *Nostoc muscorum* predominating, contributes a substantial part of the nitrogen input to the juvenile lava field (Englund 1978).

1.3 Environmental factors affecting nitrogen fixation

Light

It is self-evident that light is the most important single factor in the natural distribution of photosynthetic micro-organisms though its effect on growth, photosynthesis, or nitrogen fixation is not always simple.

The observation that nitrogen fixation in natural waters or in soils is light-dependent was generally considered to indicate that photosynthetic micro-organisms are the agents responsible (Dugdale and Dugdale 1962; Goering and Nees 1964; Horne and Fogg 1970; Stewart 1965; Jurgensen and Davey 1968; Granhall and Henriksson 1969). Though nitrogen fixation in photosynthetic organisms is principally, but not exclusively, dependent on the energy and reductant generated in the photochemical process, the nitrogenase reaction *per se* is independent of light (Fay 1965; Cox 1966; Cox and Fay 1969). Many blue–green algae can grow and fix nitrogen in complete darkness in the presence of suitable organic substrates (Allison, Hoover, and Morris 1937; Fay 1965; Watanabe and Yamamoto 1967; Khoja and Whitton 1975). This ability appears to be more pronounced in terrestrial than in planktonic forms (Goering and Nees 1964) though nitrogen fixation even in obligate phototrophic species, like *Anabaena cylindrica*, will continue in the dark at the expense of endogenous substrates produced in the previous light period, and of ATP and reductant generated in respiratory processes (Cox 1966). Planktonic populations and soil algae continue to fix nitrogen in the dark for a limited period; such dark nitrogen fixation being greater and lasting longer in algae previously exposed to bright light for a prolonged period (Fig. 1.1) (Dugdale and Dugdale 1962; Goering and Nees 1964; Horne and Fogg 1970; Granhall and Lundgren 1971; Stewart, Mague, Fitzgerald, and Burris 1971; Fay 1976; Burris and Peterson 1978; Stewart *et al.* 1978). Light may further

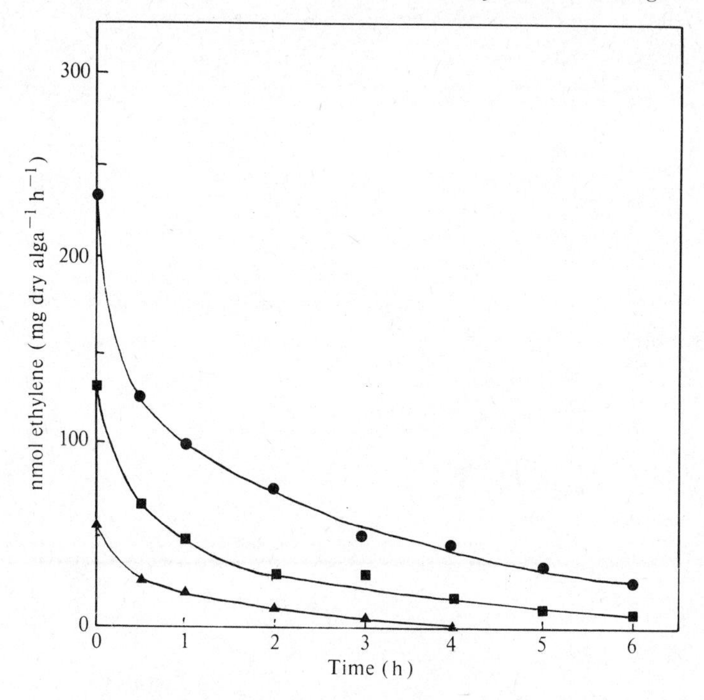

FIG. 1.1. Effect of light pretreatment of 1500 (▲), 3000 (■), and 6000 (●) lux intensity on nitrogenase activity of *Anabaenopsis circularis* in the dark. (After Fay (1976)).

enhance assimilation of organic substances, and this in turn will promote the uptake of elemental nitrogen in algae which exhibit heterotrophic abilities (Fay 1965; Watanabe and Yamamoto 1967; see also p. 22).

Nitrogenase activity, like the rate of photosynthesis, increases linearly in photosynthetic micro-organisms with the increase of light intensity (Fay 1970) (Fig. 1.2), but it may become depressed at the full intensity of solar radiation to which algae are eventually exposed at the surface of natural waters. Maximum rates of nitrogen fixation in planktonic blue–green algae measured in the subsurface layers (about 2 m below the surface), follow those of photosynthesis (Horne and Fogg 1970; Burris and Peterson 1978). Blue–green algae are more sensitive to high light intensities than diatoms or green algae (Brown and Richardson 1968). They grow best and fix nitrogen at higher rates in tropical rice fields at intensities less than 10 per cent of the full incident light (Reynaud and Roger 1978). Rates of nitrogen fixation measured at the surface of temperate lakes are only about 50 per cent of the maximum recorded in subsurface samples; rates fall off rapidly below the

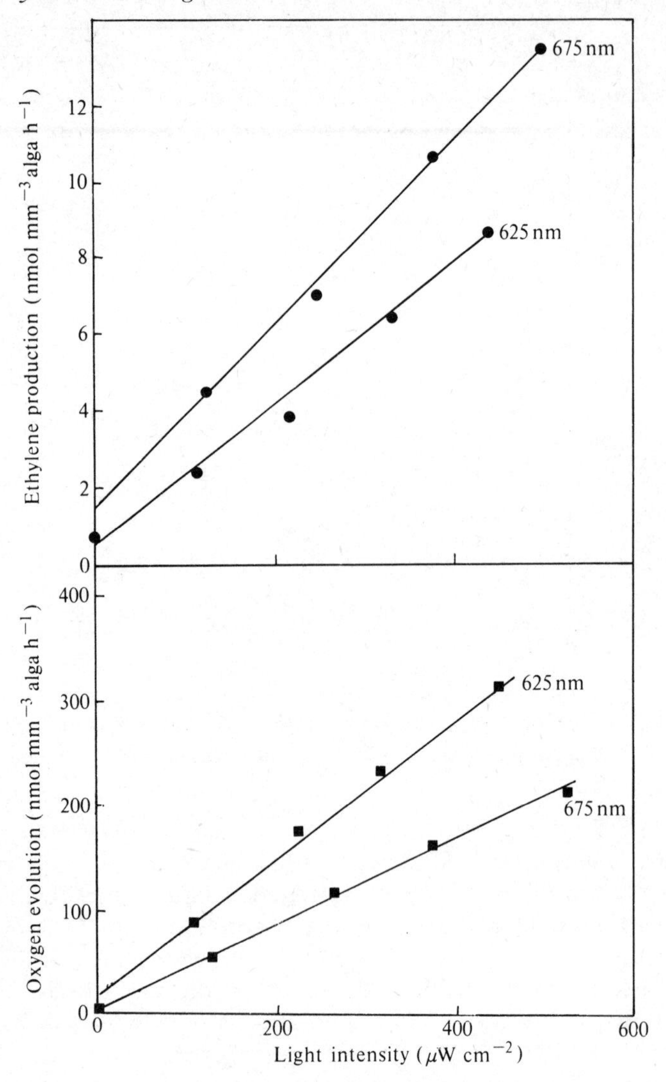

FIG. 1.2. Rates of acetylene reduction (ethylene production) (●) and photosynthetic oxygen evolution (■) by *Anabaena cylindrica v.* light intensity at 625 and 675 nm. (After Fay (1970).)

subsurface zone and become insignificant at a depth of about 5 m below the surface (Horne and Fogg 1970; Granhall and Lundgren 1971). A similar activity profile was obtained from experiments performed in the marine environment with *Oscillatoria* (*Trichodesmium*) or with the *Rhizosolenia–Richelia* symbiosis dominating the plankton

(Goering *et al.* 1966; Mague *et al.* 1974). However, the active zone extended down to a depth of 30–75 m, evidently because of light penetrating deeper in sea water.

Diurnal cycles

It is conceivable that diurnal variations of illumination in the natural environment will influence the rates of nitrogen fixation in photosynthetic micro-organisms. Notwithstanding, changes in the rate of nitrogenase activity rarely follow closely those of light intensity.

Nitrogenase activity in planktonic populations of temperate freshwater lakes was shown to be highest in the morning hours (Horne and Fogg 1970; Stewart *et al.* 1971). It declines thereafter while the intensity of sunlight increases to a peak at midday, and a slow decline continues in the early afternoon. Nitrogenase activity can remain appreciable during the night.

Intense sunlight was shown to inhibit nitrogen fixation by blue–green algae in rice fields (Reynaud and Roger 1978). While in the control unshaded plots nitrogenase activity was markedly depressed when light intensity was at midday maximum, no similar decline was noticed on shaded plots or on a hazy day. The observations point to the importance of plant cover for growth and nitrogen fixation of blue–green algae in paddy fields.

The diurnal variation of nitrogenase activity showed a different pattern in a subarctic lake (Lundgren 1978). Rates of acetylene reduction increased, first rapidly then more slowly, during the morning hours until midday, reaching a maximum in the early afternoon. In a similar way, no appreciable decrease of nitrogenase activity in blue–green algae was noticed over midday in the littoral–sublittoral zone of the Baltic Sea (Hübel and Hübel 1974).

Results of a detailed investigation, by Burris and Peterson (1978), of diurnal variation in acetylene reduction in natural populations of blue–green algae in Lake Mendota, Wisconsin, direct attention to the complexity of natural situations. Their findings suggest that nitrogenase activity is influenced not only by light intensity but by several other factors, like the rate of photosynthesis, concentration of dissolved oxygen, species composition, and vertical migration of the planktonic population. Owing to their buoyancy, planktonic blue–green algae rise to the subsurface layers of lake water in the early morning hours, and there they photosynthesize and fix nitrogen at increasing rates as light intensity gradually increases (Figs. 1.3 and 1.4). Towards midday, however, the gas-vacuolate algae begin to migrate to deeper waters, owing to the partial collapse of gas vesicles (see Walsby 1978). Consequently rates of photosynthesis and acetylene reduction decline. Later in the afternoon, as light intensity decreases, the algae may display

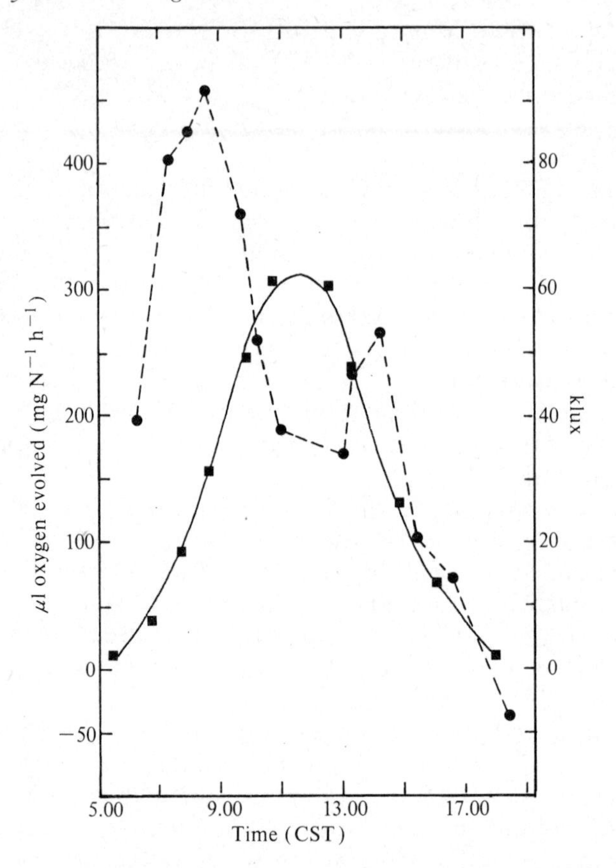

FIG. 1.3. Diurnal variation in the rate of photosynthetic oxygen evolution (●) by planktonic blue–green algae *v.* changes in light intensity (■) in surface waters of Lake Mendota, Wisconsin on 21 July 1975. (After Burris and Peterson (1978).)

smaller secondary peaks of activity. Maximum rates of acetylene reduction were recorded in the early morning hours at relatively low light intensities, before photosynthesis was appreciable and when concentration of dissolved oxygen was at its lowest level (Fig. 1.4).

Temperature

Nitrogen-fixing blue–green algae are common in temperate region, abundant in the tropics, and also present in the extreme environments of polar lakes and hot springs. Although this apparently wide range of temperate tolerance may be true for the group as a whole it is less applicable to the individual species which appear well adapted to their

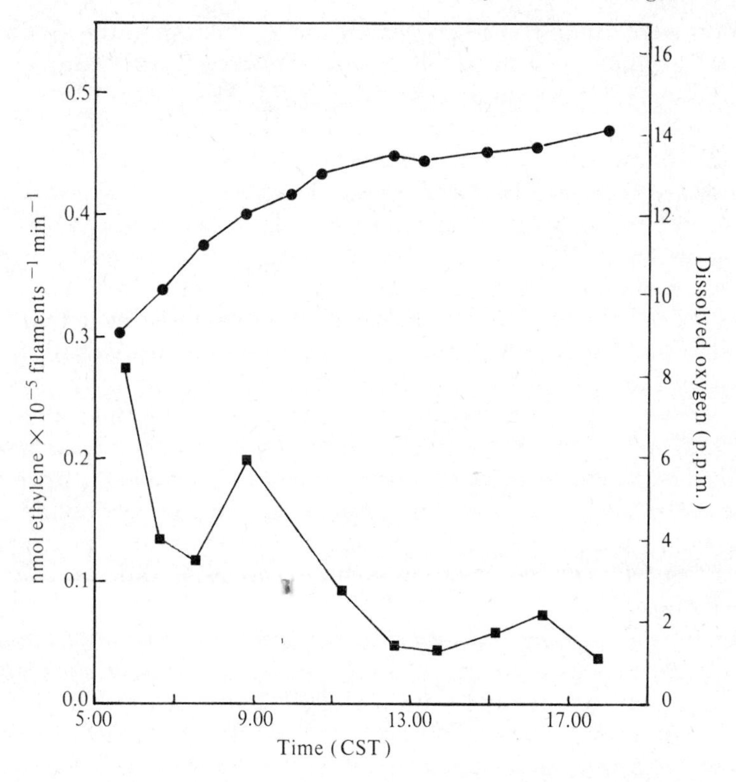

FIG. 1.4. Diurnal variation in the rate of acetylene reduction (ethylene production) (■) by planktonic blue–green algae *v.* changes in the concentration of dissolved oxygen (●) in surface waters of Lake Mendota, Wisconsin on 21 July 1975. (After Burris and Peterson (1978).)

more specific habitats (Brock 1970). Nitrogen fixation in laboratory cultures and in the field is influenced by temperature (Fogg and Than-Tun 1960; Fogg and Stewart 1968), and the temperature coefficient (Q_{10}) was found to be between 3 and 6 (Goering and Nees 1964; Horne and Fogg 1970). Nitrogen-fixing blue–green algae appear more abundant in soils of Asia and Africa south of 30° latitude than north of this latitude (Watanabe and Yamamoto 1971).

Arctic and antarctic blue–green algae can tolerate long periods of exposure to subzero temperatures, and may survive diurnal fluctuations of temperature causing repeated freezing and thawing (Holm-Hansen 1963, 1964; Fogg and Stewart 1968). Although significant [^{15}N]-nitrogen fixation was detected during *in situ* experiments in natural samples of the icy lakes of Signy Island (Horne 1972), rates were relatively low at around 0 °C. Optimum temperatures for nitrogen fixation were found to be between 10 and 20 °C even in the arctic and antarctic blue–green algae (Fogg and Stewart 1968; Granhall and Henriksson 1969; Alexander and Schell 1973; Alexander 1975).

Nitrogen fixation could not be detected during winter in the sublittoral–littoral marine zone though blue–green algae were present in abundance (Stewart 1964). Nor was nitrogenase activity measurable during winter in Scottish soils, but samples taken to the laboratory, which contained *Nostoc* and *Cylindrospermum* species, reduced acetylene at 0 °C, (maximum nitrogenase activity was observed between 15 and 20 °C; Stewart *et al.* 1978). Unexpectedly, nitrogen-fixing activity was still significant at the ecologically irrelevant temperature of 40 °C. On the other hand, samples from a Brazilian soil incorporating *Stigonema* have shown peak activity at 30 °C and a sharp decline at higher temperatures. Soil crust samples from Nigeria containing *Scytonema* reduced acetylene at an increasing rate with an increasing temperature up to 40 °C (Fig. 1.5). These findings indicate a complex relationship between the temperature range of the natural habitat and the temperature range of enzyme activity in a native species, and deserve further investigation.

Heterocyst-forming blue–green algae (predominantly *Mastigocladus* and *Calothrix* species) are abundant in thermal springs, and are found to proliferate at temperatures around 50 °C possibly because of the absence of grazers (Brock, Wiegert, and Brock 1969). Nitrogen-fixing activity is appreciable between 25–54 °C (Stewart 1970) and still detectable at 64 °C. In thermal streams of Yellowstone Park dominated by *Calothrix* species, nitrogenase activity was significant between 28 and 46 °C and highest at 37 °C (Stewart 1970). *Mastigocladus* exhibited peak activity at 43 °C, and the upper temperature limit for acetylene reduction was about 54 °C.

Moisture

Proliferation of terrestrial algae takes place during cool wet periods of moderate light intensity (see Alexander 1977). Intense solar radiation and drought cause soil desiccation and bring active metabolism in micro-organisms to a standstill. Blue–green algae in general are well equipped, by means of their mucilagenous sheaths, to tolerate long periods of desiccation, and the dehydrated cells are capable of absorbing large amounts of moisture upon rewetting. Heterocystous species produce resistant dormant structures, a type of spore called an akinete, under conditions of a prolonged drought (see Fogg *et al.* 1973). The distinct spatial relationship between heterocysts and akinetes may indicate that sporulation is induced by some physiological change related to nitrogen assimilation (Rother and Fay 1979).

It is scarcely surprising that nitrogen fixation, like other biochemical

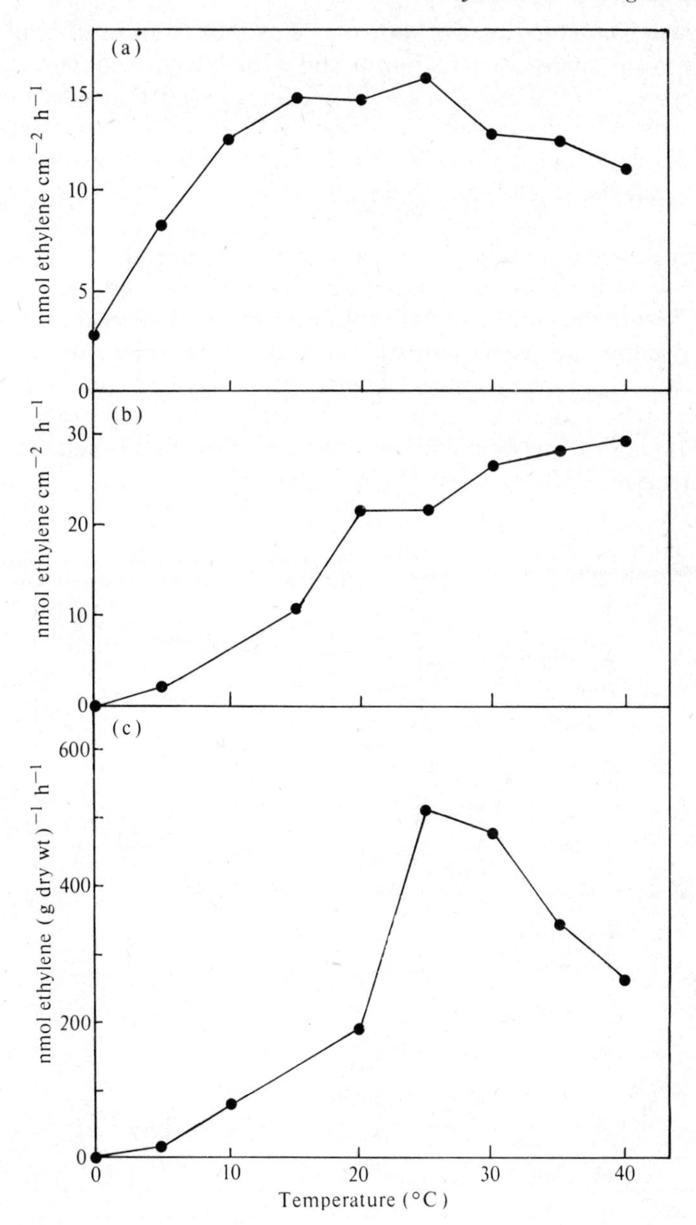

FIG. 1.5. Effect of temperature on acetylene reduction (ethylene production) by (a) soil-core samples from Scotland (containing *Nostoc* and *Cylindrospermum*), (b) *Scytonema* crusts from Nigeria, and (c) *Stigonema panniforme* from Brazil. (After Stewart *et al.* (1978).)

processes, is affected by the state of hydration. High rates of nitrogen fixation were measured in damp soil (Jurgensen and Davey 1968; Henriksson 1971; Paul *et al.* 1971), while desiccation was found to depress nitrogen fixation of blue–green algae in various habitats, e.g. on rocky shores (Stewart 1967*a*), in antarctic soils (Fogg and Stewart 1968), or at the side of hot streams of Yellowstone Park (see Fogg *et al.* 1973).

Light-dependent nitrogenase activity can develop 24 hours after rewetting of dry soil samples, but activity declines more slowly during the process of drying. An equally important factor which limits nitrogen fixation is the relative humidity of the atmosphere (Fig. 1.6) (Stewart *et al.* 1978). In Morocco *Nostoc* colonies reduced acetylene early in the morning when relative humidity was high, and ceased to reduce acetylene by midday when algae were desiccated by intense solar radiation (Stewart *et al.* 1978).

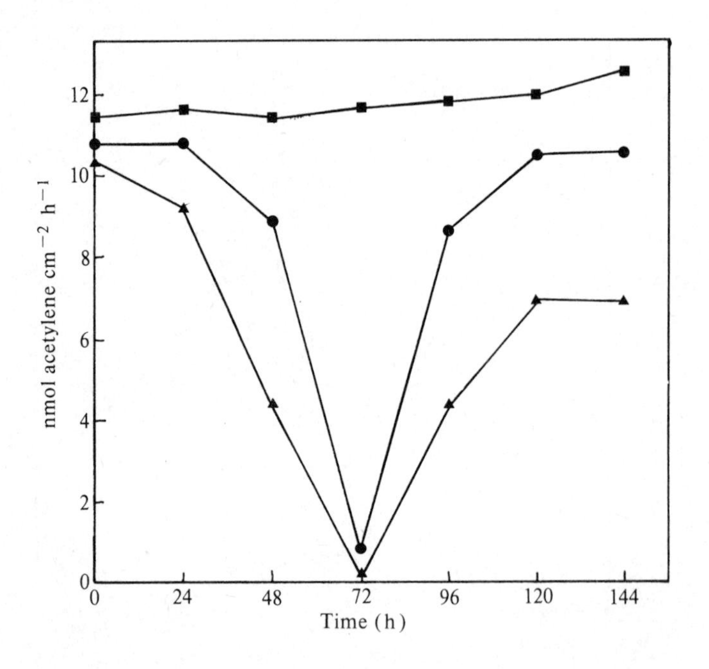

FIG. 1.6. Effect of relative humidity on acetylene reduction (ethylene production) by Scottish soil algae (*Nostoc* and *Anabaena*). Treatments: ■, incubation at relative humidity of 97.5 per cent throughout; ●, incubation for 72 h at relative humidity of 87 per cent, thereafter at 97.5 per cent; ▲, incubation for 72 h at relative humidity of 75 per cent, then at 97.5 per cent. (After Stewart *et al.* (1978).)

Oxygen

Owing to the extreme oxygen-sensitivity of nitrogenase (see Burris 1971), nitrogen fixation can take place only in an anaerobic environment, or when reducing conditions are adequate to prevent oxygen inactivation of the enzyme (Fay and Cox 1967). No mechanism to protect nitrogenase appears to be present in the photosynthetic bacteria. Nitrogen fixation even in a facultative aerobe, like *Rhodospirillum rubrum*, is suppressed in the presence of 4000 Pa oxygen (Pratt and Frenkel 1959). In blue–green algae, compartmentation of nitrogen fixation within the heterocyst appears to provide the particular micro-environment for efficient enzyme function (see Fay 1973). Most non-heterocystous nitrogen-fixing species, however, depend on external reducing conditions for nitrogenase synthesis and active nitrogen fixation (see Stewart 1977).

A classical case of nitrogen fixation being restricted by oxygen concentration in natural populations of photosynthetic bacteria was reported by Stewart (1968) who measured nitrogen-fixing activity in a sharply segragated layer of a Norwegian lake. The lake, which was once isolated from the sea, retained a bottom stratum of sea water overlayered with fresh water. The two layers were also well-demarcated with regard to the concentration of dissolved oxygen (absent in the hypolimnion; about 8 mg l^{-1} in the epilimnion) and of dissolved sulphide (absent in the epilimnion; between 18–27 mg l^{-1} in the hypolimnion). Light penetrating through the aerobic epilimnion to the anaerobic hypolimnion permits photosynthesis and nitrogen fixation of photosynthetic bacteria, predominantly *Pelodictyon*, which proliferate in the uppermost zone of the anaerobic hypolimnion.

While sulphide is essential in the maintenance of reducing conditions required for nitrogen fixation and growth of photosynthetic bacteria, its concentration could be critical. Green sulphur bacteria are exceptional in tolerating very high levels (4 to 8 mM) of hydrogen sulphide, but purple sulphur bacteria only grow well at 0.8 to 4 mM sulphide in the medium, and purple non-sulphur bacteria can tolerate no more than about 2.0 mM of sulphide (Pfennig 1975). Differences in tolerance to sulphide often produce a distinct vertical zonation of photosynthetic bacteria in the water column and within the sediments.

Stewart and Pearson (1970) have shown that growth and nitrogen fixation of heterocystous blue–green algae (*Anabaena* and *Nostoc*) in laboratory culture is inhibited by high concentrations of dissolved oxygen and that nitrogenase activity is highest under micro-aerobic conditions. High levels of dissolved oxygen also inhibit photosynthesis (Lex, Silvester, and Stewart 1972). Fogg (1969) suggested that the positive correlation consistently observed between the amounts of dissolved organic matter and the growth of planktonic (heterocystous) blue–green algae (see p. 22) may, at least partly, be explained on the basis of associated deoxygenation in eutrophic waters which results from increased oxygen consumption by the proliferating bacterial populations.

Burris and Peterson (1978) have observed an interesting discrepancy between oxygen inhibition of photosynthesis and that of nitrogenase activity in native populations of blue–green algae occupying the surface and subsurface (2 m deep) waters of Lake Mendota, Wisconsin. While photosynthesis is inhibited by oxygen concentrations (in equilibrium with 20–40 per cent oxygen) usually prevailing in the lake, acetylene reduction was depressed only at much higher concentrations of oxygen, corresponding to 40 per cent and higher concentrations of oxygen in the gas phase (Fig. 1.7). It was suggested that elevated oxygen concentra-

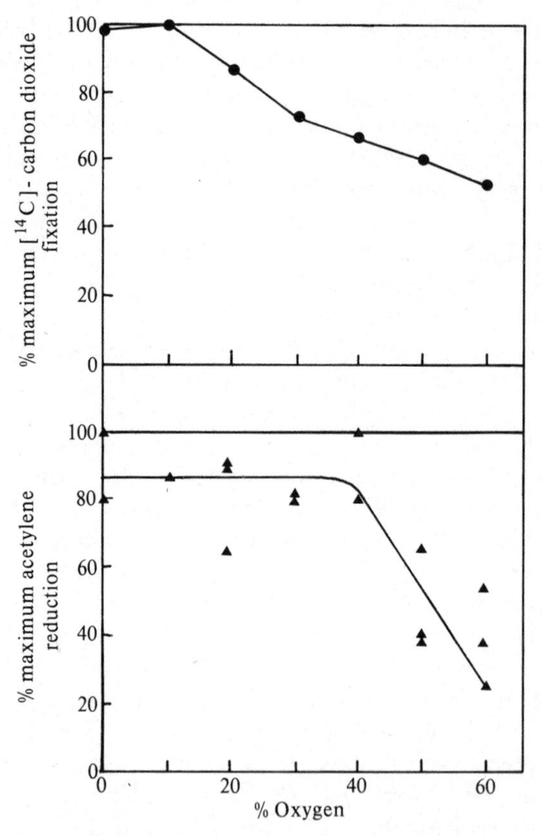

FIG. 1.7. Effect of oxygen concentration on rates of [^{14}C]-carbon dioxide fixation (●) and acetylene reduction (▲) by planktonic blue–green algae (*Aphanizomenon* and *Anabaena*) from Lake Mendota, Wisconsin, at saturating light intensity. (After Burris and Peterson (1978).)

tions could enhance photorespiration which would deplete the pool of reductant over a period of several hours (Lex *et al.* 1972) but may not affect nitrogenase activity in a 30-minute assay.

Carbon dioxide and hydrogen ion concentration

Hydrogen ion concentration and the concentration of carbon dioxide are closely interrelated factors which effect the growth of micro-organisms in the natural environment, and it is therefore convenient to consider the two together.

Blue–green algae grow better under an atmosphere enriched in carbon dioxide (Allen and Arnon 1955). Carbon-enriched algae, produced by pre-incubating *Anabaenopsis circularis* under a gas phase of 1 per cent (v/v) carbon dioxide in air, showed (by a factor of 8) increased nitrogenase activity, during subsequent light and dark assay periods, when compared with the control samples which were pre-incubated under ambient atmosphere (Fay 1976) (Table 1.1).

TABLE 1.1
Effect of carbon dioxide fixation in the light upon nitrogenase activity of Anabaenopsis circularis *in the light and in the dark*

Pretreatment* (gas phase)	C_2H_4 produced (nmol mg dry wt^{-1} h^{-1}) in light (1500 lx)	in dark
Air (control)	39	26
1% CO_2 in air	320	209

* Incubation in light (1500 lx) for 24 h.
From Fay (1976).

It is noteworthy that carbon dioxide concentrations in the subsurface atmosphere of well aerated microbially active soils may exceed the atmospheric levels by factors of 10 to 100 (Alexander 1977).

Photosynthetic bacteria and blue–green algae can tolerate a wide range of pH conditions, though they grow best in the range of pH 7.0–8.5 in culture (Pfennig 1967; Fogg *et al.* 1973). Most blue–green algae fix nitrogen best at pH around neutrality. Hence the alkaline reaction of sea water should be favourable for nitrogen fixation (Fogg 1978). pH has an important influence on the distribution of nitrogen-fixing soil micro-organisms. Blue–green algae are abundant in alkaline soils, rare in acid forest soils, and almost absent in soils with pH below 5 (Jurgensen and Davey 1968; Granhall and Henriksson 1969; Dooley and Houghton 1973). A significant positive correlation was observed between soil pH and the abundance of nitrogen-fixing blue–green algae in rice fields (Reynaud and Roger 1978). The beneficial effect of lime treatment on crop yield is attributed to a stimulatory effect on nitrogen-fixing activity in paddy soils (Subrahmanyan, Relwani, and Manna 1964). Blue–green algae in tropical soils, unlike those in temperate soils, appear to be less sensitive to low pH conditions and may display appreciable nitrogenase activity even at a pH as low as 4 (Stewart *et al.* 1978).

The abundance of blue–green algae in nutrient-rich eutrophic waters is attributed among others (see pp. 15 and 20–3) to their efficiency in obtaining carbon dioxide at low concentrations and to their ability of utilizing bicarbonate as a source of carbon dioxide. By regulating pH, the concentration of dissolved carbon dioxide and of essential nutrients (phosphate, nitrate), Shapiro (1973) was able to influence the development and composition of planktonic populations of Lake Emily, Minnesota. Increasing carbon dioxide concentration and/or lowering the pH caused a distinct shift in the composition of plankton from blue–green algae. Addition of nutrients promoted growth which in turn decreased the amounts of free carbon dioxide and raised pH. Blue–green algae can prosper under these conditions.

Combined nitrogen

Combined nitrogen has long been known to affect nitrogen fixation (Fogg 1942; Burris and Wilson 1946) but it was only in 1958 that Pengra and Wilson established the action of ammonia and to a lesser extent of nitrate in preventing nitrogenase synthesis in *Klebsiella* (*Aerobacter*). This was confirmed later with *Nostoc muscorum* (Stewart *et al.* 1968) and with *Rhodospirillum rubrum* (Munson and Burris 1969). It was noticed that repression of nitrogenase synthesis is only partial at low concentrations of ammonium-nitrogen or nitrate-nitrogen (Daesch and Mortenson 1968; Strandberg and Wilson 1968; Stewart *et al.* 1968; Dalton and Postgate 1969).

The range of concentration of combined nitrogen required to inhibit nitrogen fixation under natural conditions appears to differ from that observed in laboratory cultures. In a study of nitrogen fixation in a polluted canal system, Pearson and Taylor (1978) concluded that concentrations of ammonium-nitrogen (maximum 3 mg l^{-1}) and of nitrate-nitrogen (maximum 9 mg l^{-1}), which were sufficient to suppress nitrogenase activity in natural populations of blue–green algae *in situ*, were 10–20 times lower than those required to inhibit nitrogenase synthesis in culture (Strandberg and Wilson 1968; Stewart *et al.* 1968).

Fogg (1971) suggested that ammonium-nitrogen or nitrate-nitrogen may become locally depleted within a dense natural population of blue–green algae, despite the fact that their concentrations may be considerable in the surrounding medium. Considering further that dissolved elemental nitrogen is available in natural waters at high concentrations and that its diffusion coefficient in water is higher than that of either ammonium chloride or sodium nitrate, it is conceivable that nitrogen fixation in natural waters may take place in the presence of combined nitrogen (Dugdale and Dugdale 1962; Goering *et al.* 1966).

In general, however, nitrogen fixation in the field is significant only under conditions of low concentrations of combined nitrogen (Dugdale and Dugdale 1962; Goering and Nees 1964; Horne and Fogg 1970; Granhall and Lundgren 1971; Horne and Goldman 1972). Rates of nitrogen fixation in the English lakes Esthwaite and Windermere were highest when concentrations of nitrate were below 0.3 mg l^{-1} (Horne and Fogg 1970). Nitrogen-fixing activity appears to decline during destratification of temperate lakes in autumn when fresh supplies of combined nitrogen from the hypolimnion become available (Dugdale and Dugdale 1962; Granhall and Lundgren 1971; Horne and Goldman 1972). In spite of the large amounts of combined nitrogen present in the oceans, its concentration in the photic zone is in general limiting for the growth of algae, and hence the ability to fix nitrogen consitutes a considerable ecological advantage (Goering *et al.* 1966; Fogg 1978).

Phosphate

Biological nitrogen fixation has an absolute requirement for ATP (see Burris 1971). The generally observed light stimulation of nitrogen fixation in photosynthetic micro-organisms is apparently through the generation of ATP in photosynthetic phosphorylation (Pratt and Frenkel 1959; Bulen, Burns, and LeComte 1965; Cox and Fay 1969; Fay 1970). Available phosphorus is thus an essential factor which regulates growth of photosynthetic micro-organisms under nitrogen-fixing conditions.

Anabaena flos-aquae grown in phosphorus-limited culture developed high levels of alkaline phosphatase activity (Bone 1971). Addition of phosphate to phosphorus-starved cells of heterocystous blue–green algae markedly enhanced nitrogenase activity (Stewart, Fitzgerald, and Burris 1970). Planktonic populations of heterocystous blue–green algae have shown a similar response to added phosphate (Stewart and Alexander 1971), though the effect may vary according to the size of the cellular pool of phosphorus. Blue–green algae possess a marked ability to store phosphorus in the form of polyphosphate granules (Talpasayi 1963). The addition of orthophosphate or phosphorus-containing commercial detergents can equally stimulate nitrogenase activity (Stewart and Alexander 1971). This may explain an earlier finding by Horne and Fogg (1970) who measured maximum nitrogen fixation in the south basin of Lake Windermere which showed the characteristics of the early stage of eutrophication.

The influence of phosphorus loading on the rate of nitrogen fixation by planktonic and periphytic blue–green algae was tested in *in situ* experiments by Lean, Liao, Murphy, and Painter (1978) within large enclosures located in the eutrophic Bay of Quinte of Lake Ontario. They estimated a yearly total input of 60 g N from nitrogen fixation in enclosures receiving phosphate (0.88 g P m^{-2} a^{-1}), against 12 and 6 g N per year, respectively, in enclosures which received no additions or were supplemented with both phosphate and nitrate (Fig. 1.8). Phosphorus fertilization brought about a similar (tenfold) increase of nitrogenase activity in a subarctic oligotrophic lake maintaining a periphytic population dominated by heterocystous blue–green algae (Lundgren 1978).

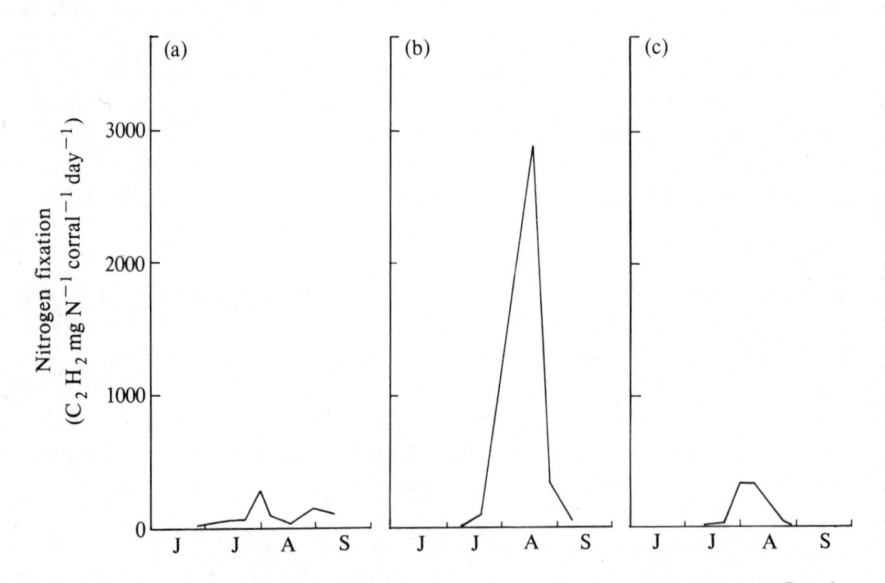

FIG. 1.8. Effect of phosphorus loading on the rate of nitrogen fixation (acetylene reduction) by planktonic algae of Lake Ontario, maintained in large enclosures during the summer of 1974. (a) No nutrients added; (b) 14.14 μg P l^{-1} wk^{-1} added; (c) 14.14 μg P and 184 μg N l^{-1} wk^{-1} added. (After Lean *et al.* (1978).)

Mineral elements

Photosynthetic bacteria and blue–green algae require a variety of mineral nutrients for growth but only a few appear specifically indispensible for growth under nitrogen-fixing conditions.

Molybdenum

Among the trace elements required, molybdenum is distinguished by being an integral part of two systems, nitrogenase and nitrate reductase, involved in inorganic nitrogen metabolism. But higher concentrations (0.2 p.p.m.) of molybdenum are needed for optimum growth of *Anabaena cylindrica* on elemental nitrogen than when the alga is grown on nitrate, and molybdenum is not necessary in the presence of ammonium-nitrogen (Wolfe 1954). Molybdenum deficiency causes such symptoms as the loss of pigmentation and increased heterocyst production, which are characteristic of nitrogen-starved algae. Instead of increasing, nitrogenase activity diminishes gradually (Fay and De Vasconcelos 1974). When the supply of molybdenum is restored, nitrogenase activity increases at a remarkable speed and the alga recovers rapidly.

Primary productivity studies in oligotrophic lakes of New Zealand and North America have shown molybdenum to be the limiting factor for growth of planktonic blue–green algae, which in turn effects the growth of zooplankton and livestock (Goldman 1964). Molybdenum may equally limit nitrogen fixation in terrestrial habitats. Light-dependent nitrogenase activity has increased considerably when sodium molybdate (0.1 p.p.m. Mo) was supplied to Scottish soils containing *Anabaena* and *Nostoc* (Stewart *et al.* 1978). Activity was saturated at 0.5 p.p.m. molybdenum (Fig. 1.9).

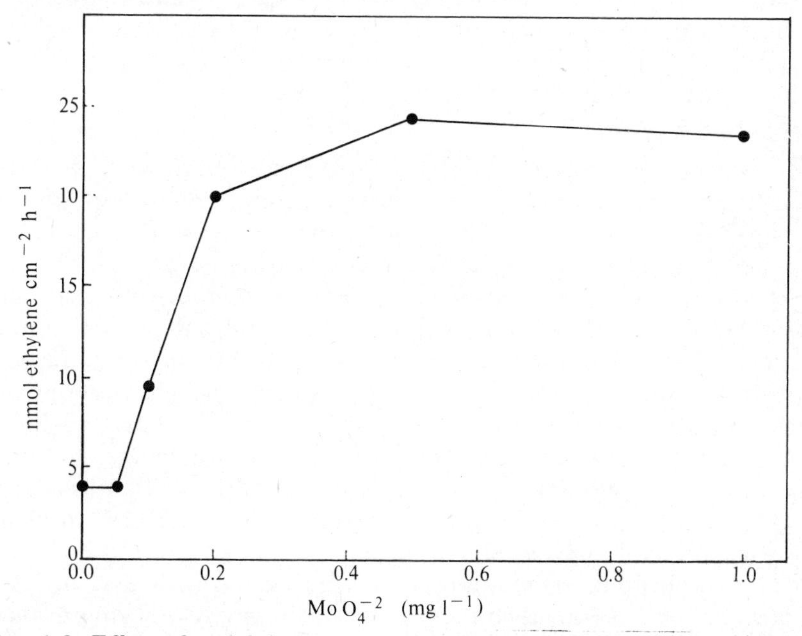

FIG. 1.9. Effect of molybdenum concentration on light-dependent acetylene reduction (ethylene production) of Scottish soil cores. (After Stewart *et al.* (1978).)

Iron

Iron is a component of both protein fractions of nitrogenase as well as of other non-haem proteins and cytochromes. It is routinely provided in a complex form, chelated with citrate, EDTA (ethylene-diamino tetra acetic acid) or other chelating agents, to prevent its precipitation at the slightly alkaline reaction of most culture media used.

Iron may become depleted in natural waters following long periods of intense nitrogen fixation associated with the abundance of nitrogen-fixing blue–green algae (Lean *et al.* 1978). Iron deficiency induces in blue–green algae the liberation of siderochromes, powerful iron-specific hydroxamate chelators. This ability confers on blue–green algae a considerable competitive advantage over other groups of algae.

Calcium

Calcium was considered by early workers to be required particularly for nitrogen fixation in *Azotobacter* (Burk 1937) and in blue–green algae (Allison *et al.* 1937; Allen and Arnon 1955; Fay 1962), but its function is still poorly understood (Jakobsons, Zell, and Wilson 1962).

Gallon (1978) has recently suggested that calcium requirement by blue–green algae is related to the presence and the action of chelators. Accordingly, chelators may cause the removal of calcium and magnesium ions from the cells of blue–green algae, which could depress nitrogenase activity especially in species, like *Gloeocapsa*, which are apparently dependent on calcium for growth. The inhibitory effect is relieved upon addition of calcium.

Organic substances

Although photosynthesis is the principal mode of nutrition in photosynthtic bacteria and blue–green algae, a great number of species can assimilate simple organic substances, sugars, and organic acids, in light-dependent reactions (see Pfennig 1967; Fogg *et al.* 1973; Stanier and Cohen-Bazire 1977). Photoassimilation of organic substances can promote growth at light intensities which are limiting photosynthesis. At saturating light intensities, however, photoassimilation of organic substrates cannot increase growth rate above that which is obtained by photosynthetic carbon dioxide fixation alone (see Fogg 1969). A number of blue–green algae can grow and fix nitrogen in complete darkness when suitable organic substrates are provided (Allison *et al.* 1937; Fay 1965; Watanabe and Yamamoto 1967) but growth in the dark is invariably slower than in the light.

The importance of organic compounds for growth and nitrogen fixation of photosynthetic micro-organisms in the natural environment and of the heterotrophic potential in interspecies competition is not easy to assess because of the multiplicity of factors and interactions which appear to be involved.

There is ample evidence for a relationship between organic pollution and the abundance of planktonic blue–green algae (Pearsall 1932; Brook 1959; Vance 1965), and a similar correlation between the rate of nitrogen fixation and the concentration of dissolved organic nitrogen (Dugdale and Dugdale 1962; Goering and Nees 1964; Horne and Fogg 1970). Multiple regression analysis of a large number of observations from English lakes showed that this positive correlation is statistically significant (Horne and Fogg 1970). Among the various direct and indirect effects of organic compounds, Fogg (1969) attached great importance to the complex-forming properties of organic substances which help to maintain iron and essential trace metals in solution. He suggested that natural chelating agents may play a notable part in promoting the growth of planktonic blue–green algae, particularly in alkaline lakes.

The ecological advantage gained through photoassimilation of organic substances may be appreciable in dimly-lit situations, like at the bottom of the photic zone in natural waters, or in soils, where organic substrates could provide a source of maintenance energy during conditions of inadequate photosynthesis. Facultative chemoheterotrophy appears to be more characteristic of terrestrial species (like *Nostoc muscorum*, *Tolypothrix tenuis*, *Chlorogloea fritschii*, *Anabaenopsis circularis*), but even in these algae the rate of dark assimilation of organic substrates is relatively low. They cannot compete with heterotrophic bacteria which in general are incomparably more efficient in utilizing organic substrates present often at very low concentrations in natural waters (Hobbie and Wright 1965).

Nevertheless, there are indications that heterotrophic nitrogen fixation could be important in the field both under light and dark conditions. Fay (1976) showed that glucose greatly stimulates acetylene reduction by *Anabaenopsis circularis* in the light and it supports prolonged nitrogenase activity in the dark (Fig. 1.10). Activity is increased up to 25-fold when glucose is assimilated in the light (Table 1.2). During the light period and in the presence of glucose the alga can accumulate adequate carbon reserves to allow subsequent nitrogen fixation (acetylene reduction) in the dark for more than 60 hours, even when organic substrates are not available during the dark period.

Soil type

Soils differ in mineral composition, particle size, and content of organic matter which in turn determine the texture, relative surface area, pore size, aeration, and moisture as well as pH and ion exchange capacity of soils (see Alexander 1977). The complex set of physical-chemical characteristics of a soil constitutes the microenvironment in which soil micro-organisms function and eventually fix nitrogen.

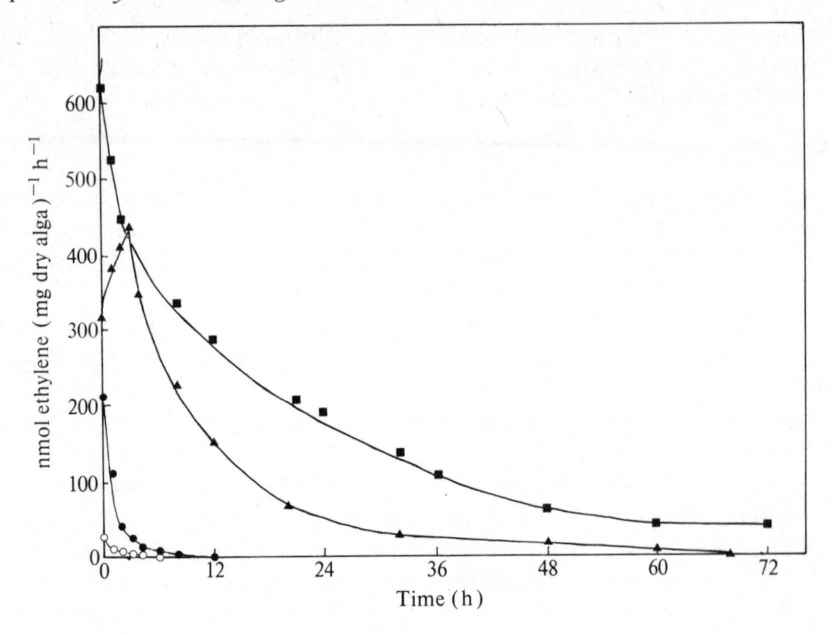

FIG. 1.10. Effect of carbon assimilation in the light on the rates and duration of acetylene reduction (ethylene production) by *Anabaenopsis circularis* in the dark. Pretreatment in light (1500 lx) for 24 h. Incubation: ○, under air throughout: ●, under 1 per cent carbon dioxide in air throughout; ■, under air with 0.03 M glucose throughout; ▲, under air with 0.03 M glucose in the light period only. (After Fay (1976).)

TABLE 1.2

Effect of glucose assimilation in the light upon nitrogenase activity of Anabaenopsis circularis *in the light and in the dark*

Pretreatment* (carbon source)	C_2H_4 produced (nmol mg dry wt^{-1} h^{-1}) in light (1500 lx)	in dark
Air (control)	39	26
0.03 M glucose	957	620

* Incubation in light (1500 lx) for 24 h.
From Fay (1976).

Investigating the relationship between soil type and the occurrence of nitrogen-fixing blue–green algae, Granhall (1975) showed that in average almost half of the soils in temperate regions contained nitrogen-fixing blue–green algae, and that they were particularly abundant in calcareous and clay soils (Table 1.3).

Clay soils are neutral or alkaline, high in water content, and able to support abundant growth of nitrogen-fixing blue–green algae. Calcareous soils are alkaline, rich in minerals (including phosphate), and contain without exception nitrogen-fixing species. Sandy and moraine soils are often acid, have little organic content, and accommodate few nitrogen-fixing blue–green algae.

TABLE 1.3
Relationship between soil type and the occurrence of blue–green algae

Soil type	Soils with blue–green algae (%)	Soils with nitrogen-fixing algae (%)	Occurrence of heterocystous filaments (%)	Total no. of genera	No. of nitrogen-fixing genera
Sand	50	17	0	4	1
Moraine	57	0	0	4	0
Clay	100	81	31	13	7
Calcareous	100	93	40	10	4
Forest	25	10	0	3	1

From Granhall (1975).

1.4 Concluding remarks

Photosynthesis and nitrogen fixation are key biological processes which generate chemically stored energy and organic matter, and upon which all forms of life depend. The combination of these two fundamental processes in a single organism, or within a symbiotic association of two distinct organisms, represents a highly productive and efficient biological system. It seems likely that such a combination has first developed some billion years ago in an ancient photosynthetic prokaryote (Olson 1970) under reducing conditions and in the absence of free oxygen in the Earth's atmosphere. Photosynthetic bacteria may have preserved this earlier stage in evolution which limits their distribution in the present oxidizing environment. They are found in restricted and less accessible habitats which could be one of the reasons for the paucity of information concerning their activities and of their role as primary producers in the natural environment. It is hoped that future research will penetrate these unexplored areas.

Blue–green algae were probably the first photosynthetic micro-organisms to have acquired a second photosystem to extract electrons from water with a concomitant discharge of free oxygen. They became well adapted to the aerobic environment initially created as a consequence of their own photosynthetic activities, and have developed specialized cells to protect nitrogenase.

Research with blue–green algae (Fay 1976) (see pp. 6 and 17) and with leguminous plants (Hardy and Havelka 1975) indicate that the major factor which limits nitrogen fixation in photosynthetic organisms is the inadequate supply of photosynthates to the nitrogen-fixing system. This may arise from a shortage of any basic requirements for photosynthesis, like light intensity or the concentration of carbon dioxide. Although blue–green algae are extremely efficient in absorbing carbon dioxide at low concentrations, higher than ambient carbon-dioxide concentrations or the supply of organic nutrients may enhance considerably their nitrogen-fixing potential (see Tables 1.1 and 1.2). The exceptionally high rates of nitrogen fixation measured in the *Azolla–Anabaena* symbiotic association (Becking 1975) are most likely attributable to an ample supply of photosynthate by the host *Azolla* to the nitrogen-fixing phycobiont, *Anabaena azollae*.

References

ALEXANDER, M. (1977). *Introduction to soil microbiology*, 2nd edn. John Wiley, New York.
ALEXANDER, V. (1975). In *Nitrogen fixation by free-living microorganisms* (ed. W. D. P. Stewart) p. 175. Cambridge University Press.
—— and SCHELL, D. M. (1973). *Arctic Alpine Res.* **5**, 77.
ALLEN, M. B. (1963). In *Symposium on marine microbiology* (ed. C. H. Oppenheimer) p. 85. C. C. Thomas, Springfield, Illinois.
—— and ARNON, D. I. (1955). *Pl. Physiol., Lancaster* **30**, 366.
ALLISON, F. E., HOOVER, S. R., and MORRIS, H. J. (1937). *Bot. Gaz.* **98**, 433.
BECKING, J. H. (1975). In *Symbiotic nitrogen fixation in plants* (ed. P. S. Nutman) p. 539. Cambridge University Press.
BONE, D. H. (1971). *Archs Mikrobiol.* **80**, 147.
BROCK, M. L., WIEGERT, R. G., and BROCK, T. D. (1969). *Ecology* **50**, 192.
BROCK, T. D. (1970). *A. Rev. ecol. System.* **1**, 191.
BROOK, A. J. (1959). *Proc. Soc. Water Treat. Exam.* **8**, 133.
BROWN, T. E. and RICHARDSON, F. L. (1968). *J. Phycol.* **4**, 38.
BULEN, W. A., BURNS, R. C., and LECOMTE, J. R. (1965). *Proc. natn. Acad. Sci. U.S.A.* **53**, 532.
BUNT, J. S., COOKSEY, K. E., HEEB, M. A., LEE, C. C., and TAYLOR, B. F. (1970). *Nature, Lond.* **227**, 1163.
BURK, D. (1937). *Biochimica* **2**, 312.
BURRIS, R. H. (1971). In *The chemistry and biochemistry of nitrogen fixation* (ed. J. R. Postgate) p. 105. Plenum Press, London.
—— (1976). *Aust. J. Pl. Physiol.* **3**, 41.
—— and PETERSON, R. B. (1978). In *Environmental role of nitrogen-fixing blue–green algae and asymbiotic bacteria. Ecol. Bull., Stockh.* **26**, 28.
—— and WILSON, P. W., (1946). *Bot. Gaz.* **108**, 254.
CAPONE, D. G., TAYLOR, D. L., and TAYLOR, B. F. (1977). *Mar. Biol.* **40**, 29.
CARPENTER, E. J. (1972) *Science, N.Y.* **178**, 1207.
—— and COX, J. L. (1974). *Limnol. Oceanogr.* **19**, 429.
—— and MCCARTHY, J. J. (1975). *Limnol. Oceanogr.* **20**, 389.

CARR, N. G. and WHITTON, B. A. (eds.) (1973). *The biology of blue–green algae.* Blackwell, Oxford.
CASTENHOLZ, R. W. (1976). *J. Phycol.* **12**, 54.
—— (1977). *Microbial Ecol.* **3**, 79.
COHEN, Y., PADAN, E., and SHILO, M. (1975). *J. Bact.* **123**, 855.
COX, R. M. (1966). *Archs Mikrobiol.* **53**, 263.
—— and FAY, P. (1969) *Proc. R. Soc.* **B172**, 357.
DAESCH, G. and MORTENSON, L. E. (1968). *J. Bact.* **96**, 346.
DALTON, H. and POSTGATE, J. R. (1969). *J. gen. Microbiol.* **56**, 307.
DOOLEY, F. and HOUGHTON, J. A. (1973). *Br. phycol. J.* **8**, 289.
DUGDALE, R. C., MENZEL, D. W., and RYTHER, J. H. (1961). *Deep Sea Res.* **7**, 297.
—— DUGDALE, V. A., NEES, J., and GOERING, J. (1959). *Science, N.Y.* **130**, 859.
DUGDALE, V. A. and DUGDALE, R. C. (1962). *Limnol. Oceanogr.* **7**, 170.
DURRELL, L. W. (1964). *Trans. Am. microsc. Soc.* **83**, 79.
ENGLUND, B. (1976). *Oikos* **27**, 428.
ENGLUND, B. (1978). *Oecologia, Berl.* **34**, 45.
FAY, P. (1962). Ph.D. thesis, University of London.
—— (1965). *J. gen. Microbiol.* **39**, 11.
—— (1970). *Biochim. biophys. Acta* **216**, 353.
—— (1973). In *The biology of blue–green algae* (eds. N. G. Carr and B. A. Whitton) p. 238. Blackwell, Oxford.
—— (1976). *Appl. Environm. Microbiol.* **31**, 376.
—— and COX, R. M. (1967). *Biochim. biophys. Acta* **143**, 562.
—— and DE VASCONCELOS, L. (1974). *Archs Microbiol.* **99**, 221.
FOGG, G. E. (1942). *J. exp. Biol.* **19**, 78.
—— (1951). *J. exp. Bot.* **2**, 117.
—— (1969). *Proc. R. Soc.* **B173**, 175.
—— (1971). *Pl. Soil* Special Volume, p. 393.
—— (1974). In *Algal physiology and biochemistry* (ed. W. D. P. Stewart) p. 560. Blackwell, Oxford.
—— (1978). In *Environmental role of nitrogen-fixing blue–green algae and asymbiotic bacteria. Ecol. Bull., Stockh.* **26**, 11.
—— and STEWART, W. D. P. (1968). *Br. Antarct. Surv. Bull.* **15**, 39.
—— and THAN-TUN (1960). *Proc. R. Soc.* **B153**, 111.
—— STEWART, W. D. P., FAY, P., and WALSBY, A. E. (1973). *The blue–green algae.* Academic Press, London.
GALLON, J. R. (1978). In *Environmental role of nitrogen-fixing blue–green algae and asymbiotic bacteria. Ecol. Bull., Stockh.* **26**, 60.
GARLICK, S., OREN, A., and PADAN, E. (1977). *J. Bact.* **129**, 623.
GOERING, J. J. and NEES, J. C. (1964). *Limnol. Oceanogr.* **9**, 530.
—— DUGDALE, R. C., and MENZEL, D. W. (1966). *Limnol. Oceanogr.* **11**, 614.
GOLDMAN, C. R. (1964). *Verh. int. Ver. theor. angew. Limnol.* **15**, 365.
GRANHALL, U. (1975). In *Nitrogen fixation by free-living microorganisms* (ed. W. D. P. Stewart) p. 189. Cambridge University Press.
—— (ed.) (1978). *Environmental role of nitrogen-fixing blue–green algae and asymbiotic bacteria. Ecol. Bull., Stockh.* **26**, p. 189.
—— and HENRIKSSON, E. (1969). *Oikos* **20**, 175.
—— and LUNDGREN, A. (1971). *Limnol. Oceanogr.* **16**, 711.
HARDY, R. W. F. and HAVELKA, U. D. (1975). *Science, N.Y.* **188**, 633.
HENRIKSSON, E. (1971). *Pl. Soil* Special Volume, p. 415.
HIRANO, M. (1965). In *Biogeography and ecology in Antarctica* (eds. J. van Mieghem and P. van Oye) *Monogr. Biol.* **15**, 127.

HOBBIE, J. E. and WRIGHT, R. T. (1965). *Mem. Ist. Ital. Idrobiol.* **18** Suppl., 175.
HOLM–HANSEN, O. (1963). *Physiologia Pl.* **16**. 530.
—— (1964). *Phycologia* **4**, 43.
HORNE, A. J. (1972). *Br. Antarct. Surv. Bull.* **27**, 1.
—— and FOGG, G. E. (1970). *Proc. R. Soc.* **B175**, 351.
—— and GOLDMAN, C. R. (1972). *Limnol. Oceanogr.* **17**, 678.
—— and VINER, A. B. (1971). *Nature, Lond.* **232**, 417.
HÜBEL, von H. and HÜBEL, M. (1974). *Arch. Hydrobiol.* Suppl. **46**, 39.
ISHIZAWA, S., SUZUKI, T., and ARARAGI, M. (1975). In *JIBP synthesis: nitrogen fixation and nitrogen cycle* (ed. H. Takahashi) p. 41. University of Tokyo Press.
JAKOBSONS, A., ZELL, E. A., and WILSON, P. W. (1962). *Archs Mikrobiol.* **41**, 1.
JORDAN, D. C., MCNICOL, P. J., and MARSHALL, M. R. (1978). *Can. J. Microbiol.* **24**, 643.
JURGENSEN, M. F. and DAVEY, C. B. (1968). *Can. J. Microbiol.* **14**, 1179.
—— —— (1971). *Pl. Soil* **34**, 341.
KHOJA, T. M. and WHITTON, B. A. (1975). *Br. phycol. J.* **10**, 139.
KNOWLES, R. (1976). In *A treatise on dinitrogen fixation* (eds. R. W. F. Hardy and A. H. Gibson) Section IV, p. 33. John Wiley, New York.
KOBAYASHI, M. (1975). In *JIBP synthesis: nitrogen fixation and nitrogen cycle* (ed. H. Takahasi) p. 69. University of Tokyo Press.
—— and HAQUE, M. Z. (1971). *Pl. Soil* Special Volume, p. 443.
—— TAKAHASHI, E., and KAWAGUCHI, K. (1967). *Soil Sci.* **104**, 113.
KONDRAT'EVA, E. N. (1965). *Photosynthetic bacteria.* Translation from Russian by Israel Programme for Science Translations.
LEAN, D. R. S., LIAO, C. F. H., MURPHY, T. P., and PAINTER, D. S. (1978). In *Environmental role of nitrogen-fixing blue–green algae and asymbiotic bacteria. Ecol. Bull., Stockh.* **26**, 41.
LEX, M., SILVESTER, W. B., and STEWART, W. D. P. (1972). *Proc. R. Soc.* **B180**, 87.
LUNDGREN, A. (1978). In *Environmental role of nitrogen-fixing blue–green algae and asymbiotic bacteria. Ecol. Bull., Stockh.* **26**, 52.
MACRAE, I. C. and CASTRO, T. F. (1967). *Soil Sci.* **103**, 277.
MAGUE, T. H. (1976). In *A treatise on dinitrogen fixation* (eds. R. W. F. Hardy and A. H. Gibson) Section IV, p. 95. John Wiley, New York.
—— WEARE, N. M., and HOLM–HANSEN, O. (1974). *Mar. Biol.* **24**, 109.
MARUYAMA, Y. (1975). In *JIBP synthesis: nitrogen fixation and nitrogen cycle* (ed. H. Takahasi) p. 79. University of Tokyo Press.
—— TAGA, N., and MATSUDA, O. (1970). *J. Oceanogr. Soc. Japan* **26**, 30.
MATERASSI, R. and BALLONI, W. (1965). *Annls Inst. Pasteur, Paris* Suppl. **3**, 218.
MOELLER, E. and ROSKOSKI, J. P. (1978). *Hydrobiologia* **60**, 13.
MUNSON, T. O. and BURRIS, R. H. (1969). *J. Bact.* **97**, 1092.
OGAWA, R. E. and CARR, J. F. (1969). *Limnol. Oceanogr.* **14**, 342.
OLSON, J. M. (1970). *Science, N.Y.* **168**, 438.
PAUL, E. A., MYERS, R. J. K., and RICE, W. A. (1971). *Pl. Soil* Special Volume, p. 495.
PEARSALL, W. A. (1932). *J. Ecol.* **20**, 241.
PEARSON, H. W. and TAYLOR, R. (1978). In *Environmental role of nitrogen-fixing blue–green algae and asymbiotic bacteria. Ecol. Bull. Stockh.* **26**, 69.
PENGRA, R. M. and WILSON, P. W. (1958). *J. Bact.* **75**, 21.
PFENNIG, N. (1967). *A. Rev. Microbiol.* **21**, 285.
—— (1975). *Pl. Soil* **43**, 1.
—— (1977). *A. Rev. Microbiol.* **31**, 275.
PRATT, D. C. and FRENKEL, A. W. (1959). *Pl. Physiol., Lancaster* **34**, 333.
REYNAUD, P. A. and ROGER, P. A. (1978). In *Environmental role of nitrogen-fixing blue–green algae and asymbiotic bacteria. Ecol. Bull., Stockh.* **26**, 148.

ROTHER, J. A. and FAY, P. (1979). *Freshwater Biol.* **9**, 369.
SHAPIRO, J. (1973). *Science, N.Y.* **179**, 382.
SINGH, R. N. (1961). *Role of blue–green algae in nitrogen economy of Indian agriculture.* Indian Council of Agricultural Research, New Delhi.
STANIER, R. Y. and COHEN-BAZIRE, G. (1977). *A. Rev. Microbiol.* **31**, 225.
STEWART, W. D. P. (1964). *J. gen. Microbiol.* **36**, 415.
—— (1965). *Ann. Bot.* **29**, 229.
—— (1967*a*). *Ann. Bot.* **31**, 385.
—— (1967*b*). *Nature, Lond.* **214**, 603.
—— (1968). In *Algae, man and environment* (ed. D. F. Jackson) p. 53. University Press, Syracuse, New York.
—— (1970). *Phycologia* **9**, 261.
—— (1971). In *Fertility of the sea* (ed. J. D. Costlow) p. 537. Gordon and Breach, London.
—— (1973). *A. Rev. Microbiol.* **27**, 283.
—— (1977). In *A treatise on dinitrogen fixation* (eds. R. W. F. Hardy and W. S. Silver) Section III, p. 63. John Wiley, New York.
—— and ALEXANDER, G. (1971). *Freshwater Biol.* **1**, 389.
—— and PEARSON, H. W. (1970). *Proc. R. Soc.* **B175**, 293.
—— FITZGERALD, G.P., and BURRIS, R. H. (1968). *Arch. Mikrobiol.* **62**, 336.
—— —— —— (1970). *Proc. natn. Acad. Sci. U.S.A.* **66**, 1104.
—— MAGUE, T., FITZGERALD, G. P., and BURRIS, R. H. (1971). *New Phytol.* **70**, 497.
—— SAMPAIO, M. J., ISICHEI, A. O., and SYLVESTER–BRADLEY, R. (1978). In *Limitations and potentials for biological nitrogen fixation in the tropics* (eds. H. J. Döbereiner *et al.*) p. 41. Plenum Press, New York.
STRANDBERG, G. W. and WILSON, P. W. (1968). *Can. J. Microbiol.* **14**, 25.
SUBRAHMANYAN, R., RELWANI, L. L., and MANNA, G. B. (1964). *Curr. Sci.* **33**, 485.
TALPASAYI, E. R. S. (1963). *Cytologia* **28**, 76.
TAYLOR, B. F., LEE, C. C., and BUNT, J. S. (1973). *Arch. Mikrobiol.* **88**, 205.
VANCE, B. D. (1965). *J. Phycol.* **1**, 81.
VENKATARAMAN, G. S. (1975). In *Nitrogen fixation by free-living micro-organisms* (ed. W. D. P. Stewart) p. 207. Cambridge University Press.
VENRICK, E. L. (1974). *Limnol. Oceanogr.* **19**, 437.
WALSBY, A. E. (1978). In *Relationship between structure and function in the prokaryotic cell.* 28th Symp. Soc. Gen. Microbiol., p. 327. Cambridge University Press.
WÄRMLING, P. (1973). *Bot. mar.* **16**, 237.
WATANABE, A. and YAMAMOTO, L. Y. (1967). *Nature, Lond.* **214**, 738.
—— —— (1971). *Pl. Soil* Special Volume, p. 403.
WIEBE, W. J., JOHANNES, R. E., and WEBB, K. L. (1975). *Science, N.Y.* **188**, 257.
WOLFE, M. (1954). *Ann. Bot.* **18**, 299.
WOLK, C. P. (1973). *Bact. Rev.* **37**, 32.
WYNN-WILLIAMS, D. D. and RHODES, M. E. (1974). *J. appl. Bact.* **37**, 217.

2 Heterotrophic micro-organisms

V. Jensen

2.1 Introduction

The heterotrophic nitrogen-fixing micro-organisms all belong to the kingdom *Prokaryotae*, division II: *The Bacteria.* Previous claims of nitrogen fixation by various filamentous fungi and yeasts have been ruled out by thorough reinvestigations using more sophisticated methods (Millbank 1969). Nitrogen-fixing bacteria do not constitute a homogenous taxonomic group however. In the eighth edition of *Bergey's Manual* (1975) the bacteria are divided into nineteen major parts, and nitrogen-fixing species occur in at least eight of these parts.

The nitrogenase system is remarkably uniform in all species hitherto examined, but apart from this the nitrogen-fixing bacteria have little or nothing in common, except of course properties that are common to all bacteria. The ability to fix nitrogen is found in both Gram-positive and Gram-negative species, in cocci, rods, and spirilla, in both spore-forming and non-spore-forming, in chemo-organotrophic, chemolithotrophic, photo-organotrophic, and photolithotrophic forms, and in both aerobic, anaerobic, facultatively anaerobic, and micro-aerophilic species.

The number of genera and species cannot be stated exactly because of continuous changes caused by description of new species and revision of previous taxa, but heterotrophic nitrogen-fixing bacteria are found within at least twenty different genera and comprise at least twice as many species.

Until recently heterotrophic free-living nitrogen-fixing bacteria were considered to contribute only insignificantly to the nitrogen economy of natural and in particular agricultural systems. More recent studies using improved techniques, e.g. the acetylene reduction assay, have proved this to be an underestimate of their importance. This, together with recent investigations of nitrogen-fixing rhizosphere associations, have evoked a rush of renewed interest and excitement about heterotrophic nitrogen-fixing organisms resulting in the production of a flood of literature, of which a large part is concerned with environmental influences.

Within a limited space it is therefore not possible to give a comprehensive review of all the existing literature concerning the effects of the environment on heterotrophic nitrogen-fixing bacteria. It has been necessary to make a selection among the existing material. Further details can be found in other more specialized reviews, to which references will be given throughout the text.

2.2 Environmental factors influencing the natural occurrence of nitrogen-fixing bacteria

In the following the effects of the environment on nitrogen-fixing bacteria will be discussed at three different levels, (1) effects on the occurrence and natural distribution of the individual species of nitrogen-fixing bacteria, (2) effects on growth and general activities of the nitrogen-fixing organisms in their environments, and (3) effects on the regulatory mechanisms in the cells controlling the nitrogenase system. The main environmental factors are discussed one at a time, but of course it should be kept in mind that the individual factors do not act independently. Numerous interactions occur, which make the whole situation quite complex and it may often be difficult to distinguish between the effects of different environmental factors acting simultaneously.

Temperature

All known heterotrophic nitrogen-fixing bacteria belong to the mesophilic group. No truly psychrophilic or thermophilic species have been described. It should be emphasized, however, that specially adapted strains sometimes are able to grow and fix nitrogen at temperatures outside the normal range of the species. Table 2.1 shows that nitrogen-fixing bacteria can be found over the whole range from 0 to 60 °C, and that occurrence of nitrogen-fixing bacteria is excluded by temperature alone, only from the permanently hottest of the naturally occurring ecological niches (thermal springs).

The natural distribution of the individual species, however, may be affected to some degree by local or regional temperature conditions. Several species have rather high minimum temperatures (10–20 °C) and are excluded, therefore, from the permanently cold regions of the world, i.e. the arctic and antarctic regions and the larger part of the marine environment, where the temperature is always below 5 °C. In terrestrial environments within the temperate, subtropical, and tropical zones, on the other hand, temperature conditions seem to have

TABLE 2.1
Temperature requirements of nitrogen-fixing bacteria

Species	Minimum (°C)	Optimum (°C)	Maximum (°C)	Reference
Aquaspirillum fasciculus	20	30	40	Strength *et al.* (1976)
A. peregrinum	10	—	30	Pretorius (1963)
Azomonas spp	—	20–30	—	*Bergey's Manual* (1974)
Azotobacter spp	10	30	40–45	Jensen (1954)
	5–20	—	40–50	*Bergey's Manual* (1974)
Bacillus macerans	approx. 0	25–30	40–42	Mishustin and Shil'nikova (1971)
B. polymyxa	5–10	—	35–45	—
	—	30	—	Dalton (1974)
Beijerinckia spp	<16	25–30	37	Becking (1961*a*)
	10	20–30	35	*Bergey's Manual* (1974)
Clostridium spp	—	25–30	37–45	Mishustin and Shil'nikova (1971)
	—	27–39	—	Dalton (1974)
C. butyricum	—	25–37	—	*Bergey's Manual* (1974)
C. pasteurianum	—	37	—	—
Derxia gummosa	15	25–35	42	Jensen, Petersen, De, and Bhattacharya (1960)
Desulfotomaculum orientis	30	30–37	42	*Bergey's Manual* (1974)
D. ruminis	30	37	48	—
Desulfovibrio sp	0	25–30	44	—
Klebsiella spp	—	35–37	—	*Bergey's Manual* (1974)
Spirillum lipoferum	17	32–40	42	Day and Döbereiner (1976)

little influence on the distribution of the individual species of nitrogen-fixing bacteria.

According to Döbereiner (1968) at least six species of nitrogen-fixing bacteria, namely three *Beijerinckia*, two *Derxia*, and one *Azotobacter* species, are restricted to tropical environments, but this viewpoint can no longer be maintained in the case of *Beijerinckia*, which has been shown to be quite common in soils of southern Australia (Thompson 1968), and to occur as far north as Canada (Anderson 1966). Recently it was reported to occur even in the High Arctic (Devon Island) by Jordan and McNicol (1978). Furthermore, Becking (1961*a*) demonstrated that the distribution of *Beijerinckia* is determined by factors other than temperature. With regard to *Derxia* the observed restrictions in occurrence may be due to the use of unsuitable isolation methods as suggested by Döbereiner and Campelo (1971), and may have little to do with its temperature requirements.

The only well documented example of a nitrogen-fixing species strictly confined to the tropical zone is *Azotobacter paspali*, but this is caused by its strict dependence upon a tropical host plant, *Paspalum notatum* (Bahia grass), and not to a direct temperature effect (Döbereiner 1966, 1968, 1970; Döbereiner and Campelo 1971).

Mishustin and Yemtsev (1975) observed differences in distribution of nitrogen-fixing clostridia in northern and southern soils within the USSR, the most common species in northern soils being *Clostridium pasteurianum* followed by *C. butyricum*, while *C. acetobutylicum* was the most common species in the southern soils. Again this was probably not a result of differences in temperature, but rather due to differences in type and nutritional status of the northern and southern soils.

Humidity

Humidity is often a limiting factor for micro-organisms in terrestrial environments both within the soil and on aerial plant surfaces, and it may also limit the occurrence of nitrogen-fixing bacteria. According to Fedorow (1960) and Mishustin and Shil'nikova (1969, 1971) species of *Azotobacter* are very exacting with respect to soil moisture, and the absence of *Azotobacter* from soils, otherwise favourable for their development, can often be explained by too low moisture contents.

Similar observations were made by Döbereiner and Campelo (1971) for the genera *Beijerinckia* and *Derxia*, both showing increasing frequency with increasing soil moisture. Karaguishieva (1972), on the other hand, found that the nitrogen-fixing mycobacteria are capable of functioning with a comparatively low soil moisture content.

The very prolific development of micro-organisms, including nitrogen-fixing species, in the phyllosphere of plants growing in humid tropical environments can also be explained partly by the permanently favourable moisture conditions (Ruinen 1965, 1971, 1974). Outside the humid tropics the moisture conditions of the phyllosphere are extremely variable with periodic drying, and nitrogen-fixing species occur much more sparsely (Bessems 1973; Silver 1977). Baines and Millbank (1978) observed that the occurrence of nitrogen-fixing bacteria in decaying wood was limited by moisture conditions.

Hydrogen ion concentration

The optimum pH values and the tolerance ranges of a number of nitrogen-fixing bacteria are recorded in Table 2.2. All the recorded species have their pH optimum close to neutrality like most other bacteria, but the tolerance ranges show some differences. These differences may in part be responsible for the different natural distribution of some species, especially in soil.

One of the most pH-sensitive nitrogen-fixing genera is *Azotobacter*, which does not occur in soils with pH below 5.6–5.8 (Mishustin and Shil'nikova 1969). Several authors have reported increasing percentages of *Azotobacter*-positive soils with increasing pH up to pH 7.5–8.0 (Jensen 1965; Becking 1961*a*; Mishustin and Shil'nikova 1971; Ibrahim 1974). *Azotobacter beijerinckii* seems to tolerate slightly lower pH values than *A. chroococcum*, and particularly tolerant strains have been isolated (Tchan 1953). *Azotobacter paspali*, on the other hand, is probably the most exacting of the *Azotobacter* species, showing good development only within a very narrow pH range (Döbereiner 1970).

Beijerinckia and *Derxia* are more acid-tolerant than *Azotobacter*. Both have been observed in soils with pH as low as 4.0–4.5 (Becking 1961*a*; Thompson 1968; Döbereiner and Campelo 1971). The highest frequency was found around pH 5.5–6.5, and at higher pH values *Beijerinckia* was replaced with *Azotobacter*. Nitrogen-fixing clostridia are probably the most ubiquitous of all nitrogen-fixing soil bacteria, and they seem to be very little affected by soil pH (Di Menna 1966; Mishustin and Shil'nikova 1969; Brouzęs, Lasik, and Knowles 1969; Jurgensen and Davey 1970). Nitrogen-fixing mycobacteria which have been isolated from Russian turf-podzolic soils, are also rather acid-tolerant (Dalton 1974).

Spirillum lipoferum shows pH requirements rather similar to those of the *Azotobacter* group. Being a rhizosphere inhabitant, however, its natural occurrence is determined by the pH of the rhizosphere rather than the pH of the surrounding soil. It has been found to occur in the rhizosphere of plants growing in soils with pH below 5.0 (Day 1976). This is also true of *Azotobacter paspali*.

The closely related *Aquaspirillum* species have similar pH requirements (Strength, Isani, Linn, Williams, Vandermolen, Laughon, and Krieg 1976), but they are strictly aquatic organisms, and the pH of their

TABLE 2.2
Optimum pH and pH ranges of nitrogen-fixing bacteria

Species	Minimum	Optimum	Maximum	Reference
Aquaspirillum fasciculus	5.5	7.0	8.5	Strength *et al.* (1976)
A. peregrinum	5.0	—	8.0	Pretorius (1963)
Azomonas spp	4.5	7.0–7.5	9.0	*Bergey's Manual* (1974)
Azotobacter spp	5.5–6.0	7.5	9.0–10.0	Jensen (1954)
	4.5	7.0–8.0	9.0	Mishustin and Shil'nikova (1971)
Azotobacter paspali	—	6.7–7.0	—	Döbereiner (1970)
Beijerinckia spp	3.0	4.0–6.0	>9.0	Becking (1961*a*, 1974)
	3.0	—	9.5–10.0	*Bergey's Manual* (1974)
Clostridium spp	4.7	6.9–7.3	>8.5	Mishustin and Shil'nikova (1971)
Derxia gummosa	5.0	5.6–6.2	9.0	Jensen *et al.* (1960)
Desulfovibrio	—	7.0–8.0	—	Dalton (1974)
Klebsiella spp	—	7.2	—	*Bergey's Manual* (1974)
Spirillum lipoferum	5.5	6.8–7.8	9.0	Day and Döbereiner (1976)

natural environment will almost always be close to their optimum range. The pH of sea water is normally between 7.5 and 8.0, and most freshwater environments have pH values above 7.0, so the natural distribution of nitrogen-fixing bacteria in aquatic ecosystems will almost always be restricted by factors other than pH.

Salinity

Effects of salinity on the occurrence of nitrogen-fixing bacteria have been studied extensively in the USSR (see reviews by Fedorow (1960) and Mishustin and Shil'nikova (1969, 1971)). In general, *Azotobacter* was found to occur only sparsely in saline soils, even if particularly halotolerant strains sometimes could be isolated. Abd-el-Malek (1971) reported similar results in Egypt. *Azotobacter* counts in saline alkaline soils were very low compared to neighbouring drained and washed soils. Leaching of these saline soils resulted in significant increases in *Azotobacter* counts. Ibrahim (1974) compared Egyptian soils with varying contents of soluble salts and found a significant, negative correlation between *Azotobacter* counts and salt content. The Russian observation that strains isolated from saline soils often were more halotolerant than strains from non-saline soils was also confirmed. Mahmoud, El-Sawy, Ishac, and El-Safty (1978) found that an *Azotobacter* strain isolated from a strongly saline soil could grow with sodium chloride concentrations up to 0.75 M, but showed maximum nitrogen fixation with 0.14 M sodium chloride. An isolate from a non-saline soil showed maximum nitrogen fixation in the absence of sodium chloride.

Regular occurrence of *Azotobacter* in all depths of water and in the sediments in the Black Sea was reported by Psenin (1963), but later attempts on isolating *Azotobacter* from seawater have failed (Wood 1965; Wynn-Williams and Rhodes 1974; Werner, Evans, and Seidler 1974). Herbert (1975) isolated *Azotobacter* from shallow estuarine sediments, but they were probably of terrestrial origin, since they were unable to fix nitrogen in the presence of salt concentrations corresponding to sea water.

Other species of nitrogen-fixing bacteria, however, are able to tolerate the salinity of sea water. Werner, Evans, and Seidler (1974) isolated a strain of *Klebsiella pneumoniae* and a strain of *Enterobacter aerogenes*, which could grow and fix nitrogen in presence of sea water. Herbert (1975) found that five strains of nitrogen-fixing clostridia were completely unaffected by sodium chloride in the medium up to 0.5 M, and that twelve strains of *Desulfovibrio* had an obligate salt requirement for nitrogen fixation. Maximum rates occurred at salt concentrations between 0.2–0.4 M NaCl (see also Herbert, Brown, and Stanley 1977; Brown and Johnson 1977; Fogg 1978).

Patriquin (1978) isolated from roots of *Spartina* growing in salt-marshes a *Spirillum*-like nitrogen-fixing organism which showed good growth over the range 0.0 to 0.5 M NaCl, and a facultative anaerobe, which could not grow without added salt, but grew well in media containing 0.03 to 1.20 M NaCl.

Toxic substances

There are certain indications that the natural distribution of individual species of nitrogen-fixing bacteria may be affected by differences in resistance to toxic elements in the environment. Becking (1961*b*) found that *Beijerinckia* is far more tolerant to the high levels of iron, aluminium, and manganese characteristic of acid lateritic soils than *Azotobacter*, and this fact can partly explain the absence of *Azotobacter* and the presence of *Beijerinckia* in such soils. According to Mishustin and Shil'nikova (1969) the toxicity of aluminium to *Azotobacter* may also be one of the reasons for its absence from acid podzolic soils.

Several pesticides have been shown to be particularly toxic to nitrogen-fixing micro-organisms, e.g. Phoxim (Eisenhart 1975) and Dinoseb (Vlassak, Heremans, and Van Rossen 1976). Therefore, the possibility that uncritical use of pesticides in large doses may result in reduction or even eradication of populations of nitrogen-fixing species, should not be overlooked.

2.3 Environmental factors influencing growth and general activity of nitrogen-fixing bacteria

The mere demonstration of presence or absence of a nitrogen-fixing bacterium in a given environment is usually relatively simple. Measurements of *in situ* growth and activity of such organisms is a much more complicated matter, and often impossible without severely disturbing the natural system. Most information on the effects of the environment on growth and activity of nitrogen-fixing bacteria, therefore, is based on indirect evidence obtained from pure-culture studies or from measurements of activity in samples under laboratory conditions, and this should be kept in mind when the data are evaluated. The advent of the acetylene reduction assay has resulted in enormous advantages with regard to direct measurements of nitrogenase activity, but *in situ* measurements are still attended with great difficulties and uncertainty.

Temperature

While the ambient temperature apparently has little influence on the mere occurrence of most nitrogen-fixing micro-organisms (see p. 31), it is one of the major factors determining the immediate activity of the nitrogen-fixing populations present at a given time, both with regard to general metabolism and growth, and to nitrogenase activity (Balandreau, Millier, Weinhard, Ducerf, and Dommergues 1977). However, the response of nitrogen-fixing organisms to variations in temperature is rather complex.

First, a considerable strain variation is found, apparently depending upon adaptation to the ambient temperature. Strains from tropical environments often require higher growth temperatures than strains of the same species isolated from colder regions (Becking 1961*a*; Mishustin and Shil'nikova 1969; Döbereiner 1978). Second, cultures supplied with combined nitrogen often can grow at higher temperatures than cultures growing in nitrogen-deficient medium. On the other hand, cells grown at the optimum temperature and then tested by the acetylene reduction assay at both lower and higher temperatures have shown considerable nitrogenase activity outside the range, where the cultures could grow without combined nitrogen (Postgate 1974; Day and Döbereiner 1976). No biochemical explanation has been given for this behaviour.

When naturally occurring, mixed populations of nitrogen-fixing micro-organisms have been tested for nitrogenase activity at different temperatures, rather different responses have been obtained, as illustrated by the following examples. Brouzes and Knowles (1973) examined a Canadian sandy loam soil using an anaerobic incubation, and they found nitrogenase activity within the range from 10 to 45 °C with optimum activity at 37 °C. Blasco and Jordan (1976) working with a muskeg soil from northern Canada found optimum activity around 20 °C, while Waughman (1976) found that optimum activity in an English reotrophic peat occurred about 30 °C, and that nitrogenase was inactivated by exposure to temperatures above 37 °C. The activity was very low at temperatures below 15 °C. Granhall and Lindberg (1978) studied a Swedish coniferous forest soil and found no activity at 0 °C, a low but rather constant activity from 5 to 15 °C, and then a strong increase until a maximum activity was reached between 25 and 30 °C.

Nitrogenase activity in the rhizosphere of *Digitaria sanguinalis* showed an optimum temperature around 30–37 °C, and no activity at 10 and 45 °C (Barber and Evans 1976), while nitrogenase activity associated with Florida mangroves was highest at 28 °C and declined toward 37 °C, and some activity was observed at 8 °C (Zuberer and Silver 1978). Washed excised roots of *Spartina alterniflora* from the Atlantic coast of Nova Scotia exhibited maximum nitrogenase activity in the region of 25–30 °C, and no activity could be detected at 0 and 45 °C (Patriquin 1978).

There is no obvious trend in these results. Also some of the observed variation may be due to differences in the laboratory methods employed.

Humidity

Humidity is a critical factor in terrestrial environments, not only for the mere occurrence of nitrogen-fixing organisms as discussed above (p. 33), but perhaps even more for the growth and activity of these organisms in the habitats where they occur. This aspect has been studied especially in connection with nitrogen fixation in soil.

According to Abd-el-Malek (1971) and Mahmoud and Ibrahim (1971) the optimum moisture level for growth and activity of *Azotobacter* in Egyptian soils is between 50 and 75 per cent of the water holding capacity, with decreasing activity both at lower and higher water contents. In several different types of Australian soils increasing nitrogenase activity with increasing moisture contents was observed by Okafor and MacRae (1973) and Stefanson (1973).

In Canadian grassland Vlassak, Paul, and Harris (1973) found a highly significant correlation between the naturally occurring soil moisture contents and the nitrogen-fixing capacity. Kapustka and Rice (1976) likewise found a significant correlation between soil moisture and acetylene reduction in three types of grassland in central Oklahoma, and Koch and Oya (1974) observed a profound effect of soil moisture contents in Hawaiian pasture soils, both on rates of acetylene reduction and on numbers of nitrogen-fixing bacteria present, both showing maxima in the rainy season of December–March.

Balandreau *et al.* (1977) attempted to model a *Zea mays* rhizosphere system and found that soil-water content limited the rate of acetylene reduction up to a certain threshold value. The existence of such a threshold value has not been suggested in any of the previous references, but, of course, it must be assumed that a direct positive effect of increasing moisture content can occur only up to a certain moisture level. At still higher moisture contents indirect effects will predominate, caused by interactions between water content and other factors, particularly the rate of gas exchange between soil and atmosphere.

Oxygen tension

Since the discovery of the extreme oxygen sensitivity of the nitrogenase enzyme complex, innumerable studies have been performed on the effects of molecular oxygen on nitrogen fixation, including studies on enzyme preparations and intact cells, on pure and mixed cultures, and on more or less natural systems. The subject has been reviewed several times, most recently by Postgate (1974), Yates (1977), and La Rue (1977).

The sensitivity of purified nitrogenase preparations seems to be rather similar irrespective of the origin, and differences between individual species in regard to response to varying oxygen tensions depend chiefly upon more or less well developed mechanisms of protection of the nitrogenase system against oxygen damage.

The heterotrophic nitrogen-fixing bacteria include (i) obligate aerobes, which only grow and fix nitrogen in a normal atmosphere, (ii) micro-aerophiles, which require oxygen, but tolerate only low oxygen tensions when fixing nitrogen, (iii) facultative anaerobes, which can grow aerobically with combined nitrogen, but fix nitrogen only anaerobically or at low oxygen tensions, and (iv) obligate anarobes, which may be tolerant to low oxygen tensions, but never utilize molecular oxygen (see Table 2.3). These groups, however, are not very clearly separated from each other, and the response to oxygen of the individual species may depend to a large extent upon other environmental factors.

TABLE 2.3

Optimum oxygen tension required for nitrogen fixation in some species of aerobic nitrogen-fixing bacteria

Species	Optimum pO_2	Reference
Azotobacter chroococcum, N-limited	around 0.1	Drozd and Postgate (1970)
A. chroococcum, N-limited, grown at pO_2 0.09	0.05	Hill *et al.* (1972)
A. chroococcum, N-limited, grown at pO_2 0.55	0.2	
A. chroococcum, C-limited	0.025	Dalton and Postgate (1969)
A. paspali growing on Paspalum roots	0.04	Döbereiner *et al.* (1972)
Beijerinckia sp	0.15	Spiff and Odu (1973)
Corynebacterium autotrophicum	0.0036	Berndt *et al.* (1976)
Derxia gummosa	0.05	Hill (1971)
Mycobacterium flavum	0.05	Biggins and Postgate (1969)
	0.015–0.025	Yates (1977)
Spirillum lipoferum	0.01–0.02	Day and Döbereiner (1976)
	0.005–0.007	Burris, Okon, and Albrecht (1

The *Azotobacter* group is the most oxygen-tolerant of nitrogen-fixing bacteria. They have a very highly developed protection mechanism including two lines of defence—respiratory protection and conformational protection. The respiratory protection consists in an intensified respiratory activity consuming large amounts of oxygen. This mechanism requires a rich supply of available carbon compounds, and it does not function, therefore, in carbon-limited populations. The conformational protection consists in a 'switch off' from the normal, active, and oxygen-sensitive conformation of the enzyme to an inactive and insensitive conformation, from which it can be 'switched on' again. This second defence line seems to work in situations where the respiratory protection is insufficient to avoid oxygen damage. More details about these mechanisms can be found in the reviews by Postgate (1974), Yates and Jones (1974), and Yates (1977).

The group of micro-aerophilic nitrogen-fixing bacteria is a much more diverse group, including *Derxia gummosa, Mycobacterium flavum, Corynebacterium autotrophicum*, the nitrogen-fixing spirilla, the nitrogen-fixing methane oxidizers, certain strains of *Rhizobium*, and the unclassified bacterium of Jensen and Holm (1975). Characteristic of this group of organisms is lack of either oxygen protection mechanisms or very inefficient protection mechanisms capable of coping with only low concentrations of oxygen. Most, if not all, can grow normally in air when supplied with combined nitrogen, but they can fix nitrogen only at reduced oxygen tensions.

The facultative anaerobes, including the nitrogen-fixing species of the enterobacteria and of the genus *Bacillus*, and the obligate anaerobes, including species of *Clostridium, Desulfovibrio*, and *Desulfotomaculum* all grow best at completely anaerobic conditions when supplied with molecular nitrogen as the nitrogen source, although some of them can tolerate low oxygen tensions; they are probably devoid of any oxygen protection mechanisms.

There can be little doubt that nitrogen fixation in natural systems is almost always favoured by low oxygen tensions. Measurements of nitrogenase activity in samples (e.g. of soil) incubated at different oxygen tensions have given rather diverse results however (Brouzes, Mayfield, and Knowles 1971; Döbereiner, Day, and Dart 1972, 1973; Spiff and Odu 1972; Okafor and MacRae 1973; Gotto and Taylor 1976; Waughman 1976; Zuberer and Silver 1978). This probably depends upon the varying ability of the systems studied of creating or maintaining a favourable internal oxygen tension independent on the surrounding atmosphere. The favourable effect of high water contents, evt. waterlogging, of soil on its nitrogenase activity is probably also a consequence of the creation of partially or totally anaerobic conditions within the soil.

Hydrogen ion concentration

While many investigations have been concerned with the effects of pH on the natural occurrence of nitrogen-fixing bacteria (see p. 34) and on the pH requirements of pure cultures (Table 2.2), little information is available on the effects of pH on the activity of nitrogen-fixing bacteria when growing in their natural environment. Nor is such information easily obtained, since the pH of the soil, for example, interacts strongly with other factors, such as availability of various nutrients, and it is always very difficult to distinguish direct effects of pH from indirect effects caused by these interactions.

On basis of experiences from pure-culture studies one should expect a neutral to slightly alkaline reaction to be optimal for nitrogenase activity in most environments. Liming of acid soil has also been found to stimulate nitrogen-fixing bacteria and to increase nitrogen fixation (Moore 1966; Carneiro and Döbereiner 1968), but it is uncertain whether it is a direct or indirect effect of the increased pH. O'Toole and Knowles (1973) found that high concentrations of glucose under aerobic conditions suppressed nitrogenase activity, and they argue that this effect was caused by the lowering of pH from the original 6.8 to 5.6.

Carbon and energy nutrition

A wide range of low molecular weight carbon compounds can be utilized as carbon and energy substrates by heterotrophic nitrogen-fixing bacteria. Practically all of them can utilize some organic acids, e.g. pyruvate, lactate, or malate, and simple alcohols such as ethanol. Mono- or disaccharides can also be utilized by most species, but with some important exceptions—the nitrogen-fixing spirilla (Okon, Albrecht, and Burris 1976*a*; Strength *et al.* 1976), *Desulfovibrio* (Postgate 1974), *Mycobacterium flavum*, and related mycobacteria (Biggins and Postgate 1969).

Simple aromatic compounds, e.g. benzoate, can be utilized as a carbon source by *Azotobacter chroococcum* (Jensen 1954), and the range of carbon substrates utilized for nitrogen fixation also includes hydrocarbons: both gaseous such as methane (Davis, Coty, and Stanley 1964; De Bont and Mulder 1974) and long-chain alkanes like hexadecane (Rivière, Oudot, Jonquères, and Gatellier 1974). Polysaccharides and other high molecular weight compounds are generally not utilized by nitrogen-fixing bacteria, with one important exception, namely starch which is utilized by *Azotobacter chroococcum* (Jensen 1954), but generally not by *A. beijerinckii* (Jensen and Petersen 1955).

All the metabolizable carbon compounds are easily decomposable substances, which also can be utilized and decomposed by numerous non-nitrogen-fixing organisms. Consequently, these compounds are usually found only in very low concentrations in natural environments, and their availability will very often, therefore, be a limiting factor for nitrogen fixation both in terrestrial and aquatic ecosystems (Jensen 1965; Stewart 1969; Burris 1977; Brown and Johnson 1977; Herbert *et al.* 1977).

Also, it has often been demonstrated experimentally that addition of of decomposable carbon substances usually results in considerable increases in nitrogenase activity in widely different natural systems (Fehr, Pang, Hedlin, and Cho 1972; Okafor and MacRae 1973; Spiff and Odu 1972; Murphy 1975; Blasco and Jordan 1976; Pearson and Taylor 1978; Zuberer and Silver 1978). Of course, exceptions from this general rule have also been observed. Thus Hanson (1977*b*) found that glucose addition to a Spartina salt marsh system suppressed nitrogen fixation on a long-term basis, but this was explained as an indirect effect caused by impoverished nitrogen nutrition of the plants resulting in suppressed productivity and root exudation.

Combined nitrogen

Presence of combined nitrogen in the environment affects the activity of nitrogen-fixing bacteria in different ways: (i) by interfering directly with the regulatory mechanisms controlling synthesis and activity of nitrogenase (see pp. 46 and 47), (ii) by allowing the nitrogen-fixing bacteria to grow under conditions where they are unable to fix molecular nitrogen, or (iii) by interfering with the competitive ability of the nitrogen-fixing bacteria in relation to non-nitrogen-fixing organisms.

It is an old experience that nitrogen-fixing bacteria, when given the choice, will make use of assimilable combined nitrogen in preference to molecular nitrogen, so that nitrogen fixation will take place only when the level of utilizable combined nitrogen is below a certain threshold value. This threshold value, however, may be rather different for different organisms and under different circumstances.

Hüser (1965) found in experiments with beech litter and humus, amended with glucose and ammonium chloride, that nitrogen fixation decreased strongly when the ratio between available carbon and available combined nitrogen was below 70, and it stopped completely when this ratio reached 26. Spiff and Odu (1972) studied effects of up to 6 mM ammonium nitrogen added to two Nigerian soils and observed a 50 per cent inhibition of nitrogenase activity in one soil, but no effect on the other. Rinaudo (1973) showed a depressive effect of added ammonium nitrogen in paddy soils, but only at concentrations higher than 3 to 6 mM.

Kalininskaya, Rao, Volkova, and Ippolitov (1973) examined the effects of various combinations of straw and ammonium sulphate on nitrogenase activity in a rice field soil. Straw alone caused an increase, and ammonium sulphate alone a decrease in activity. A combination with a C/N ratio of about 50 caused a transient decrease followed by an increase, and a combination with C/N ratio of about 100 gave the highest activity of all treatments, higher than the same amount of straw alone. Rather similar results were obtained by Murphy (1975) using combinations of glucose and ammonium nitrate. He found that 3 mM nitrogen was inhibitory in the presence of 1 per cent glucose in all soils, whereas 1.5 mM nitrogen was stimulatory in some soils and inhibitory in others. These results confirm the statement of Mishustin and Shil'nikova (1969) that small doses of combined nitrogen may have a stimulatory effect on nitrogen fixation.

Knowles and Denike (1974) found that the threshold value for suppression of nitrogenase activity in a sandy loam soil was 2 mM ammonium nitrogen in the presence of 1 per cent glucose, but only 0.2 mM in presence of 0.05 per cent glucose, and they did not observe any stimulatory effects. Hanson (1977*a*, *b*), working with a *Spartina* salt marsh ecosystem, showed that addition of organic nitrogen compounds had very little influence on nitrogenase activity. Inorganic nitrogen, both ammonia and nitrate, on the other hand, inhibited nitrogenase activity, but only at concentrations many times higher than those normally found in the salt marsh system studied. On a long-term basis, however, nitrogenase activity was increased by monthly additions of ammonium nitrate, probably owing to increased plant productivity and root exudation. Most of these reported results seem to support the view of Stewart (1969) that levels of combined nitrogen in natural, unfertilized ecosystems are rarely high enough to inhibit nitrogenase activity.

The rather different levels of mineral nitrogen found to be inhibitory in different environments, may be explained at least partially by differences in ammonium-fixation capacity of the different soils, as suggested by Hanson (1977*a*). The stimulatory effect of low nitrogen doses observed in some cases, is probably related to a similar phenomenon encountered in pure cultures of micro-aerophilic or facultatively anaerobic nitrogen-fixing bacteria. Initiation of growth of these bacteria in media with oxygen tensions above optimal is greatly facilitated by small amounts of combined nitrogen, as this allows the bacteria to grow and consume oxygen, so that the oxygen tension is reduced to a level more favourable to nitrogen fixation.

It seems to be a general rule that the tolerance range with regard to various environmental factors is wider when the bacteria utilize combined nitrogen than when they fix molecular nitrogen. This is particularly conspicuous with regard to oxygen tension, but it has also been found that some organisms, e.g. *Azotobacter* can grow at a lower pH value with combined than with molecular nitrogen (Burk, Lineweaver, and Horner 1934), and that several nitrogen-fixing bacteria can grow at higher temperatures when supplied with combined nitrogen than with molecular nitrogen (Postgate 1974).

Other mineral nutrients

Nitrogen-fixing bacteria have a specific requirement for molybdenum because of the content of this element in the Mo–Fe-component of nitrogenase. Molybdenum is also necessary for assimilation of nitrate, but in much smaller amounts (Mulder 1948), and it is apparently not required at all when the bacteria are supplied with reduced nitrogen compounds (ammonia). Becking (1962) observed a large variation in the demand for molybdenum in different species of *Azotobacter* and *Beijerinckia*. It was high for *A. chroococcum*, very low for *A. vinelandii*, and with the *Beijerinckia* species at various levels in between. The molybdenum requirement of *Beijerinckia* was larger at low pH than at a neutral reaction.

The molybdenum supply is probably sufficient to secure optimum growth of nitrogen-fixing organisms in most natural environments, but it may be low in acid, lateritic soils. Döbereiner (1977) found that leaf spraying with ammonium molybdate increased nitrogenase activity in the rhizosphere of field-grown maize, but this seems to be the only direct observation of the effect of molybdenum fertilization on free-living heterotrophic nitrogen-fixing organisms.

Iron is found in both components of nitrogenase, as well as in ferredoxin, and it is required in larger amounts when the bacteria are fixing nitrogen than when they utilize combined nitrogen (Postgate 1974). Iron is probably seldom or never a limiting factor for nitrogen fixation in natural environments.

The calcium requirements of nitrogen-fixing bacteria have been investigated in detail, e.g. by Jensen (1948), Norris and Jensen (1958), Becking (1961*b*), and Jakobson, Zell, and Wilson (1962). All the common *Azotobacter* species, namely *A. chroococcum, A. vinelandii*, and *A. beijerinckii*, have been found to need calcium for growth both in presence and absence of combined nitrogen. This was also true for *Azomonas insignis*. *A. macrocytogenes*, *A. agilis*, and *Beijerinckia indica*, on the other hand, have no detectable calcium requirements. This may be one reason for the good development of *Beijerinckia* and the absence of *Azotobacter* in the calcium-deficient lateritic soils (Becking 1961*b*, 1974).

There is also a conspicuous difference between *Azotobacter* and *Beijerinckia* with regard to phosphate requirements. *Azotobacter* is generally considered to be very exacting in its requirement for phosphate; this is probably the reason for its absence from some soils otherwise favourable for its development (Jensen 1965; Mishustin and Shil'nikova 1969, 1971). Phosphate fertilization has also been shown in some cases to stimulate growth and nitrogen fixation by *Azotobacter* in soil (Mishustin and Shil'nikova 1969).

Beijerinckia is much more tolerant towards large variations in the phosphate level. It can grow very well in very low phosphate concentrations, but it can also tolerate concentrations as high as 2 per cent phosphate in the medium (Becking 1962).

The specific requirement for sodium chloride of some of the marine species of nitrogen-fixing bacteria is mentioned above (p. 36).

2.4 Environmental factors controlling the nitrogenase system

The physiological and genetic control of the nitrogenase system involves rather complex mechanisms, which are not yet fully understood (see, e.g. Yates 1977). The mechanisms controlling synthesis of nitrogenase and those controlling the activity of already formed nitrogenase will be discussed separately in the two following sections.

Regulation of nitrogenase synthesis

The synthesis of nitrogenase is influenced by several factors in the environment including ammonia and some other nitrogen compounds, molecular oxygen, and molybdenum (Yates 1977). Perhaps the most clear-cut example is the effect of the ammonium ion (NH_4^+), which has been shown to repress nitrogenase synthesis in all types of nitrogen-fixing organisms (Hill, Drozd, and Postgate 1972; Postgate 1974). The degree of repression caused by a particular concentration of ammonia, however, depends upon the population density of the nitrogen-fixing organisms. At low densities, only low concentrations of ammonia are needed for complete repression. At low ammonia concentrations nitrogen fixation and ammonia assimilation may occur at the same time. The intracellular pool of ammonia formed during nitrogen fixation will also partly repress nitrogenase synthesis, and the highest concentration of nitrogenase has been found in cultures growing in a nitrogen-free atmosphere with ammonia in limiting concentration as the nitrogen source (Dalton and Postgate 1969; Dalton and Mortenson 1972).

The regulatory protein involved in ammonia repression is probably glutamine synthetase, which is adenylylated in the presence of ammonia. The non-adenylylated form induces nitrogenase synthesis, and the conversion into the adenylylated form causes repression of nitrogenase synthesis (Shanmugam and Valentine 1975; Yates 1977).

Other nitrogen sources, e.g. nitrate, may cause a similar repression as ammonia, at least in some species, but it is not clear whether they act directly or only after conversion to ammonia (Drozd, Tubb, and Postgate 1972; Burns and Hardy 1975).

Nitrogenase is destroyed by molecular oxygen, and nitrogenase synthesis in the presence of destructive concentrations of oxygen would be a wasteful process. It might be expected, therefore, that a mechanism preventing such waste would exist, and evidence has been presented that molecular oxygen actually functions as a repressor of nitrogenase synthesis by a mechanism independent of ammonia repression (Brill, Steiner, and Shah 1974; St. John, Shah, and Brill 1974; Yates 1977; Eady, Issack, Kennedy, Postgate, and Ratcliffe 1978).

It has been suggested that molecular nitrogen should be necessary as an inducer of nitrogenase synthesis, but it has never been proved, and such mechanisms seem very unlikely because natural environments completely devoid of molecular nitrogen hardly exist (Dalton and Mortenson 1972). Molybdenum, on the other hand, has been shown to be necessary for induction of nitrogenase synthesis (Brill *et al.* 1974; Cardenas and Mortenson 1975; Yates 1977), so this element is required not only as a direct component of nitrogenase, but also as an inducer of its synthesis.

Regulation of nitrogenase activity

Ammonia has been assumed to act not only as a repressor of nitrogenase synthesis but also as an inhibitor of nitrogenase activity, but the evidence presented is not very convincing (Drozd *et al.* 1972; Postgate 1974; Brown, MacDonald-Brown, and Meers 1974; Yates 1977). The only physiological compound which has been proved to control nitrogenase activity, at least *in vitro*, is ADP. It seems that at least ten times more ATP than ADP is necessary for maximal activity of nitrogenase, and the activity is completely inhibited when the ratio of ATP to ADP is 0.5. Therefore, it is impossible for nitrogenase to consume more than 60–70 per cent of the ATP supplied, and when the ATP supply within the cell is low, the remaining ATP can be diverted to more critical cell functions (Dalton and Mortenson 1972; Yates 1977).

A number of non-physiological substrates of nitrogenase, such as carbon monoxide and acetylene, can inhibit nitrogen fixation competitively, and this may play a role in certain natural environments (Dalton and Mortenson 1972).

2.5 The biotic environment

In nature, microbial populations seldom live alone. Other micro-organisms, plants, and animals will usually be present, and the conditions for a given organism will be affected not only by the non-living components of the ecosystem, but also by the other organisms composing the biotic environment. The interactions between the inhabitants of an ecosystem may be direct as in parasitism and predation, but most interactions are indirect: one organism changes, by its metabolic activities, the environment so that it becomes either a richer or a poorer environment for another organism.

Interactions with other micro-organisms

A large number of papers have appeared reporting commensalistic or mutalistic interactions between nitrogen-fixing bacteria and other micro-organisms (see review by Jensen and Holm 1975). The micro-organisms involved include both other bacteria, yeasts, filamentous fungi, and protozoa, and such associations may be of great ecological significance. In most cases the stimulatory effect of the associative organisms can be ascribed to (i) modification of an otherwise unavailable substrate, (ii) production of growth-promoting substances, or (iii) removal of inhibitory substances.

The best example of an unavailable carbon substrate which is made available by microbial action, is cellulose (Jensen 1940, 1941*a*, *b*). Other examples have been reported by Richards (1939) and Fedorow and Kalininskaya (1961). Stimulatory effects caused by excretion of growth-promoting substances have been described by Dommergues and Mutaftschiev (1965), Sandrak (1971), and Jensen and Holm (1975) amongst others.

Inhibitory substances, which can be removed by associated micro-organisms, include molecular oxygen (Jensen 1941*b*; Mulder, Lie, and Woldendrop 1969; Evans, Campbell, and Hill 1972; Darbyshire 1972; Line and Loutit 1973; Jensen and Holm 1975). Another possibility consists of removal of nitrogenous substances, which would otherwise inhibit nitrogenase activity either directly or indirectly (Mulder *et al.* 1969; Döbereiner 1974).

Antagonistic effects on nitrogen-fixing bacteria have not attracted the same attention as the stimulatory effects, but it is obvious that competition for carbon and energy substrates must be a very important factor for the activity of nitrogen-fixing organisms. In general, these organisms will be competitive only in environments where combined nitrogen is a limiting factor.

Antagonism caused by production of inhibitory substances, e.g. antibiotics, is probably also important in natural environments, and it has been suggested as an explanation for the restricted occurrence of nitrogen-fixing bacteria in the rhizosphere of temperate plants (Fedorow and Savkina 1961; Döbereiner 1974). Antagonistic effects on *Azotobacter* caused by lowering the pH below the minimum value have also been reported (Chan, Basavanand, and Liivak 1970).

The rhizosphere of higher plants

The rhizoplane–rhizosphere region of higher plants is an environment which is significantly different from the surrounding soil. It is characterized in general by a strongly increased microbial activity, but the individual types of micro-organisms may be stimulated to a very different degree, some of them even occurring in lower numbers in the rhizosphere than in the surrounding soil (or completely absent).

The differences between the rhizoplane–rhizosphere and the surrounding soil are caused mainly by (i) excretion of root exudates and sloughing off of dead root tissue, (ii) selective absorption of plant nutrients by the root, and (iii) root respiration resulting in decreased oxygen tension and increased concentration of carbon dioxide. A secondary result of these processes will often be an altered pH.

Selective stimulation of nitrogen-fixing bacteria in the rhizosphere region has been found to occur in some cases and not in others, apparently depending upon the nature of the root exudates. These again are determined chiefly by the plant species, the developmental stage of the plant, and the climatic conditions (see reviews by

Döbereiner (1974) and Knowles (1977)).

Both numbers and activity of nitrogen-fixing bacteria are generally low in the rhizosphere of plants growing in temperate regions, and the associations between plants and bacteria seem rather non-specific (Nelson, Barber, Tjepkema, Russell, Powelson, Evans, and Seidler 1976; Neira and Döbereiner 1977; Barber, Tjepkema, and Evans 1978). The most common types of nitrogen-fixing bacteria are facultatively anaerobic members of the enterobacteria and the genus *Bacillus*. *Azotobacter* is usually not stimulated by the rhizosphere environment of temperate plants, and it is not established easily in the rhizosphere even after inoculation (Döbereiner 1974).

The rhizosphere of subtropical and tropical plants appears to be a much more favourable environment for nitrogen-fixing bacteria than that of temperate plants, owing to the high photosynthetic activity of these plants, which results in large amounts of carbon-rich and nitrogen-poor root exudates. This is especially true of plants possessing the C_4-dicarboxylic acid photosynthetic pathway like most tropical grasses (Döbereiner and Day 1975; Neyra and Döbereiner 1977).

Plant–bacterium associations also seem to be more specific under tropical conditions, the most specific being the *A. paspali–P. notatum* association. *A. paspali* is easily grown in pure culture, but in nature it seems to be strictly confined to the rhizosphere of a few ecotypes of *P. notatum* (Döbereiner 1970, 1977). The rhizosphere of sugar-cane, which is characterized by a high content of sucrose, has been found to stimulate selectively other members of the *Azotobacter* family. In Brasil, the predominant nitrogen-fixing bacterium under sugar-cane is *Beijerinckia* sp (Döbereiner 1961), whereas this genus was absent from sugar-cane fields in Egypt, being replaced here by *Azotobacter*, especially *A. vinelandii* (Eid 1978).

In most other tropical plants *Spirillum lipoferum* is the predominant nitrogen-fixing organism in the rhizosphere. This species seems to be especially well adapted to the prevailing conditions in the environment, i.e. the low oxygen tension, the neutral pH, and the rich supply of carboxylic acids as the carbon source (Döbereiner, Marriel, and Nery 1976; Döbereiner 1978).

The phyllosphere of higher plants

The aerial surfaces of higher plants, the phyllosphere or phylloplane, represents an environment which may harbour large populations of micro-organisms, especially in the humid tropical regions where moisture conditions are favourable most of the time. In temperate regions the development of phyllosphere populations may be severely hampered by periodic drying (Ruinen 1974, 1975; Silver 1977; Knowles 1977).

The supply of nutrients to phyllosphere micro-organisms is mainly based on leachates and exudates from the leaves, and the composition of this material is probably not very different from that of root exudates. In fact, a close analogy exists between the phyllosphere and rhizosphere habitats. Very large numbers of *Azotobacter* and *Beijerinckia* cells have been found on the surface of *Citrus* and *Cacao* leaves in Surinam (Ruinen 1965, 1975) and on mulberry leaves in India (Vasantharajan and Bhat 1968). The sheath liquid of tropical grasses is also a habitat rich in nitrogen-fixing bacteria (Ruinen 1971; Bessems 1973). In contrast, the nitrogen-fixing microflora in the phyllosphere of temperate plants usually is restricted to small numbers of enterobacteria or bacilli.

Animal habitats

The intestinal tract of mammals harbours a very rich microbial community, fermenting large amounts of organic matter, and it is not at all surprising to find putatative nitrogen-fixing bacteria among the many bacterial species occurring in this environment. It is an anaerobic habitat, often with a very low redox potential, so only obligate or facultative anaerobes can be active. The most common nitrogen-fixing organisms in this environment are members of the enterobacteria, but nitrogen-fixing species of *Bacillus*, *Clostridium*, or *Desulfotomaculum* may also occur (Bergersen and Hipsley 1970; Elleway, Sabine, and Nicholas 1971; Jones and Thomas 1974).

Growth conditions in the intestinal tract are favourable for nitrogen-fixing bacteria as far as carbon nutrition, pH, temperature, and moisture are concerned, but an important obstacle to their activity is the presence of combined nitrogen, especially ammonia, in rather high concentrations. According to Hobson (1976) the ammonia concentration in the rumen is usually in excess of the requirements of the bacteria. Therefore, nitrogenase synthesis must be assumed to be completely repressed, and the low activities which have been measured directly on rumen contents by acetylene reduction assays are probably due to bacteria present in the ingested feed (Granhall and Ciszuk 1971; Hobson, Summers, Postgate, and Ware 1973; Jones and Thomas 1974).

While the nitrogen-fixation process in the intestinal tract of mammals thus seems quite insignificant as far as the nitrogen nutrition of the animal is concerned, conditions are different in certain lower animals, where apparently a much more favourable environment for nitrogen fixation can be established. Rather high rates of acetylene reduction have been measured on a number of different genera and species of wood-eating termites both in America and Australia, and the activities seem high enough to account for a significant proportion of the nitrogen needed by the animal (Benemann 1973; Breznak, Brill, Mertins, and Coppel 1973). The nitrogenase activity was also shown to be closely related to the nitrogen content of the diet. Nitrogen fixation takes place in the hindgut of the termites, which functions as a fermentation chamber rather analogous to the rumen, and the bacteria responsible have been identified as members of the enterobacteria (French, Turner, and Bradbury 1976; Potrikus and Breznak 1977; Eutick, O'Brien, and Slaytor 1978).

Still higher activities have been measured on four species of marine shipworms, which also use wood as their main or sole nutrient source (Carpenter and Culliney 1975). In one case, the activity was high enough to result in a doubling of the nitrogen content of the animal in only 1.4 days. However, the most remarkable feature was perhaps the ability of the organism responsible (a facultatively anaerobe, *Spirillum*-like bacterium) to hydrolyse cellulose.

Citernesi, Neglia, Seritti, Lepidi, Filippi, Bagnoli, Nuti, and Galluzzi (1977) have isolated nitrogen-fixing bacteria, including species of *Bacillus*, *Clostridium*, *Enterobacter*, and *Klebsiella*, from the intestinal tract of a number of different soil animals, but it was not established whether these bacteria were permanent inhabitants of the animals, or just soil bacteria accidentally passing through the intestinal tract with the ingested food.

2.6 Concluding remarks

The effects of the total environment on heterotrophic nitrogen-fixing bacteria are highly complex because of interactions between the individual environmental factors. Examples of such interactions have been given in the preceding paragraphs. Because of this complexity it is impossible as yet to analyse in detail the effects of the total environment. Much more research is needed before it becomes possible to formulate accurate mathematical models, which could describe these relationships and make predictions feasible.

In most cases, at least in terrestrial environments, a high activity of heterotrophic nitrogen-fixing bacteria is considered desirable, and this can be achieved only by (i) a pH not far from neutrality, (ii) a rather high moisture level and a correspondingly low oxygen tension, (iii) a rich supply of carbon and energy sources, and (iv) with a low concentration of available combined nitrogen.

Most of these prerequisites can be fulfilled by suitable measures. In many environments, however, especially in the intestinal tract of mammals or in highly productive agricultural soils, the main obstacle to a high activity of heterotrophic nitrogen-fixing bacteria is the presence of combined nitrogen in amounts sufficient to repress the synthesis of nitrogenase.

Certainly, it is possible by selection of the proper mutants to obtain strains of nitrogen-fixing bacteria whose control mechanisms are non-functional, and which would be able, therefore, to fix nitrogen in environments rich in combined nitrogen. The main difficulty in this connection is that a genetic change of this sort almost certainly would interfere with the competitive ability of these bacteria in relation to other micro-organisms, and it would probably prove impossible to establish stable populations of such bacteria in natural environments.

Another way out of the dilemma of how to maintain high plant production and a high activity of asymbiotic heterotrophic nitrogen-fixing bacteria, would be to try to separate the nitrogen nutrition of the plants from that of the bacteria, either in space or time. This might be possible, for example by supplying nitrogen fertilizers to the aerial parts of plants growing in a nitrogen-deficient soil, or perhaps by creating a closer association between nitrogen-fixing bacteria and plant roots with the bacteria growing inside the root tissue separated from the surrounding soil environment.

A different approach would be to try to produce a nitrogen source which could be utilized only by the plants and not by the nitrogen-fixing bacteria. Such a compound could be present in the soil in high concentrations without interfering directly with the nitrogen fixation process, and it would also be useful from other points of view.

These speculations may seem fanciful and non-realizable, and perhaps the problem about how to create favourable environments for nitrogen-fixing bacteria will be solved in a quite different and yet unforeseen manner. The problem, however, is important enough to justify intensive research. Its solution would represent a valuable contribution to a world, deficient in both energy and food.

References

Abd-el-Malek, Y. (1971). *Pl. Soil* Special Volume, 423–42.

Anderson, G. R. (1966). *J. Bact.* **91**, 2105–6.

Baines, E. F. and Millbank, J. W. (1978). *Ecol. Bull., Stockh.* **26**, 193–8.

Balandreau, J. P., Millier, C. R., Weinhard, P., Ducerf, P., and Dommergues, Y. R. (1977). In *Recent developments in nitrogen fixation* (ed. W. Newton, J. R. Postgate, and C. Rodriguez-Barrueco) pp. 523–9. Academic Press, London.

Barber, L. E. and Evans, H. J. (1976). *Can. J. Microbiol.* **22**, 254–60.

—— Tjepkema, J. D., and Evans, H. J. (1978). *Ecol. Bull., Stockh.* **26**, 366–72.

Becking, J. H. (1961*a*). *Pl. Soil* **14**, 49–81.

—— (1961*b*). *Pl. Soil* **14**, 297–322.

—— (1962). *Pl. Soil* **16**, 171–201.

—— (1974). *Soil Sci.* **118**, 196–212.

Benemann, J. R. (1973). *Science, N.Y.* **181**, 164–5.

Bergerson, F. J. and Hipsley, E. H. (1970). *J. gen Microbiol.* **60**, 61–5.

Bergey's manual of determinative bacteriology (1974). 8th edn. Williams and Wilkins, Baltimore.

Berndt, H., Ostwal, K.-P., Lalucat, J., Schumann, C., Mayer, F., and Schlegel, H. G. (1976). *Archs Mikrobiol.* **108**, 17–26.
Bessems, E. P. M. (1973). *Agric. Res. Repts.* 786. [Doctoral Thesis, Wageningen.]
Biggins, D. R. and Postgate, J. R. (1969). *J. gen Microbiol.* **56**, 181–93.
Blasco, J. A. and Jordan, D. C. (1976). *Can. J. Microbiol.* **22**, 897–907.
Breznak, J. A., Brill, W. J., Mertins, J. W., and Coppel, H. C. (1973). *Nature, Lond.* **244**, 577–9.
Brill, W. J., Steiner, A. L., and Shah, V. K. (1974). *J. Bact.* **118**, 986–9.
Brouzes, R. and Knowles, R. (1973). *Soil Biol. Biochem.* **5**, 223–9.
—— Lasik, J., and Knowles, R. (1969). *Can. J. Microbiol.* **15**, 899–905.
—— Mayfield, C. I., and Knowles, R. (1971). *Pl. Soil* Special Volume, 481–94.
Brown, C. M. and Johnson, B. (1977). *Adv. aquat. Microbiol.* **1**, 49–114.
—— MacDonald-Brown, D. S., and Meers, J. L. (1974). *Adv. microb. Physiol.* **11**, 1–52.
Burk, D., Lineweaver, H., and Horner, C. K. (1934). *J. Bact.* **27**, 325–40.
Burns, R. C. and Hardy, R. W. F. (1975). *Nitrogen fixation in bacteria and higher plants.* Springer Verlag, Berlin.
Burris, R. H. (1977). In *Genetic engineering for nitrogen fixation* (ed. A. Hollaender) pp. 9–17. Plenum Press, New York.
—— Okon, Y., and Albrecht, S. L. (1977). In *Genetic engineering for nitrogen fixation* (ed. A. Hollaender) pp. 445–50. Plenum Press, New York.
Cardenas, J. and Mortenson, L. E. (1975). *J. Bact.* **123**, 978–84.
Carneiro, A. M. and Döbereiner, J. (1968). *Pesq. Agropec. Bras.* **3**, 151–7.
Carpenter, E. J. and Culliney, J. L. (1975). *Science, N.Y.* **187**, 551–2.
Chan, E. C. S., Basavanand, P., and Liivak, T. (1970). *Can. J. Microbiol.* **16**, 9–16.
Citernesi, U., Neglia, R., Seritti, A., Lepidi, A. A., Filippi, C., Bagnoli, G., Nuti, M. P., and Galuzzi, R. (1977). *Soil Biol. Biochem.* **9**, 71–2.
Dalton, H. (1974). *C. R. C. Crit. Rev. Microbiol.* **3**, 183–220.
—— and Mortenson, L. E. (1972). *Bact. Rev.* **36**, 231–60.
—— and Postgate, J. R. (1969). *J. gen. Microbiol.* **56**, 307–19.
Darbyshire, J. F. (1972). *Soil Biol. Biochem.* **4**, 371–6.
Davis, J. B., Coty, V. F., and Stanley, J. P. (1964). *J. Bact.* **88**, 468–72.
Day, J. M. (1976). In *Biological nitrogen fixation in farming systems of the tropics* (ed. A. Ayanaba and P. J. Dart) pp. 273–88. Wiley, New York.
—— and Döbereiner, J. (1976). *Soil Biol. Biochem.* **8**, 45–50.
De Bont, J. A. M. and Mulder, E. G. (1974). *J. gen. Microbiol.* **83**, 113–21.
Di Menna, M. E. (1966). *N. Z. Jl agric. Res.* **9**, 218–26.
Döbereiner, J. (1961). *Pl. Soil* **15**, 211–17.
—— (1966). *Pesq. Agropec. Bras.* **1**, 357–65.
—— (1968). *Pesq. Agropec. Bras.* **3**, 1–6.
—— (1970). *Zbl. Bakt.* **124**, 224–30.
—— (1974). In *The biology of nitrogen fixation* (ed. A. Quispel) pp. 86–120. North-Holland, Amsterdam.
—— (1977). In *Recent developments in nitrogen fixation* (ed. W. Newton, J. R. Postgate, and C. Rodriguez-Barrueco) pp. 513–22. Academic Press, London.
—— (1978). *Ecol. Bull., Stockh.* **26**, 343–52.
—— and Campelo, A. B. (1971). *Pl. Soil* Special Volume, 457–70.
—— and Day, J. M. (1975). In *Nitrogen fixation by free-living micro-organisms* (ed. W. D. P. Stewart) pp. 39–56. Cambridge University Press.
—— —— and Dart, P. J. (1972). *J. gen Microbiol.* **71**, 103–16.
—— —— —— (1973). *Soil Biol. Biochem.* **5**, 157–9.
—— Marriel, I. E., and Nery, M. (1976). *Can. J. Microbiol.* **22**, 1464–73.

DOMMERGUES, Y. and MUTAFTSCHIEV, S. (1965). *Annls Inst. Pasteur, Paris* **109** (Suppl. No. 3), 112–20.
DROZD, J. and POSTGATE, J. R. (1970). *J. gen. Microbiol.* **60**, 427–9.
—— TUBB, R. S., and POSTGATE, J. R. (1972). *J. gen. Microbiol.* **73**, 221–32.
EADY, R. E., ISSACK, R., KENNEDY, C., POSTGATE, J. R., and RATCLIFFE, H. D. (1978). *J. gen Microbiol.* **104**, 277–85.
EID, M. A. (1978). Studies on the free N_2-fixing bacteria in the rhizosphere of sugar cane. M.Sc. Thesis, Faculty of Agriculture, Cairo University.
EISENHART, A. R. (1975). *Tidsskr. PlAvl* **79**, 254–58.
ELLEWAY, R. F., SABINE, J. R., and NICHOLAS, D. J. D. (1971). *Archs Mikrobiol.* **76**, 277–91.
EUTICK, M. L., O'BRIEN, R. W., and SLAYTOR, M. (1978). *Appl. Environ. Microbiol.* **35**, 823–8.
EVANS, H. J., CAMPBELL, N. E. R., and HILL, S. (1972). *Can. J. Microbiol.* **18**, 13–21.
FEDOROW, M. W. (1960). *Biologische Bindung des Atmosphärischen Stickstoffs*. VEB Deutschen Verlag der Wissenschaften, Berlin.
—— and KALININSKAYA, T. A. (1961). *Mikrobiology [eng. transl.]* **30**, 833–40.
—— and SAVKINA, E. A. (1961). *Mikrobiology [eng. transl.]* **29**, 622–5.
FEHR, P. I., PANG, P. C., HEDLIN, R. A., and CHO, C. M. (1972). *Agron. J.* **64**, 251–4.
FOGG, G. E. (1978). *Ecol. Bull., Stockh.* **26**, 11–19.
FRENCH, J. R. J., TURNER, G. L., and BRADBURY, J. F. (1976). *J. gen. Microbiol.* **95**, 202–6.
GOTTO, J. W. and TAYLOR, B. F. (1976). *Appl. Environ. Microbiol.* **31**, 781–3.
GRANHALL, U. and CISZUK, P. (1971). *J. gen. Microbiol.* **65**, 91–3.
—— and LINDBERG, T. (1978). *Ecol. Bull., Stockh.* **26**, 178–92.
HANSON, R. B. (1977*a*). *Appl. Environ. Microbiol.* **33**, 596–602.
—— (1977*b*). *Appl. Environ. Microbiol.* **33**, 846–52.
HERBERT, R. A. (1975). *J. exp. mar. Biol. Ecol.* **18**, 215–25.
—— BROWN, C. M., and STANLEY, S. O. (1977). In *Aquatic microbiology* (ed. F. A. Skinner and J. M. Shewan) Soc. appl. Bact. Symp. No. 6, pp. 161–77. Academic Press, London.
HILL, S. (1971). *J. gen. Microbicl.* **67**, 77–83.
—— DROZD, J. W., and POSTGATE, J. R. (1972). *J. appl. Chem. Biotechnol.* **22**, 541–58.
HOBSON, P. N. (1976). *The microflora of the rumen (Patterns of progress)*. Meadowfield Press, Shildon, Co. Durham.
—— SUMMERS, R., POSTGATE, J. R., and WARE, D. A. (1973). *J. gen. Microbiol.* **77**, 225–26.
HUSER, R. (1965). *Pl. Soil* **23**, 236–46.
—— (1970). *Z. PflErnähr. Düng Bodenk.* **127**, 49–56.
IBRAHIM, A. N. (1974). *Acta agron. hung.* **23**, 113–18.
JAKOBSON, A., ZELL, E. A., and WILSON, P. W. (1962). *Arch. Mikrobiol.* **41**, 1–10.
JENSEN, H. L. (1940). *Proc. Linn. Soc. N.S.W.* **65**, 543–56.
—— (1941*a*). *Proc. Linn. Soc. N.S.W.* **66**, 89–106.
—— (1941*b*). *Proc. Linn. Soc. N.S.W.* **66**, 239–49.
—— (1948). *Proc. Linn. Soc. N.S.W.* **72**, 299–310.
—— (1954). *Bact. Rev.* **18**, 195–214.
—— (1965). In *Soil nitrogen* (ed. W. W. Bartholomew and F. E. Clark), Amer. Soc. Agron. Monogr. Vol. 10, pp. 436–80.
—— PETERSEN, E. J., DE, P. K., and BHATTACHARYA, R. (1960). *Arch. Mikrobiol.* **36**, 182–95.
JENSEN, V. and HOLM, E. (1975). In *Nitrogen fixation by free-living micro-organisms* (ed. W. D. P. Stewart) pp. 101–19. Cambridge University Press.

—— and Petersen, E. J. (1955). *Royal Veterinary and Agricultural College Yearbook 1955*, 107–26.
Jones, K. and Thomas, J. G. (1974). *J. gen. Microbiol.* **85**, 97–101.
Jordan, D. C. and McNicol, P. J. (1978). *Appl. Environ. Microbiol.* **35**, 204–5.
Jurgensen, M. F. and Davey, C. B. (1970). *Soils Fertil.* **33**, 435–46.
Kalininskaya, T. A., Rao, V. R., Volkova, T. N., and Ippolitov, L. T. (1973). *Microbiology* [*eng. transl.*] **42**, 426–9.
Kapustka, L. A. and Rice, E. L. (1976). *Soil Biol. Biochem.* **8**, 497–503.
Karaguishieva, D. (1972). *Mikrobiology* [*eng. transl.*] **43**, 136–7.
Knowles, R. (1977). In *A treatise on dinitrogen fixation.* Sect. IV (ed. R. W. F. Hardy and A. H. Gibson) pp. 33–83. Wiley, New York.
—— and Denike, D. (1974). *Soil Biol. Biochem.* **6**, 353–8.
Koch, B. L. and Oya, J. (1974). *Soil Biol. Biochem.* **6**, 363–7.
La Rue, T. A. (1977). In *A treatise on dinitrogen fixation.* Sect. III (ed. R. W. F. Hardy and W. S. Silver) pp. 19–62. Wiley, New York.
Line, M. A. and Loutit, M. W. (1973). *J. gen. Microbiol.* **74**, 179–80.
Mahmoud, S. A. Z., el-Sawy, M., Ishac, Y. Z., and el-Safty, M. M. (1978). *Ecol. Bull., Stockh.* **26**, 99–109.
—— and Ibrahim, A. N. (1971). *Acta agron. hung.* **20**, 157–63.
Millbank, J. R. (1969). *Arch. Mikrobiol.* **68**, 32–9.
Mishustin, E. N. and Shil'nikova, V. K. (1969). In *Soil biology. Reviews of research*, pp. 65–124. Natural Resources Research IX. UNESCO.
—— —— (1971). *Biological fixation of atmospheric nitrogen.* Macmillan, London.
—— and Yemtsev, V. T. (1975). In *Nitrogen fixation by free-living microorganisms* (ed. W. D. P. Stewart) pp. 29–38. Cambridge University Press.
Moore, A. W. (1966). *Soils Fertil.* **29**, 113–29.
Mulder, E. G. (1948). *Pl. Soil* **1**, 94–119.
—— Lie, T. A., and Woldendorp, J. W. (1969). In *Soil biology. Reviews of research*, pp. 163–208. Natural Resources Research IX. UNESCO.
Murphy, P. M. (1975). *Proc. R. Irish Acad. B* **75**, 453–64.
Nelson, A. D., Barber, L. E., Jepkema, J., Russell, S. A., Powelson, R., Evans, H. J., and Seidler, R. J. (1976). *Can. J. Microbiol.* **22**, 523–30.
Neyra, C. A. and Döbereiner, J. (1977). *Adv. Agron.* **29**, 1–38.
Norris, J. R. and Jensen, H. L. (1958). *Arch. Mikrobiol.* **31**, 198–205.
Okafor, N. and MacRae, I. C. (1973). *Soil Biol. Biochem.* **5**, 181–6.
Okon, Y., Albrecht, S. L., and Burris, R. H. (1976*a*). *J. Bact.* **127**, 1248–54.
—— —— —— (1976*b*). *J. Bact.* **128**, 592–7.
O'Toole, P. and Knowles, R. (1973). *Soil Biol. Biochem.* **5**, 789–97.
Patriquin, D. G. (1978). *Ecol. Bull., Stockh.* **26**, 20–7.
Pearson, H. W. and Taylor, R. (1978). *Ecol. Bull., Stockh.* **26**, 69–82.
Postgate, J. R. (1974). In *The biology of nitrogen fixation* (ed. A. Quispel) pp. 663–86. North Holland, Amsterdam.
—— and Campbell, L. L. (1966). *Bact. Rev.* **30**, 732–8.
Potrikus, C. J. and Breznak, J. A. (1977). *Appl. Environ. Microbiol.* **33**, 392–9.
Pretorius, W. A. (1963). *J. gen. Microbiol.* **32**, 403–8.
Psenin, L. N. (1963). In *Symposium on marine microbiology* (ed. C. H. Oppenheim) pp. 383–91. C. C. Thomas, Springfield, Ill.
Richards, E. H. (1939). *J. agric. Sci., Camb.* **29**, 302–5.
Rinaudo, G. (1973). *Rev. Ecol. Biol. Sol.* **11**, 149–68.
Rivière, J., Oudot, J., Jonquères, J., and Gatellier, J. (1974). *Ann. Agron.* **25**, 633–44.
Ruinen, J. (1965). *Pl. Soil* **22**, 375–94.

—— (1970). *Pl. Soil* **33**, 661–71.
—— (1971). In *Ecology of leaf surface microorganisms* (ed. T. F. Preece and C. H. Dickinson) pp. 567–79. Academic Press, London.
—— (1974). In *The biology of nitrogen fixation* (ed. A. Quispel) pp. 121–67. North Holland, Amsterdam.
—— (1975). In *Nitrogen fixation by free-living microorganisms* (ed. W. D. P. Stewart) pp. 85–100. Cambridge University Press.
SANDRAK, N. A. (1971). *Mikrobiology [eng. transl.]* **40**, 603–7.
SHANMUGAM, K. T. and VALENTINE, R. C. (1975). *Science, N.Y.* **187**, 919–24.
SILVER, W. S. (1977). In *A treatise on dinitrogen fixation*, Sect. III (ed. R. W. F. Hardy and W. S. Silver) pp. 153–84. Wiley, New York.
SPIFF, E. D. and ODU, C. T. I. (1972). *Soil Biol. Biochem.* **4**, 71–7.
STEFANSON, R. C. (1973). *Soil Biol. Biochem.* **5**, 869–80.
STEWART, W. D. P. (1969). *Proc. R. Soc.* **B172**, 367–88.
ST. JOHN, R. T., SHAH, V. K., and BRILL, W. J. (1974). *J. Bact.* **119**, 266–9.
STRENGTH, W. J., ISANI, B., LINN, D. M., WILLIAMS, F. D., VANDERMOLEN, G. E., LAUGHON, B. E., and KRIEG, N. R. (1976). *Int. J. Syst. Bact.* **26**, 253–68.
TCHAN, Y. T. (1953). *Proc. Linn. Soc. N.S.W.* **78**, 83–4.
THOMPSON, J. P. (1968). *IX Int. Cong. Soil Sci.* Vol. 2, pp. 129–39.
VANCURA, V., ABD-EL-MALEK, Y., and ZAYED, M. N. (1965). *Folia Microbiol.* **10**, 224–9.
VASANTHARAJAN, V. N. and BHAT, J. V. (1968). *Pl. Soil* **28**, 258–67.
VLASSAK, K., HEREMANS, K. A. H., and VAN ROSSEN, A. R. (1976). *Soil Biol. Biochem.* **8**, 91–3.
—— PAUL, E. A., and HARRIS, R. E. (1973). *Pl. Soil* **38**, 637–49.
WERNER, D., EVANS, H. J., and SEIDLER, R. J. (1974). *Can. J. Microbiol.* **20**, 59–64.
WAUGHMAN, G. J. (1976). *Can. J. Microbiol.* **22**, 1561–6.
WOOD, E. J. F. (1965). *Marine microbial ecology*. Chapman and Hall, London.
WYNN-WILLIAMS, D. D. and RHODES, M. E. (1974). *J. appl. Bact.* **37**, 203–16.
YATES, M. G. (1977). In *Recent developments in nitrogen fixation* (ed. W. Newton, J. R. Postgate, and C. Rodriguez-Barrueco) pp. 219–70. Academic Press, London.
—— and JONES, C. W. (1974). *Adv. Microb. Physiol.* **11**, 97–135.
ZUBERER, D. A. and SILVER, W. S. (1978). *Appl. Environ. Microbiol.* **35**, 567–75.

3 Non-leguminous root-nodule symbioses with actinomycetes and *Rhizobium*

A. D. L. Akkermans and C. van Dijk

3.1 Introduction

Root nodules of non-legumes are perennial coralloid structures in which specific micro-organisms grow and fix dinitrogen symbiotically. Because of limited growth and subsequent low nitrogen consumption by the nodule tissue, the main part of the nitrogen fixed is transported to the shoots, leaves, and roots of the host plant. This process enables nodulated plants to grow in soils poor in nitrogen.

According to the nature of the microsymbionts (or endophytes), non-leguminous nitrogen-fixing root nodules can be classified into three groups, viz. *Alnus*-type nodules with *Frankia* (*Actinomycetales*), *Cycas*-type nodules with *Nostoc* or *Anabaena* (*Cyanobacteria*), and *Parasponia*-type nodules with *Rhizobium* as the microsymbiont.

The majority of the non-leguminous nodules belong to the *Alnus*-type (Allen, Allen, and Klebesadal 1964; Bond 1974), also described as an actinomycete symbiosis (Schaede 1962), while more recently the name actinorhiza has been introduced (Fessenden 1979). They are found in a variety of perennial angiosperms, including trees (e.g. *Alnus* and *Casuarina* spp), shrubs (e.g. *Hippophaë* and *Ceanothus* spp), and occasionally herbs (*Datisca* spp). The generic name *Frankia* is used for actinomycetes involved in *Alnus*-type symbioses (Becking 1970*b*; Lechevalier and Lechevalier 1979), but knowledge of these endophytes is still too restricted to justify taxonomic treatment at the species level.

The nodules are considered as infection-induced lateral roots which show restricted apical growth and profuse branching under the influence of a persistent *intracellular* symbiont (*endosymbiont*) present in the cortex region. The microsymbiont is recognized as an actinomycete by development of thin prokaryotic hyphae which follow apical growth of the nodule lobe by infection of young cortical cells. Intracellular growth of hyphae is followed by the development of club-shaped or vesicular structures at the tips of hyphae which may play an essential role in the symbiosis. Sometimes spore formation is observed. Restricted apical growth and profuse branching of the nodule lobes eventually leads to the formation of more or less spherical, multi-lobed organs of loose (open coralloid) to very compact (closed coralloid) structures.

Similar coralloid structures are also found on roots of *Parasponia* spp and *Cycadales*. Contrary to *Alnus*-type and *Parasponia*-type symbioses, the blue–green algae in cycads generally develop *intercellularly (ectosymbiontic)*. *Cycas*-type nodules are induced even prior to infection, although longevity of these structures apparently depends on symbiotic activity.

Coralloid nodule structures are not specific to these nitrogen-fixing symbioses but are also found among mycorrhizal roots of some *Pinus* species (Zak 1973) and among heterotrophic orchids such as *Didimoplexis minor* (Burgeff 1932). The wide range of host–endophyte combinations which give rise to the coralloid nodule structure is an illustration of its non-specific nature.

As compared with the *Rhizobium*–legume symbioses, non-leguminous nodule symbioses have received relatively little attention in spite of the important role which these species play—for example as pioneers in natural ecosystems.

Present knowledge on ecological and physiological aspects of non-leguminous root-nodule symbiosis has been reviewed by Trappe, Franklin, Tarrant, and Hansen (1968); Bond (1971, 1974, 1976*a*); Becking (1970*a*, 1975, 1977); Tarrant and Trappe (1971); Millbank (1974, 1977): Quispel (1974*a*,*b*); Silvester (1976); Akkermans (1978a); and Torrey and Tjepkema (1979). In this chapter special attention is paid to the occurrence of *Alnus*-type and *Parasponia*-type nodules, their ecological importance and the extra-nodular behaviour of *Frankia*.

3.2 Survey of nodule-bearing plant species

Although the occurrence of nodulated non-leguminous plants has been known for a long time, new discoveries in the last 20 years have almost doubled the number of genera in which nodulated species were found to occur (Bond 1976*a*). Furthermore it can be expected that intensive investigations, especially among pioneer plants on nitrogen-poor soils in tropical regions, will be fruitful in the discovery of new nodule-bearing species.

Cycas-type nodules

Root-nodule symbioses with cyanobacteria are only known to occur in the *Cycadales*, which are considered as primitive gymnosperm relicts of an ancient tropical flora. Cycads (seed ferns) have had a widespread distribution in Triassic, Jurassic, and Cretaceous periods. Many species are known as fossils, and there are good reasons to believe that this type of nodule symbioses has played an important role in the nitrogen cycle of tropical areas in the past. The present-day flora in the tropics and subtropics contains about 90 species of the *Cycadales*, classified in nine genera (Table 3.1). Most of these species occur in low numbers in tropical ecosystems. Only a few, e.g. *Macrozamia* spp in Australia occur as pioneers in quantities large enough to play an important role in the nitrogen cycle. *Cycas*-type nodules have received little attention and many questions concerning the infection process and existence of host–endophyte specificity are still unsolved. Review papers on *Cycas*-type nodules have been presented by Millbank (1974, 1977); Silvester (1976); Becking (1977); and Akkermans (1978*a*).

Parasponia-type nodules

The general assumption that *Rhizobium*-induced nodules are restricted to legumes was maintained for many years, despite considerable lack of systematic research of non-leguminous root systems. This omission became evident when Trinick (1973) discovered that nodules in certain Ulmaceae are induced by a *Rhizobium* sp. So far as is known, this *Rhizobium* symbiosis with a non-legume seems to be restricted to *Parasponia* spp (Table 3.1). It should be noted, however, that only few Ulmaceae have been screened for nodulation. Of particular interest is the apparent absence of nodules in *Trema* spp which are taxonomically closely related to *Parasponia* spp. Until now three species of *Parasponia* have been found to bear nodules, viz. *P. rugosa*, *P. parviflora*, and *P. andersonii*, all native to the south-eastern part of Asia. Other examples of *Rhizobium*—non-legume symbioses have also been reported, e.g. in certain Zygophyllaceae (Sabet 1946; Mostafa and Mahmood 1951; Athar and Mahmood 1972). These observations need reinvestigation, despite the criticism on it in literature (Allen and Allen 1950; Becking 1977).

Alnus-type nodules

Nodule symbioses with actinomycetes occur in at least 140 plant species which belong to 8 families and 17 genera (Table 3.2). Most of these species grow in nitrogen-poor soils. The geographic distribution of actinorhizal plant species is widespread, especially in the temperate regions. In the tropics, actinorhizas are generally restricted to higher altitudes, with a few exceptions, e.g. *Casuarina* spp.

In certain families, viz. Casuarinaceae, Coriariaceae, Elaeagnaceae, and Myricaceae, all species are found to be nodulated, while in other families nodulation seems to be restricted to certain taxa (Bond 1976*a*).

TABLE 3.1
Occurrence of Cycas-*type and* Parasponia-*type root nodules*[a]

Order	Family	Tribe	Genus[b]	Total no. of species	No. of species examined for nodulation: with nodules	No. of species examined for nodulation: without nodules
Cycas-type root nodules						
Cycadales	Cycadaceae		**Bowenia**	2	2	0
			Ceratozamia	4	4	0
			Cycas	±20	±20	0
			Dioon	3–4	3–4	0
			Encephalartos	15–20	15–20	0
			Macrozamia	16	16	0
			Microcycas	1	1	0
			Stangeria	1	1	0
			Zamia	30–40	30–40	0
Parasponia-**type root nodules**						
Urticales	Ulmaceae			±230		
		Ulmeae	*Phyllostylon*		0	0
			Holoptelea		0	1
			Planera		0	0
			Ulmus[c]	20–25	0	1

TABLE 3.1 (*cont.*)

Order	Family	Tribe	Genus[b]	Total no. of species	No. of species examined for nodulation: with nodules	No. of species examined for nodulation: without nodules
		Celtideae	*Celtis*		0	3
			Pteroceltis		0	0
			Ampelocera		0	0
			Zelkova		0	0
			Hemiptelea		0	0
			Trema	10–15[c]	0	3
			Parasponia	5[c]	3	0
			Aphananthe		0	0
			Mirandaceltis		0	0
			Gironniera		0	1
			Chaetacme		0	0
			Lozanella		0	0
	Cannabiaceae		2	3	0	0
	Moraceae		75	1850	0	0
	Urticaceae		49	1900	0	13[d]
	Barbeyaceae		1	1	0	0
	Eucommiaceae		1	1	0	0

[a] Taxonomy according to Hutchinson (1964, 1967) with the exception of Cycadales.
[b] Genera in bold type denote nodule-bearing species.
[c] Soepadmo (1977).
[d] Akkermans (1978*b*).

TABLE 3.2

Occurrence of Alnus-*type root nodules*[a]

Order	Family	Tribe	Total no. of genera per family	Genus[b]	Total no. of species	Number of species examined for nodulation: with nodules	Number of species examined for nodulation: without nodules
Casuarinales	Casuarinaceae		1	**Casuarina**	45[d]	18[e]	0
Coriariales	Coriariaceae		1	**Coriaria**	15[d]	13[e]	0
Fagales	Betulaceae		2	**Alnus**	35[d]	33[e]	0
				Betula	±40	0	0[b]
	Fagaceae		8	—	±400	0	0[b]
	Corylaceae		4	—	50	0	0[b]
Cucurbitales	Cucurbitaceae		126	—	±1280	0	0
	Begoniaceae		5	—	>500	0	0
	Datiscaceae		3	**Datisca**	2	2	0
				Octomelis	2	0	0
				Tetramelis	1	0	0
	Caricaceae		5	—	65	0	0
Myricales	Myricaceae		2	**Myrica**	35[d]	20[e]	0
				Comptonia	1	1	0
Rosales	Rosaceae		124	—	±3375		
		Rubieae		**Rubus**	>200	1	3
		Dryadeae		*Fallugia*	1	0	0
				Cowania	3–4	0	0
				Novosieversia	1	0	0
				Sieversia	25	0	0
				Geum	±70	0	0
				Acomastylis	15	0	0
				Orthurus	2	0	0

TABLE 3.2 (*cont.*)

Order	Family	Tribe	Total no. of genera per family	Genus[b]	Total no. of species	No. of species examined for nodulation: with nodules	No. of species examined for nodulation: without nodules
				Dryas	4	3	0
				Coluria	7	0	0
				Waldsteinia	6–7	0	0
				Purshia	2	2	0
				Chamaebatia	2	0	0
		Cercocarpeae		*Coleogyne*	1	0	0
				Potaninia	1	0	0
				Cercocarpus	20	3[e]	0
	Dichapetalaceae		4	—	±110	0	0
	Calycanthaceae		2	—	7	0	0
Rhamnales	Heteropyxidaceae		?	—	?	0	0
	Vitaceae		?	—	?	0	0
	Elaeagnaceae		3	**Elaeagnus**	45[d]	14	0
				Hippophaë	3[d]	1[e]	0
				Shepherdia	3[d]	2[e]	0
	Rhamnaceae[c]		58				
		Rhamneae[g]		*Sageretia*		0	1
				Rhamnus		0	3
				Ceanothus	55[d]	31	0
				Colubrina		0	2
				Alphitonia		0	1
				Pomaderris		0	5
				Spyridium		0	1
				Cryptandra		0	1

Continued

TABLE 3.2 (*cont.*)

Order	Family	Tribe	Total no. of genera per family	Genus[b]	Total no. of species	No. of species examined for nodulation: with nodules	No. of species examined for nodulation: without nodules
		Zizypheae[g]		*Zizyphus*		0	5
				Berchemia		0	1
				Maesopsis		0	1
		Ventilagineae		*Ventilago*		0	2
				Smythea		0	1
		Colletieae[g]		*Talguenea*		0	1
				Trevoa		1	0
				Retanilla		0	1
				Discaria	10[d]	5	2
				Colletia	17[d]	3	0
		Gouanieae		*Gouania*		0	2

[a] Taxonomy according to Hutchinson (1964, 1967) with the exception of the Elaeagnaceae (Engler and Prantl 1894) and the Rhamnaceae (Suessenguth 1953).

[b] Genera in bold type denote nodule-bearing species.

[c] Details on nodulation of the Rhamnaceae are given in Table 3.3.

[d] According to Willis (1966).

[e] According to Becking (1977).

[f] Search for nodulation not reported, but absence of nodules generally known.

[g] Genera of the Rhamnaceae, not yet searched for nodulation are:

Rhamneae—*Scutia, Oreorhamnus, Macrorhamnus, Schistocarpeae, Hovenia, Noltea, Emmenosperma, Tzellemtinia, Ampelozizyphus, Cormonema, Hybosperma, Phylica, Nesiota, Lasiodiscus, Siegfriedia, Trymalium.*

Zizypheae—*Paliurus, Sarcomphalus, Condalia, Condaliopsis, Microrhamnus, Lamellissepalum, Reynosia, Karwinskia, Auerodendron, Phyllogeiton, Doerpfeldia, Chaydaia, Berchemiella, Rhamnella, Krugiodendron, Rhamnidium, Dallachya.*

Colletieae—*Adolphia, Kentrorhamnus.*

Gouanieae—*Pleuranthodes, Reissekia, Helinus, Crumenaria.*

Fig. 3.1. Root nodules of (a) *Alnus glutinosa*, (b) *Ceanothus velutinus*, and (c) *Datisca cannabina*. (After Chaudhary (1979).)

In the Betulaceae only *Alnus* spp are found to bear nodules (Fig. 3.1(a)). Much less information is available on other plant families. Among the Rosaceae nodulated specimens of *Dryas* spp have been found in Alaska and Canada (Lawrence, Schoenike, Quispel, and Bond 1967; Bond 1976*a*). Of particular interest is the apparent absence of nodulation in *D. octopetala* from Europe and Japan. Other nodule-bearing Rosaceae include *Cercocarpus* (Vlamis, Schultz, and Biswell 1964; Hoeppel and Wollum 1971; Youngberg and Hu 1972), *Purshia* (Wagle and Vlamis 1961; Webster, Youngberg, and Wollum 1967; Krebill and Muir 1974; Bond 1976*b*; Dalton and Zobel 1977), and *Rubus* (Bond 1976*a*; Soemartono, personal communication; Becking 1979*b*). The occurrence of nodulation in *Purshia* seems to be irregular, depending on soil conditions. Since similar effects may occur in other Rosaceae not yet investigated, it is of great importance that future studies on nodulation include many specimens over a wide range of natural habitats. Special attention should be paid to *Cowania* spp (Bond 1976*a*,*b*) because *C. mexicana* var. *stansburyana* (syn. *C. stansburyana*) hybridizes with *Purshia tridentata* and *P. glandulosa* (Stutz 1972; Deitschman, Jorgensen, and Plummer 1974).

In *Rubus* only one species has yet been found to bear *Alnus*-type nodules, viz. *R. ellipticus* (Java, Indonesia). Although absence of nodulation in other *Rubus* spp has occasionally been reported (Bond 1976*a*; Becking 1979) this widespread genus needs further investigation.

In the Rhamnaceae nodulation has been confirmed for *Ceanothus* (Fig. 3.1(b)), *Trevoa*, *Discaria*, and *Colletia* spp (5 per cent of all known genera in this family) which all belong to the tribes Rhamneae and Colletieae (Tables 3.2 and 3.3). However, about 70 per cent of all genera in this family have never been examined for nodulation. Special attention should be paid to species belonging to the Colletieae. Nodulation has first been described in *Discaria toumatou* (Morrison and Harris 1958) and about 20 years later in *Colletia*, which is closely related to *Discaria* (Bond 1976*a*). These species are frequently present in botanical gardens and nodulated specimens of both genera were recently found in Argentina (Medan and Tortosa 1976), England (Bond 1976*a*), Indonesia, and in The Netherlands (Akkermans, see Table 3.3). Nitrogen-fixing activity of two *Colletia* spp from Indonesia has been confirmed (Akkermans 1978*b*). Many species of these genera have not been investigated. Absence of nodules has occasionally been reported (Table 3.3). Actinorhizas have also been observed in *Trevoa trinervis* in Chili (Rundel and Neel 1978).

Another plant family which includes nodule-bearing species is the Datiscaceae. The presence of root nodules in *D. cannabina* has first been described by Trotter (1902, 1906) and subsequently by Montemartini (1905, 1906) and Severini (1922). Surprisingly these old botanical and microbiological papers have not been reported in any of the more recent

TABLE 3.3
Occurrence of Alnus-*type root nodules in Rhamnaceae*

Tribe	Genus searched	Species searched for nodulation	Presence of nodules[a]	References[b]
Rhamneae	*Alphitonia*	*A. excelsa* (Fenzl.) Benth.	–	2
	Ceanothus	31 spp	+	2
	Colubrina	*C. asiatica* (L.) Brongn.	–	1
		C. nepalensis Don.	–	1
	Cryptandra	*C. amara Sm.*	–	2
	Pomederris	*P. apetala* Labill.	–	2
		P. elliptica Labill.	–	2
		P. kumeraho A. Cunn.	–	2
		P. lanigera (Andr.) Sims	–	2
		P. phylicifolia Lodd. ex Link	–	2
	Rhamnus	*R. frangula*	–	2
		R. pentapomica Parker	–	2,3
		R. prinoides L'Hérit.	–	2
	Sageretia	*S. filiformis* (Roth.) Don.	–	1
	Spyridium	*S. ulicinum* Benth.	–	2
Zizypheae	*Berchemia*	*B. floribunda* Wall.	–	1
	Zizyphus	*Z. horsfieldii* Bl.	–	1
		Z. jujuba Lam.	–	2,3
		Z. oenoplia (L.) Mill.	–	1
		Z. rotundifolia Lam.	–	2,3
		Z. rufula Miq.	–	1
	Maesopsis	*M. eminii* Engl.	–	1,2
Ventilagineae	*Smythea*	*S. lanceata* (Tul.) Summ.	–	1
	Ventilago	*V. madraspatana* Gaertn.	–	1
		V. oblongifolia Bl.	–	1
Colletieae	Colletia	**C. armata** Miers.	+	2,6
		C. cruciata Gill ex Hook	+	1,2,4,6
		C. spinosa	+	1,4,6
	Discaria	**D. americana** Gill et Hook	+	4
		D. articulata (Phil.) Miers	–	4
		D. nana (Clos) Weberb.	+	4
		D. pubescens Druce	–	2
		D. serratifolia (Vent.) Benth et Hook	+	4
		D. toumatou Raoul.	+	2,5
		D. trinervis (Gill.) Reiche	+	4
	Trevoa	T. trinervis	+	7
	Talguenea	*T. quinquenervis* (Gill., Hook.). Johnst.	–	7
	Retanilla	*R. ephedra* (Vent.) Brongn.	–	7
Gouanieae	*Gouania*	*G. leptostachia* DC	–	1
		G. longispicata Engl.	–	2
Total	16			

[a] Presence (+) or absence (–) of nodules reported.

[b] 1: Akkermans (1978*b*); 2: Bond (1976*a*); 3: Khan (1971); 4: Medan and Tortosa (1976); 5: Morrison and Harris (1958); 6: Akkermans (observations in botanical gardens and tree nurseries in The Netherlands); 7: Rundel and Neel (1978).

papers on symbiotic nitrogen fixation until Chaudhary (1977, 1979) rediscovered nodulation in *D. cannabina* (Fig. 3.1(c)). Other examples of rediscovery of nodulated plants are presented on page 94. The family Datiscaceae includes only three genera. The genus *Datisca* contains two species with disjunct distribution, viz. *D. cannabina* in the Himalayas and *D. glomerata* in California. Chaudhary (1979) recently discovered nitrogen-fixing nodules in the latter species, indicating that nodulation in *Datisca* is a generic character. It is still not known whether the other genera, viz. *Tetramelis* and *Octomelis*, woody species from Indonesia, are also nodulated.

3.3 *Alnus*-type root nodules

General features of nodulation

The perennial nature of *Alnus*-type nodules permits long-term survival but most nodules die within a few years. A study of the age distribution of nodule populations from 20-year-old *A. glutinosa* trees revealed a mean nodule age of 3–4 years with a maximum age of eight years (Akkermans 1971; Akkermans and van Dijk 1976). During root growth, decaying nodules are continuously replaced by new infections on young roots. Consequences of this rapid turn-over of nodule tissue will be discussed on page 74.

Various reports have demonstrated the universal nodulation of *Alnus* species in nature (loc. cit. in Bond 1976*a*). Other species, such as *Myrica gale* are much less abundantly nodulated than *Alnus* spp, if at all. In certain plant species the abundance of nodules seems to be correlated with root age, as has been demonstrated for *Hippophaë rhamnoides* (Stewart and Pearson 1967; Akkermans 1971; Oremus 1979). Young plants and stolons are generally abundantly nodulated, while relatively few nodules are found on old shrubs. Other plant species, such as *Ceanothus*, *Purshia*, and *Dryas* spp, even have a more heterogenous nodulation pattern which probably due to different environmental conditions, although differences in susceptability for nodulation by different ecotypes of a plant species may also play a part (Wollum, Youngberg, and Chichester 1968).

The formation of effective nodules is primarily affected by the distribution and infectivity of proper endophyte strains in the soil. As compared to the homologous combination, numerous but small and ineffective nodules were formed on *M. faya* inoculated with *M. gale* endophyte (Mian, Bond, and Rodriguez-Barrueco 1976). Molybdenum deficiency equally increased the number of nodules on *A. glutinosa* (Becking 1961). Other environmental factors which are known to affect nodule formation include soil pH, combined nitrogen, temperature, water-stress, and soil aeration.

In contrast with most legumes, a number of actinorhizal plant species can grow in acid soils. In north-western Europe *M. gale* is most acid tolerant, *A. glutinosa* moderately tolerant, and *H. rhamnoides* and *Coriaria myrtifolia* both intolerant to low pH (Bond 1951; Ferguson and Bond 1953; Bond, Fletcher, and Ferguson 1954; Bond, MacConnel, and McCallum 1956; Canizo, Miguel, and Rodriguez-Barrueco 1978). Of all species investigated, the nodulation process proved to be more sensitive to low pH than plant growth. Skogen (loc. cit. in Bond 1976*a*) reported that non-nodulated plants of *H. rhamnoides* grew vigorously at particular sites in a nitrogen-rich swamp at pH 4.5–5.0. The degree of acid tolerance of these species reflects their ecological distribution: *M. gale* is a colonizer of acid peat soils (pH 3.5–5), *A glutinosa* occurs on moderately acid soils (pH 4.5–6.5), and *H. rhamnoides* ssp *rhamnoides* is a pioneer of calcaereous coastal sand dunes (pH 6–9) in western Europe.

Nodule weight per plant of *A. glutinosa* and *M. gale* increased in the presence of low concentrations of nitrogen (10 mg NH_4^+-N l^{-1}) but was depressed at higher concentrations (MacConnell and Bond 1957). Nodulation of other species, e.g. *H. rhamnoides* and *Shepherdia canadensis* is much more depressed in the presence of combined nitrogen (Bond *et al.* 1954; 1956; Gardner and Bond 1957). Similar effects are known for legumes and can be explained by a depression of the carbon supply to the nodules after addition of combined nitrogen, as demonstrated by Small and Leonard (1969) with *Pisum sativum*.

Root nodules are preferentially formed at well-aerated sites. When *A. glutinosa* grows at sites with a high watertable, nodules are formed only in the upper soil layer which is not permanently waterlogged. Alder trees along streams and lakes are mainly well nodulated at the border just above the water surface. Here permanently high humidity and good aeration provide optimum conditions for nodulation. Under anaerobic soil conditions (peat bogs) roots are never nodulated.

Poor nodulation may also occur under conditions of both low soil temperature and water stress, as was observed for *Purshia tridentata* in Oregon on pumice soils (Dalton and Zobel 1977).

Nodule development and intranodular life cycle of the microsymbionts

Nodule development

Early stages in the development of root nodules have mainly been studied in *Alnus* spp (Pommer 1956; Becking 1966; Becking, de Boer, and Houwink 1964; Angulo Carmona 1974; Angulo Carmona, van Dijk, and Quispel 1976). Angulo Carmona (1974) distinguished *primary infection* and *primary nodule* as developmental stages prior to the formation of the *true nodule* in *A. glutinosa*. Primary infection is restricted to the entrance of the microsymbiont into the root hair and the development of thin actinomycete hyphae in the root-hair cell. The next stage comprises the primary nodule in which the cortex cells undergo cell division prior to infection with hyphae spreading from the root-hair cell. Most infected cortical cells show abundant clustering of hyphae which eventually are transformed by the formation of vesicles at their tips. Primary nodules may be recognized as local thickenings of the roots which turn red on illumination, at least in *Alnus* spp. The true nodule is completed by the outbreak of a lateral root which is induced close to the primary nodule and transformed immediately into a nodule lobe by persistent infection of its cortex region. Callaham and Torrey (1977) have described a similar mode of nodule initiation in *Comptonia peregrina* although the term primary nodule has been used in a different way.

In true nodules the penetration of young cortex cells by thin hyphae and the subsequent differentiation of the endophyte is essentially the same as in the primary nodule. However, in the latter proliferation of the endophyte depends on local dedifferentiation of cortex cells, while in the true nodule continuous growth of the endophyte is ensured by the production of young cortex cells from the apical nodule meristem.

The abundance and arrangement of infected cells within the cortex may vary among nodules of different plant species. Generally, less cortex cells are infected as compared to bacteroid tissue of annual leguminous nodules. In nodules of *Coriaria* spp the infected region is confined to one side of the stele (Shibata and Tahara 1917; Bond 1962), while in other *Alnus*-type nodules infected host cells are arranged concentric to the stele. In most *Myrica* species (Becking 1977) and in *Rubus ellipticus* (Bond 1976*a*) infection is confined to only one or two cell layers, but in *Alnus* infection is scattered over a wider part of the cortex region. Cross-inoculation between *Myrica gale* and *M. faya* (Mian *et al.* 1976) showed that the actual arrangement of infected cells is determined entirely by the host.

Development of hyphae and vesicles

Apical growth of the nodule lobe causes zonation of successive stages in the development of the endophyte from the apex to the base of the nodule lobe (Fig. 3.2). In the meristematic region A (Fig. 3.2) cortex cells are not susceptible to infection till cell division has ended. Once cell growth has started, the first sign of infection is observed by the presence of thin hyphae coiling around the nucleus. Eventually host cells are filled with tight clusters of hyphae (region B). In *Alnus* spp this stage covers only a narrow zone and soon spherical vesicles develop at the tip of branched hyphae near the host cell wall. Here clusters of vesicles with the appearance of bunches of grapes show metabolic activity (region C) until digestion by the host cell (region D). The

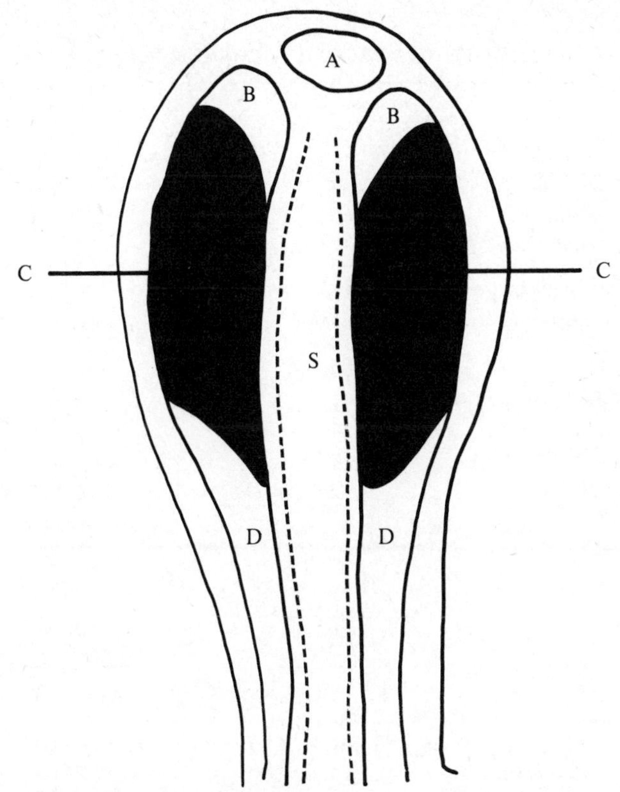

Fig. 3.2. Distribution of distinct endophyte regions in a longitudinally sectioned nodule lobe of *Alnus glutinosa*. S: stele.

description of endophyte development fits for all *Alnus*-type nodules, but shape and structure of vesicles may vary among the plant species (Table 3.4). Spherical and club-shaped vesicles ('clubs') are characterized by the presence of irregularly arranged, incomplete cell walls. Formation of complete cell walls was observed in a minority of vesicles in *Alnus* and *Elaeagnus* spp and it was suggested that these vesicles might break up into individual cells (Schaede 1933; Gardner 1965; Becking 1976). It is possible that these aberrant vesicles are terminal analogues of intercallary sporogenous bodies (see below).

Recent cross-inoculation studies with pure cultures of a *Comptonia peregrina* endophyte revealed that shape and structure of vesicles are also determined by the host. This organism forms 'clubs' in *Comptonia* nodules (Callaham, del Tredici, and Torrey 1978), but in *A. glutinosa* the same organism develops spherical vesicles (Lalonde 1979).

The complex ultrastructure of vesicles (Becking *et al.* 1964; Gatner and Gardner 1970; Lalonde and Knowles 1975; van Dijk and Merkus

TABLE 3.4
Morphological features of Alnus-*type endophytes*

Family	Genus	Microsymbiont Vesicle type[a]	Spores present[b]
Betulaceae	*Alnus*	S	+ −
Casuarinaceae	*Casuarina*	C	+
Coriariaceae	*Coriaria*	?	?
Datiscaceae	*Datisca*	?	−
Elaeagnaceae	*Elaeagnus*	S	?
	Hippophaë	S	?[c]
	Shepherdia	S	?
Myricaceae	*Comptonia*	C	+ −
	Myrica	C	+ −
Rhamnaceae	*Ceanothus*	P u	?[c]
	Colletia	S	−
	Discaria	S	−
	Trevoa	?	?
Rosaceae	*Cercocarpus*	C	?
	Dryas	C u	+
	Purshia	P u	+
	Rubus	C	?

[a] Shape of the vesicles: club-shaped (C), spherical (S), or pear-shaped (P). Not septated (u); unknown (?).
[b] Spores present (+), absent (−), or not yet described (?).
[c] Aberrant spore formation described by Gardner (1976).

1976) suggests specialization and high metabolic activity. Their function in nitrogen fixation is discussed on page 88.

Development of spores

Besides hyphae and vesicles, a third stage in the development of the endophyte has been described as bacteroids (Schaede 1933, 1962), bacteria-like cells, polyhedral-shaped cells (Becking *et al.* 1964; Becking 1970*a*), granules (Quispel 1974*a*; Akkermans and van Dijk 1976), or spores (van Dijk and Merkus 1976). The latter term was proposed in accordance with the terminology used for free-living actinomycetes. Spores are recognized as tightly packed, thick-walled cells of somewhat irregular shape and their size may vary between 0.5–1.0 μm. Clumps of spores are most abundant in the older part of region C and in region D (Fig. 3.2). The development of spores from hyphae via sporogenous bodies may start in cortex cells containing hyphae or vesicles as well as in intercellular spaces (van Dijk and Merkus 1976). In pure culture the *Comptonia peregrina* endophyte formed similar intra-axial and terminal sporogenous bodies (Callaham *et al.* 1978). The same type of spore formation was also observed in pure cultures of *Frankia* isolated from *Alnus* species (Quispel and Tak 1978; Berry and Torrey 1979; Lalonde 1979) and from *Elaeagnus umbellatus* (Baker and Torrey 1979). Although the term sporangium has been introduced to replace the term sporogenous body, there is no clear evidence that local septation of hyphae which leads to spore formation occurs in sack-like structures or spore vesicles as is required to justify the term sporangium.

In contrast with hyphae and vesicles which are always present in nitrogen-fixing actinorhizae, spore formation may be absent. When spores are present, they may persist in the old cortex region where most if not all hyphae and vesicles are disintegrated by host cell activity. Spore-free (Sp(−)) nodules and spore-rich (Sp(+)) nodules were observed in species of *Alnus*, *Comptonia*, and *Myrica* (Table 3.4) (Schaede 1933; Quispel 1974*a*; Akkermans and van Dijk 1976; van Dijk and Merkus 1976; van Dijk 1978). In various *Alnus* species Sp(+) and Sp(−) nodules are induced by different strains of the endophyte (van Dijk 1978). So far, all pure cultures of *Frankia* produce spores to some extent, but behave like Sp(−) strains within the symbiosis. Pure cultures of symbiotic Sp(+) strains of *Frankia* have not yet been obtained, which is probably due to a more complex growth requirement (Quispel and Tak 1978).

Germination of spores was observed in pure cultures of a *Comptonia peregrina* endophyte by Callaham *et al.* (1978). Arguments for the infectivity of spores were forwarded by Käppel and Wartenberg (1958) who obtained nodule formation on test plants with crude spore samples of *A. glutinosa* root nodules, and by van Hiele and van Dijk (unpublished) who used suspensions of a defined number of spores (Table 3.5). The spores were derived from spore clumps which were isolated from nodule sections with the aid of thermo-regulated micropipets (diameter suction point 5–10 μm) handled by micro-manipulation. Purity of the spore suspensions was established by light and electron microscopy. Subsequently, a dilution series of spores liberated from the clumps were used as inoculi for test plants. Up to 10 per cent of the pure spores in the inoculi gave rise to nodules (Table 3.5).

TABLE 3.5

Nodulation capacity of spores isolated from Alnus glutinosa *(Sp+) nodules*[a]

Number of spores used as inoculum	1200	300	75	18	0
Nodulation capacity[b]	83	25	8	0	0

[a] Spore suspensions at different concentrations were added to five test plants.

[b] Nodulation capacity denotes the number of nodules formed per set of five plants.

Extranodular behaviour of the microsymbionts

Although descriptions of intranodular development of endophytes are numerous, the fate of the microsymbionts free in the soil (extranodular endophyte) is hardly known. The existence of populations of extranodular endophytes is essential for establishment and maintenance of root-nodule populations of non-leguminous hosts in the field. The occurrence of endophyte populations in the soil can be explained by liberation and dissemination of *Frankia* from regularly decaying root nodules (Fig. 3.3). This view is supported by experimental evidence for survival and infectivity of the endophyte after termination of the symbiotic relationship. On the other hand, recent success in preparation of pure cultures of several strains of *Frankia*, isolated from root nodules suggests that extranodular growth of the endophyte might occur locally in the soil, although substantial evidence is not yet available.

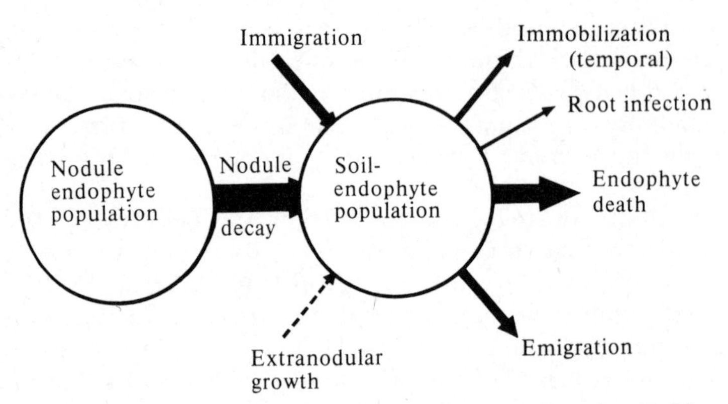

FIG. 3.3. Factors affecting *Alnus*-type endophyte populations in soil. The size of the arrows denotes the importance of the different processes.

Nodule decay and survival of extranodular endophytes

The survival and infectivity of the endophyte released from its symbiotic relationship has been repeatedly demonstrated by the use of nodule homogenates as inoculi for root-nodule production on test plants. The infectivity of diluted homogenates, although steadily decreasing, can be maintained at 6 °C for several months. Air-dried nodules of *A. glutinosa* kept over silica gel at 6 °C were highly infective (10^5–10^6 infective particles per g) after three months preservation, but infectivity was extremely low (0.01 per cent of the original infectivity) after seven years of preservation (van Dijk, unpublished). Oremus (1980) observed high infectivity of air-dried and powdered *H. rhamnoides* nodules after 18 months of preservation at 6 °C. Maintenance or even an increase of infectivity of crushed *A. glutinosa* nodule suspensions, kept at room temperature for several months, was observed by Quispel

(1954*a*, 1955, 1960) when alcoholic peat extracts or alder root extracts were added during incubation. Similar observations were made by Rogers and Wollum (1974). Akkermans and Houwers (1979) demonstrated that nodule homogenates of *A. glutinosa*, when stored at −20 °C, maintain the same grade of infectivity for at least one year. This procedure enables us to store strains of *Frankia* of well-defined origin and known infectivity as an alternative for the use of pure cultures of endophytes as inoculi, as has been proposed by Lalonde (1979).

Quispel (1954*b*) demonstrated that the relation between inoculum densities (expressed as a dilution series of a fresh nodule homogenate) and the number of nodules thus obtained in hydroponic cultures of *A. glutinosa* test plants fits a Mitscherlich equation. At low inoculum densities this relationship approximates a straight line and can be used for determination of the nodulation capacity, which represents the minimum number of infective endophyte particles per quantity of nodule or soil. Nodulation capacities of homogenates of fresh *A. glutinosa* nodules with and without spores (Akkermans and van Dijk 1976; van Dijk, in preparation) and of *H. rhamnoides* root nodules (Oremus 1980) are presented in Table 3.6.

TABLE 3.6

Nodulation capacity of nodule homogenates of Alnus glutinosa *and* Hippophaë rhamnoides[a]

Species	Fresh nodules	Decayed nodules[c]
A. glutinosa Sp(−)	10–10^2 (10^3)[b]	—
A. glutinosa Sp(−)	15	5
A. glutinosa Sp(+)	5×10^3–10^5 (10^6)[b]	—
A. glutinosa Sp(+)	10^4	10^3
H. rhamnoides[d]	10^2–10^3	—

[a] Minimum number of infective endophyte particles per mg fresh nodule weight.
[b] Maximum values, incidentally observed.
[c] One year after onset of nodule decay.
[d] After Oremus (1980).

So far data on survival of *Frankia* concerned the endophyte which was liberated mechanically from healthy nodule tissue. It is most likely that under natural conditions the endophyte will be released only from decaying root nodules. Experiments designed to pursue survival of endophyte during nodule decay and subsequent release into the soil (van Dijk, in preparation) showed that fresh nodule lobes of *A. glutinosa* burried in 200 g of soil at about 70 per cent of its water-holding capacity and incubated in the dark at 10/15 °C (12/12 h) were completely disintegrated between 25 and 40 weeks after onset of the experiment, although considerable endophyte release into soil was observed as early as 15 weeks after the start of the experiment. After an incubation period of one year the number of infective particles, initially introduced by the fresh nodule lobes, had decreased to about 30 per cent of the initial number. The difference between the nodulation capacities of Sp(+) and Sp(−) nodules, as repeatedly found for fresh nodule homogenates (Table 3.6), was reflected in the extranodular endophyte populations resulting from one year decay of Sp(+) and Sp(−) nodules. This observation suggests that the presence of spores in nodules is advantageous to the level of free endophyte in the soil but strain differences and clustering of nodule types in the field (van Dijk 1978) favour a more complicated situation.

In an attempt to quantify the annual release of endophyte from a nodule population in the field, the turnover on nodule biomass in *A. glutinosa* peat-bog vegetation was derived from the age and weight distribution of the Sp(−) nodule populations and from the nodulation capacity of Sp(−) nodules. It was calculated that in the case of a steady-state nodule population, the annual amount of decaying nodules approximated 10 g dry weight per m^2 soil surface, which accounted for about 35 per cent of the total nodule biomass. Based on the nodulation capacity of Sp(−) nodules (Table 3.6) and on the fact that the whole nodule population was restricted to the upper 15 cm of the soil layer by the permanently high watertable, the quantity of endophyte released annually from decaying nodules into the soil was calculated as 10–100 infective particles per cm^3 of soil. Comparable data on Sp(+) nodule populations in the same area showed an annual release of approximately 10^3–10^5 particles per cm^3 of soil. The higher value is due to the high nodulation capacity of Sp(+) nodules.

Surprisingly, the calculated data for annual endophyte release equalled the population density of *Frankia* at a particular moment for soil samples from the same area (10–100 infective particles per cm^3 Sp(−) soil and 30–10^5 infective particles per cm^3 of Sp(+) soil). In this case the extranodular endophyte population may be explained entirely by nodule decay, if the endophyte released survives for approximately one year or more.

It is not known exactly what stages in endophyte development participate in extranodular survival after nodule decay. However, spores appear to be either absent or to occur in hardly detectable amounts in many *Alnus*-type nodules examined (Table 3.4). It is most likely that fragments of thin hyphae are able to cause nodulation as well. The infectivity of selected microtome sections from the apical region of the nodule lobe, where spores and vesicles are still absent, supports this idea. The possibility of additional spore formation in Sp(−) nodules after onset of nodule decay is not favoured by microscopical observations or by the nodule decay experiment previously mentioned. The observation of 'endophytic bacterial cells' in the rhizosphere of *A. glutinosa* test plants and the involvement in the root-hair infection process (Lalonde 1977; Lalonde and Quispel 1977) needs confirmation.

Distribution of soil endophyte and host plant

It is obvious that free soil endophyte is present wherever nodulated host plants occur but how strict is this relation? Many workers are in favour of considerable host-plant-dependent endophyte distribution. Becking (1970*a*) examined the nodulation potential to *A. glutinosa* of 11 experimental plots of sandy soil, deprived of host plants for at least 44 years and treated annually with different fertilizer combinations. No endophyte could be demonstrated in ten plots, but the plot amended with farmyard manure had a high nodulation potential. Approximately the same results were obtained by Akkermans and Houwers (in preparation) about ten years later, although an additional clay plot regularly amended with the same farmyard manure as the sandy plot showed an extremely low nodulation potential. All soil samples provided abundant nodulation when endophyte was added. It is suggested that endophyte is either introduced together with the manure and can survive under proper conditions, or that the manure may stimulate extra-nodular growth of the endophyte.

A steady increase of the percentage of nodulated seedlings of *Ceanothus velutinus* during a four-year period after settlement of the species suggests a rapid host-plant-dependant increase of the endophyte population in the soil (Wollum *et al.* 1968).

Complete absence of *Frankia* endophytes compatible with *A. glutinosa* was observed in soil samples from a coastal sand-dune area in The Netherlands, covered with different types of plant communities. Here *H. rhamnoides* was abundant, but other actinorhizal host species were absent (van Dijk, in preparation). In the same area endophyte compatible with *H. rhamnoides* spp *rhamnoides* was most abundant in soil samples from young *Hippophae* vegetation, although low amounts of endophyte were occasionally found at sites still free from *Hippophae* (Oremus 1975).

In several cases nodulation of actinorhizal species failed, when grown far from their natural distribution area in botanical gardens and nursery soils in Japan (Uemura 1971) and Europe (Schaede 1962; Bond 1974; Becking 1977). Contrary to the previous data, several investigations are in favour of an endophyte distribution which is less dependant on the actual distribution area of compatible actinorhizal host species. Roberg (1934) obtained nodulation of *Elaeagnus* and *Hippophaë* test plants with soil samples from a meadow, while *Elaeagnus* (and *Hippophaë*?) were completely absent from the area. Nodulation of

Coriaria arborea could be obtained from New Zealand soil samples which had not supported this species for at least 100 years (Allen, Silvester, and Kalin 1966). Benecke (1969) reported high nodulation potentials to *A. viridis* of soil samples from grasslands at various sites in New Zealand deprived of *Alnus* and at an altitude where introduction of endophyte was unlikely. Some nodulation was also reported from grassland subsoils, but no significant nodulation could be attained from a *Nothofagus* forest topsoil and from screes either close to *A. viridis* or deprived of this species.

In Scotland Rodriguez-Barrueco (1968) found that 11 out of 17 soil samples collected from pastures, arable fields, gardens, and conifer plantations, all deprived of alder had low to moderate nodulation potential to *A. glutinosa* test plants. In the remaining six samples, no alder endophyte could be detected. The zero nodulation potential of a moor soil and of the conifer plantation samples could be attributed to a low pH (4.0). Curiously enough, samples from alder vegetation had high nodulation potential to *Myrica* as well. A comparable study of 100 soil samples collected from different habitats in Spain where *A. glutinosa* was occasionally present and where *M. gale* was only present at one sampling site, was carried out by Bermudes de Castro, Miguel, and Rodriguez-Barrueco (1976). All samples (71) from sites with alder had a relatively high nodulation potential to this species and 43 per cent of 29 sampling sites without alder, such as poplar, birch, oak, and conifer stands and meadows, were also infective to alder test plants although on the average to a significantly lower level than the *Alnus* sites. There was again (Rodriguez-Barrueco 1968) a remarkable and highly significant coincidence of the nodulation potentials to *M. gale* and to *A. glutinosa* at all sampling sites despite the absence of *M. gale*. In nursery soil at Tokyo, seedlings of *Ceanothus americanus* developed nodules despite the lack of known cross-inoculable species in Japan (Uemura 1971).

Some of these observations certainly suggest the existence of local populations of *Frankia* maintained independently of known actinorhizal plant species. However, definite evidence can only be obtained by exclusion of factors affecting endophyte migration. In the field, long-term survival and migration by water movement, wind, rainfall, and faunistic vectors will undoubtedly increase the distribution area of endophyte as compared to that of its host plant, but to what extent? Sp(−) endophyte from decaying *A. glutinosa* nodules showed extremely low death rate when incubated in soil samples for one year at 10–15 °C (van Dijk 1979). Wollum *et al.* (1968) demonstrated that the nodulation potential of soils to snowbrush (*Ceanothus velutinus*) is inversely proportional to the length of time that snowbrush had been absent from timber stands. As snowbrush disappears from the understory when timber stands are about 50 years old owing to the light deficiency, the relatively high nodulation potential of soil samples from timber stands up to 150 years old can be interpreted in terms of an efficient extra-nodular survival period for snowbrush endophyte of approximately 100 years. The nodulation potential of soils from timber stands older than 150 years showed a sharp decline, occasionally to the zero level. If indeed the endophyte may survive in the soil for such a long period, considerable passive migration can be expected.

Despite the absence of quantitative information about the migration rate of the endophyte in the soil, it is expected that there are similarities with the behaviour of other soil micro-organisms. Ruddick and Williams (1972) showed transport of spores of free-living actinomycetes by water movement and soil arthropodes. Significant increases of *Streptomyces* spores in air was observed following soil disturbance or rainfall. The effects were intensified by high wind velocity (Lloyd 1969). Endophyte migration rates of some hundreds of meters a year seem sufficient to explain early nodulation of *A. crispa* and *A. incana* pioneer plants on mountain slopes and barren soils in Alaska and northern America where these species play an important role in soil development and nitrogen increase (Crocker and Major 1955; Lawrence 1958; Mitchell 1968; Decker 1966; Ugolini 1968). The exposure of barren soils to rough climatic conditions stimulates considerable transport of soil dust (Bollen, Lu, Trappe, and Tarrant 1969) and adhering micro-organisms.

So far, the occurrence of extranodular endophytes in populations of compatible host species was considered as a general feature. Absence of soil endophyte in potential host-plant populations may occur when part of the ecological amplitude of the host is not suitable for nodulation or for survival of free endophyte. In *Alnus*, where nodulation is extremely widespread (Bond 1976*a*), the endophyte will cover 100 per cent of the host-plant distribution. In other nodule-bearing host-plant genera, where absence of nodules was sometimes recorded (Bond 1976*a*), lack of nodulation may be an indication of absence of endophyte, but here it is not possible to discriminate between absence of endophyte and inhibition of infection (or nodule growth) without additional research. It is likely that negative effects on nodulation at sites where host plants may survive, also influences the soil endophyte content (see pp. 68–9).

A more complicated situation occurs when several endophytes of different genetic constitution provide nodulation on the same host species. On *A. glutinosa* the Sp(+) and Sp(−) nodules obviously arise from genetically distinct endophytes and often exhibit spatial clustering per spore-type to such an extent that niche differentiation between the strains is likely (van Dijk 1978). It is speculated that differentiation among endophytes creates better possibilities for nodulation of a host-plant species adapted to a broad range of environmental conditions.

Extranodular growth of endophytes

Early attempts to isolate strains of *Frankia* on artificial substrates by von Plotho (1941), Pommer (1959), Mikola (1966), Uemura (1952, 1971), Danilewicz (1961), and Allen *et al.* (1966) were not satisfactory because the isolated organisms failed to produce nodules on suitable host plants or because nodulation experiments failed to be reproducible. In many cases *Streptomyces*-like organisms were isolated but probably as contaminants of the nodule tissue. *Actinomyces alni* Peklo, isolated by von Plotho, was recognized as *Streptomyces coelicolor* (de Vries, personal communication).

Quispel (1960) obtained increasing infectivity of *A. glutinosa* nodule homogenates incubated in artificial substrates containing alcoholic petroleum-ether extracts, but at that time the experiments were not always reproducible and definite proof of endophytic growth could not be produced.

The assumption that growth of the endophyte depends on the nodule symbiosis is supported by the nodulation response of test plants to dilution series of nodule homogenates as mentioned above. This suggests that endophyte proliferation in the rhizosphere is either extremely low or absent. However, Becking (1975) and Lalonde (1977) suggest that proliferation of *Frankia* in the rhizosphere of host plants occurs.

Indications of endophytic growth in the soil were obtained by Quispel (1955) who observed significant increase in the nodulation capacity of crushed nodule suspensions incubated in a special peat soil, and by Rossi (1964) who found migration of the *A. glutinosa* endophyte in specially designed tubes filled with sterilized soil and locally inoculated with a nodule suspension. Although Rossi explained the results in terms of endophytic growth, passive migration of the endophyte could not be excluded. A claim for the isolation of a putative *Frankia* endophyte from root nodules of *A. crispa* var. *mollis* (= *viridis* ssp *crispa*) by Lalonde, Knowles, and Fortin (1975) has been withdrawn (Lalonde, personal communication).

Recent attempts to isolate *Alnus*-type endophytes have been most encouraging. Quispel and Tak (1978) succeeded in obtaining reproducible proliferation of *A. glutinosa* Sp(−) endophyte by the use of root extracts added to the incubation medium. Attempts to cultivate the Sp(+) strain were unsuccessful. Callaham *et al.* (1978) obtained visible growth in pure cultures of a *Frankia* strain isolated from nodules of *Comptonia peregrina*. The identity of the isolated organism was confirmed by nodulation tests with *C. peregrina* host plants. Addition of root extracts to the complex growth medium was not required. Pure cultures of *Frankia* strains were also obtained from nodules of *A. viridis* spp *crispa*, *Elaeagnus umbellata* (Baker and Torrey 1979), and *A. rubra* (Berry and Torrey 1979). The isolated strains share morphological characteristics and a probably wide host range. 'Sporangia' (sporogenous bodies) and spores were produced only in pure culture; production of vesicles was found only by Lalonde and Calvert (1979). Nitrogen fixation activity could not be achieved in pure culture. The growth requirement for strains of *Frankia* in pure cultures is only poorly defined. In some cases growth was stimulated by the addition of lipid mixtures to the growth medium (Quispel and Tak 1978; Lalonde and Calvert 1979). Most peculiarly, the isolated organism is only infective when test plants were cultivated under non-sterile conditions. This suggests that the infection process is supported by other, non-infective soil micro-organisms (Knowlton, Berry, and Torrey 1979).

It is likely that specific conditions for growth of *Frankia* spp are locally present in the soil or in the rhizosphere, but the question remains to what extent such extranodular growth contributes to maintenance of the organisms under natural conditions.

Diversity among Frankia *endophytes*

Substantial evidence for diversity among *Alnus*-type endophytes was obtained by Miehe (1918), who failed to obtain nodules on a *Casuarina* sp with *A. glutinosa* nodule homogenates. Since then cross-inoculation experiments with nodule homogenates and test plants of different host species were performed to trace incompatibility as an indication of endophyte diversity (Mowry 1933; Roberg 1934, 1938; Bond 1963, 1967, 1974; Becking 1966, 1970*a*,*b*; Rodriguez-Barrueco and Bond 1968; Mackintosh and Bond 1970).

Becking (1970*b*, 1974) proposed a subdivision of *Alnus*-type endophytes into nine species with the generic name *Frankia* (Fam. Frankiaceae) based on cross-inoculation barriers between plant families. However, fragmentary knowledge of taxonomically valuable characters and recent evidence for the inconsistency of cross-inoculation barriers between plant families make this classification unsuitable for practical use. The *Comptonia peregrina* isolate CpI1 (Callaham *et al.* 1978) also induced effective nodules on *Myrica gale* and *Myrica cerifera* and on six *Alnus* species investigated (Torrey, personal communication; Lalonde 1979; Lalonde and Calvert 1979), but did not form nodules on *Casuarina*, *Elaeagnus*, and *Ceanothus* (Baker and Torrey 1979). Preliminary studies on *Frankia* isolates of *A. rubra* (ArI3) and *A. viridis* spp *crispa* (AvcI1) revealed that these isolates cover the same host range as the *C. peregrina* isolate (Berry and Torrey 1979; Baker and Torrey 1979). Successful reciprocal cross-inoculations with nodule homogenates as inoculi and members of different plant families were obtained for the combinations *M. gale*/*A. glutinosa*, *Elaeagnus angustifolia*/*A. glutinosa*, and *Coriaria myrtifolia*/*M. gale* (Miguel, Canizo, Costa, and Rodriguez-Barrueco 1978).

The discrepancies between these and earlier observations are not yet clear, but increases in successful cross-inoculation combinations and morphological similarity among the *Frankia* isolates mentioned before suggests a limited number of fairly promiscuous endophyte species. In this light, the high coincidence of nodulation potentials of soil samples to *A. glutinosa* and *M. gale* (Bermudes de Castro *et al.* 1976), can be explained by the presence of only one *Frankia* strain compatible to both plant species. In the same way Lalonde (1979) suggests that *C. peregrina* and *A. crispa* locally share the same endophyte.

On the other hand increasing evidence for endophyte diversity among closely related host species was gained by partial cross-inoculation incompatibility in combinations such as *A. glutinosa/A. jorrulensis* (Rodriguez-Barrueco 1966; Mackintosh and Bond 1970), *A. jorrulensis/A. sieboldiana* (Mackintosh and Bond 1970), *A. glutinosa/A. viridis* (Benecke 1969), and *A. crispa* var. *mollis/A. glutinosa* (Lalonde and Quispel 1977). Cross-inoculation barriers within the genus *Myrica* were discussed by Mackintosh and Bond (1970) and by Bond (1974). In both *Alnus* and *Myrica*, endophyte diversity is at least partially related to the degree of disjunction between distribution areas of host species.

At the species level, endophyte diversity is exhibited by the existence of Sp(+) and Sp(−) strains (Akkermans and van Dijk 1976; van Dijk and Merkus 1976; van Dijk 1978), both compatible to *A. glutinosa* but different in their ability to produce spores. In addition, Quispel and Tak (1978) observed different growth requirements for these strains (see p. 80).

Besides cross-inoculation experiments, there is a need for additional tools to discriminate between strains of *Frankia*. Morphological features of intranodular endophyte, such as shape and size of vesicles, are obviously determined by the host (Lalonde 1979; see p. 71) and therefore cannot be used as reliable tools in *Frankia* taxonomy. Morphological characters need to be studied in pure cultures under defined growth conditions (Callaham *et al.* 1978; Lalonde 1978). Isolation of *Frankia* endophytes also enables comparison of physiological and serological properties.

The chemical composition of actinomycete cell walls such as the presence of diaminopimelic acid (DAP) isomers, might serve to discriminate between major taxonomical groups. In endophytes of *A. glutinosa*, meso-DAP was found to be the sole isomer (Angulo Carmona *et al.* 1976; Quispel 1974*a*; Becking 1976). The same was true for Sp(+) and Sp(−) endophytes of *A. glutinosa* (van Dijk 1978), but Lalonde *et al.* (1975) reported the presence of both meso-DAP and LL-DAP in cell walls of the *A. crispa* endophyte. Neither meso-DAP nor LL-DAP could be detected in the endophytes of *H. rhamnoides*, *M. gale* (van Dijk 1978), *Casuarina* (Becking 1977), and *Shepherdia canadensis*

(Akkermans 1978*a*). Both DAP and sugar composition of the cell walls of three *Frankia* isolates refer to cell-wall type III, which is also found among the Dermatophilaceae, Actinoplanaceae, and Nocardiaceae (Lechevalier and Lechevalier 1970). A more detailed determination of the taxonomic position is not yet possible. The classification of the Euactinomyceteae by Lechevalier and Lechevalier (1970) leads directly to the Dermatophyllaceae, although a close relationship between *Frankia* and *Dermatophilus* is very unlikely. None of the other actinomycete families shows division of mycelium in all planes prior to spore formation, while introduction of the term sporangium for *Frankia* sporogenous bodies not really narrow the gap between *Frankia* and free-living sporangial actinomycetes.

Nitrogen fixation

The most striking feature of the microsymbiont is its ability to reduce nitrogen to ammonia within the nodule symbiosis. The total amount of nitrogen fixed by nodulated plants is determined by two factors, viz. the total biomass of living nodule tissue, and the nitrogenase activity of the nodules, integrated over the entire growing season. Both nodulation and nitrogen fixation are affected by the rate of photosynthesis of the plant and by environmental factors, such as soil temperature, supply of nutrients (e.g. combined nitrogen) and aeration of the soil. The effect of some of these factors will be discussed below.

As shown in Table 3.7, there is a large variation in activity among plant species, grown under different conditions. Nodules from the same tree are likewise variable in activity. Maximum value for acetylene-reducing activity of *Alnus*-type nodules was 90 μmol C_2H_4 g nodule dry wt^{-1} h^{-1} in case of *A. glutinosa* seedlings (Table 3.7). Nitrogenase activity rapidly decreased with increasing nodule age. A comparable decline has also been observed for nodule respiration of *A. glutinosa*. Oxygen consumption of 1, 1.5, and 2 years old plants amounted to 10.5, 5.1, and 1.8 ml O_2 g nodule dry wt^{-1} h^{-1} respectively (MacConnell and Bond 1957). The rapid decrease in activity with age seems to be a general feature of perennial nodules and is the main reason why *Alnus*-type nodules on the average fix less nitrogen than young nodules from annual legumes (up to 200 μmol C_2H_4 g nodule dry wt^{-1} h^{-1} (Waughman 1977).

Site of nitrogen fixation

In *A. glutinosa* and *H. rhamnoides* nitrogenase activity was found almost exclusively in the apical (2–3 mm) part of the nodule lobe (Akkermans 1971). Similar observations have been reported for *A. glutinosa* by Becking (1977) and for *Ceanothus greggi* by Kummerow, Alexander, Neel, and Fishbeck (1978). Especially in this region of the nodule lobe,

TABLE 3.7

Acetylene reduction of Alnus-*type nodules*

Species	Source	nitrogenase activity μmol C_2H_4 h^{-1} g fr wt $^{-1}$	g dry wt $^{-1}$	References
Casuarina cunninghamiana	F	0.1		Sloger (1968) *loc. cit.* Silver (1969)
	F	0.1–0.3?		Tyson and Silver (1979)
	L	1.5–3.7[a]		Bond and Mackintosh (1975)
C. equisitifolia	F	0.2–0.7		Sloger (1968) *loc. cit.* Silver (1969)
	F		0.001–0.3	Tyson and Silver (1979)
	F		75	Waughman (1977)
C. junghuniana	F	0.3–2.0		Akkermans (1978*b*)
C. rumphiana	F	13.2–15.0	47.4–54.6	Becking (1977)
C. torulosa	L	1.3–3.7[a]		Bond and Mackintosh (1975)
Coriaria myrtifolia	L	2.8–15.2		Canizo *et al.* (1978)
Alnus crispa	F	5.1		Dalton and Naylor (1975)
A. cordata	F		1–3	Pizelle (1975)
A. glutinosa	L	0.1–0.4		Gordon and Wheeler (1978)
	L	≤1.4±0.6		Mian and Bond (1978)
	L	2.5–6.5		Wheeler (1969)
	L	≤9.5		Wheeler (1971)
	L	≤9.5		Wheeler and Bowes (1974)
	L	4.4–7.4		Becking (1977)
	L		9–92	Akkermans (1971)
	L	1.7–8.8		Wheeler and Lawrie (1976)
	L	2–20		Lalonde (1979)
	F	0.4[c]		McNiel and Carpenter (1974)
	F		2–18	Pizelle (1975)
	F		1–7	Pizelle and Thiery (1977)
	L		16[d]	Stewart (1962)

TABLE 3.7 (*cont.*)

Species	Source	nitrogenase activity[e] μmol C_2H_4 h^{-1} g fr wt $^{-1}$	g dry wt $^{-1}$	References
A. glutinosa (*cont.*)	F		0.2–5	Waughman (1972)
	F		6	Waughman (1977)
	F		0.1–70	Akkermans (1971)
			(2.6–7.9)[e]	Akkermans (1971)
	F	20.4		van Straten *et al.* (1977)
	F	1–12		Akkermans *et al.* (1977)
	F	22.8		Akkermans (1978*a*)
A. incana	F		2–12	Pizelle (1975)
	F	7.7–20.0		Akkermans (1978*a*)
A. maritima	F	0.3–1.2		Akkermans (1978*b*)
A. rubra	F	1.3–1.6		Russell and Evans (1970)
	L	4.8–7.8		Russell and Evans (1970)
	F	7.0		Schubert and Evans (1976)
A. rugosa	F	4.5		Stewart, Fitzgerald, and Burris (1967)
A. tenuifolia	F	0.03–10.9		Fleschner, Delwiche, and Goldman (1976)
A. viridis	F		6.6[d]	Benecke (1970)
Datisca cannabina	F	5.5		Chaudhary (1979)
Elaeagnus angustifolia	F	0.01–0.8[c]		McNiel and Carpenter (1974)
	F	14.8		Schubert and Evans (1976)
E. ebbingi	F	3.4		Akkermans (unpublished)
E. umbellate	F	1.5[c]		McNiel and Carpenter (1974)
Hippophaë rhamnoides	F		0.1–60	Akkermans (1971)
			(8–24)	Akkermans (1971)
	F		0.2	Waughman (1972)
	F	0.7[c]		McNiel and Carpenter (1974)
	F		7	Waughman (1977)

continued

TABLE 3.7 (*cont.*)

Species	Source	nitrogenase activity μmol C_2H_4 h^{-1} g fr wt $^{-1}$	g dry wt $^{-1}$	References
	F	11.2		van Straten *et al.* (1977)
	F	6.8±5.5		Simonov *et al.* (1978)
Shepherdia argentea	F	0.3[c]		McNiel and Carpenter (1974)
S. canadensis	F	1.5[c]		McNiel and Carpenter (1974)
	F	2.9		van Straten *et al.* (1977)
Comptonia peregrina	F	0.7–2.2		Stewart *et al.* (1967)
	F		1–33	Fessenden, Knowles, and Brouzes (1973)
	L	9.6–22.2		Callaham *et al.* (1978)
	L	10		Lalonde (1978)
Myrica californica	F	1.4		Schubert and Evans (1976)
M. cerifera	F	1.6–3.4		Sloger (1968) *loc. cit.* Silver (1969)
	F	0.1–1.8[b]		Silver and Mague (1970)
M. faya	L	0.1–0.4		Miguel and Rodriguez-Barrueco (1974)
M. gale	L	6.5–14		Wheeler (1969)
	L	6		Miguel and Rodriguez-Barrueco (1974)
	F		17	Waughman (1977)
	F	0.2		van Straten *et al.* (1977)
M. javanica	F	2.4		Becking (1977)
	F	0.2–1.1		Akkermans (1978*b*)
M. pensylvanica	F	1.0		McNiel and Carpenter (1974)
	F	0.1–5.0		Morris, Eveleigh, Riggs, and Tiffny (1974)
Ceanothus americanus	F		0.7–1.4	Krochmal and McCrain (1975)
C. cuneatus	F	0.21[a]		Delwiche, Zinke, and Johnson (1965)
C. divaricatus	F	0.07[a]		Delwiche *et al.* (1965)
C. foliosus	F	0.03[a]		Delwiche *et al.* (1965)
C. gloriosus	F	0.1[a]		Delwiche *et al.* (1965)

TABLE 3.7 (*cont.*)

Species	Source	nitrogenase activity[e] μmol C_2H_4 h^{-1} g fr wt $^{-1}$	g dry wt $^{-1}$	References
C. greggi var. *perplexans*	F	0.9–1.0		Kummerov *et al.* (1978)
C. griseus	F	0.03[a]		Delwiche *et al.* (1965)
C. incanus	F	0.01[a]		Delwiche *et al.* (1965)
C. integerrimus	F	0.06[a]		Delwiche *et al.* (1965)
C. jepsonii	F	0.06[a]		Delwiche *et al.* (1965)
C. prostatus	F	0.09[a]		Delwiche *et al.* (1965)
C. thyrsiflorus	F	0.04[a]		Delwiche *et al.* (1965)
C. sorediatus	F	0.04[a]		Delwiche *et al.* (1965)
	F	0.04[a]		Delwiche *et al.* (1965)
C. velutinus	F		1–38[c]	McNabb *et al.* (1977)
	F	1.8		Schubert and Evans (1976)
Colletia cruciata	F	1.6		Bond (1976*a*)
	F	0.7–1.4		Akkermans (1978*b*)
C. spinosa	F	2.1		Akkermans (1978*b*)
Trevoa trinervis	F	0.2–0.9		Rundel and Neel (1978)
Purshia tridentata	F	1.2		Schubert and Evans (1976)
	F	0–5.1		Dalton and Zobel (1977)

[a] Expressed as mg N g total N^{-1} h^{-1}
[b] Expressed as μmol C_2H_4 mg total N^{-1} h^{-1}
[c] Expressed per nodule dry weight or fresh weight?
[d] Expressed as mg N g dry wt^{-1} day^{-1}.
[e] Average values per plant.
[f] Nodules from field-grown (F) or laboratory (L) plants.

clusters of endophytic vesicles show high metabolic activity, as was demonstrated by strong tetrazolium reduction (Akkermans 1971). The relationship between endophytic vesicles and nitrogen fixation was confirmed by significant acetylene-reducing activity of clusters of vesicles, which were isolated from nodule homogenates by filtration and supplied with ATP and dithionite (van Straten, Akkermans, and Roelofsen 1977).

When the efficiency of different nitrogen-fixing organisms is compared, nitrogenase activity should be related to the biomass of these organisms. In the case of actinorhizae the number of vesicle clusters in nodule sections (Akkermans 1971) or in nodule homogenates (Akkermans, Roelofsen, and Blom 1979) can be used as reference for the amount of endophyte responsible for nitrogen fixation. In *A. glutinosa* the amount of vesicle clusters varied from 10^6–7×10^6 per g nodule fresh weight (Akkermans 1978*a*; Akkermans *et al.* 1979). The average *in situ* nitrogenase activity expressed per vesicle cluster amounted to 0.001–0.02 nmol $C_2H_4h^{-1}$.

The relative amount of endophyte of *Alnus* actinorhizae can also be derived from the DAP content of vesicle clusters, but here impurities like dead endophyte and spores, not involved in nitrogen fixation, introduce quantitative errors. Using both methods as a reference it was shown that the large variation in activity among nodules of one tree can not be explained by differences in the amount of vesicle clusters (Akkermans *et al.* 1979).

Soil temperature

The rate of fixation is strongly affected by soil temperature, although the tolerance to extreme temperatures varies among plant species.

In *A. glutinosa*, *H. rhamnoides*, and *M. gale* nitrogen fixing activity of detached nodules is optimum at 20–25 °C, (Akkermans 1971; Waughman 1977). In *A. glutinosa* and *M. gale* activity declined rapidly above 25 °C, but not so in *H. rhamnoides* and *Casuarina* spp (Akkermans 1971; Mackintosh and Bond 1970; Waughman 1977). The tolerance to high temperature of the latter species has ecological relevance because these plants grow under conditions where soil temperature may exceed 30 °C. Low soil temperatures can limit the fixation rate significantly and is one of the reasons of low activity in *P. tridentata* in Oregon (Dalton and Zobel 1977).

Water stress

Many actinorhizal plant species which grow in dry areas are significantly affected by a temporal shortage of water, as has been demonstrated for *P. tridentata* (Dalton and Zobel 1977). This species is part of the understory in open Ponderosa pine forests on dry soils in Oregon

(Fig. 3.4(c); p. 93). In late July, nodule activity declined sharply, which corresponded with water stress. Nitrogenase activity ceased in plants with xylem pressure potentials below −25 bars. Here, massive decay of nodules generally occurs in dry summer periods.

Water logging

Growth and nitrogen-fixing activity of submerged nodules is often very low, due to insufficient gas transport through the water phase. Since the amount of oxygen consumed by nodules is 5–10 times lower than the amount of nitrogen fixed (see, e.g. Table 3.8), one may assume that the fixation rate under waterlogged conditions is primarily limited by respiration. The importance of air spaces inside the nodules is illustrated by rapid decrease of respiration and nitrogen fixation when the gas content of *Alnus* nodules is replaced by water (Table 3.8).

TABLE 3.8
Acetylene reduction, respiration, and hydrogen uptake of root nodules of Alnus glutinosa *as affected by water saturation of intercellular pore space*[a]

'reatment	Incubation time (min)	Acetylene reduction (μmol g nodule dry wt⁻¹)	Oxygen uptake (μmol g nodule dry wt⁻¹)	Hydrogen uptake (μmol g nodule dry wt⁻¹)
ore space ʼaterlogged	6	0.3	0	0
	18	0.8	11.3	0.3
	45	2.0	33.5	1.0
	60	3.1	44.7	1.6
	90	5.6	64.3	2.2
'ore space naffected	6	2.3	0	0
	18	6.9	21.7	1.2
	45	16.7	53.9	2.8
	60	21.1	66.6	3.5
	90	28.0	90.1	4.3

[a] Aliquots of 2.5 g fresh weight of nodules were submerged in water in 20-cm³ tubes. The air from he nodules was removed by evacuation of the gas phase above the solution for one minute. After ubsequent introduction of air above the solution, water was pressed into the intercellular spaces of he nodule. Untreated nodules were used as controls. In preliminary experiments it was shown that vacuation of dry nodules did not affect the gas uptake. Nodules were incubated in 16-cm³ Hungate ubes with an initial gas phase of acetylene (10 per cent), oxygen (20 per cent), hydrogen (0.7 per ent) in nitrogen. Gasses were analysed by gas chromatography.

In this experiment it was also shown that oxygen-dependent hydrogen uptake is simultaneously depressed. The gas volume in the nodule, calculated from differences in weight before and after waterlogging, amounted to 5 per cent of the total volume. Pore space of field-grown *A. rubra* nodules, determined by Wheeler, Gordon, and Ching (1979), ranged from 3.0–11.5 per cent. Under well-aerated conditions oxygen pressure of the atmosphere is about optimum for nitrogen fixation. Further increases in the oxygen partial pressure steadily decreases nitrogen fixation activity in *A. rubra* nodules (Wheeler *et al.* 1979).

Certain plant species have special structures such as lenticells and nodule rootlets which are supposed to enhance gas diffusion through the nodule tissue. Lenticells are formed at the nodule tips, especially under wet growth conditions. Most *Myrica* and *Casuarina* species form rootlets at the tips of the lobes. These rootlets have large air spaces and might enhance oxygen supply of the nodule tissue (Bond 1952; Torrey and Callaham 1978; Tjepkema 1978). When grown in liquid culture, the rootlets are negatively geotropic. However, nodule roots of field-grown nodules are generally short and often lack definite negative geotropic growth. In some cases nodules of *Casuarina* from Indonesia were free from rootlets (Akkermans 1978*b*). Moreover, we have found that *Casuarina* spp grown on extremely dry and stony soils also form nodule rootlets which makes the function of these structures disputable.

Seasonal and diurnal variations

In perennial *Alnus* nodules nitrogenase is induced in early spring when leaf emergence starts (Akkermans 1971; Pizelle and Thiery 1977; Mian and Bond 1978). Nitrogenase activity increases with increasing soil temperature and light intensity during the growing season, as was also found in *Ceanothus velutinus* (McNabb, Geist, and Youngberg 1977; McNabb 1979), *H. rhamnoides* (Akkermans 1971; Simonov, Zhiznevskaya, Khailova, Ilyasova, Kudryatseva, and Tibilov 1978; Oremus, personal communication), and in *P. tridentata* (Dalton and Zobel 1977). The total amount of nitrogen fixed per hectare may vary largely because of differences in the composition of the vegetation, heterogeneous distribution of nodules in the soil, and differences in nitrogenase activity from site to site. Although many authors have presented data on nitrogenase activity (Table 3.7), only few of them have investigated variation of the activity in the field. This makes extrapolation of activity on an annual basis impossible. Moreover, quantitative information on nodule populations in natural vegetations is still incomplete (Table 3.9). Although different estimates of annual fixation rates by actinorhizal species have been reported (Silvester

TABLE 3.9

Nodule biomass and nitrogen fixation of actinorhizal species in the field

Plant species	Age of plants (years)	Total surface examined (m^2)	Nodule dry weight ($g\ m^{-2}$)	Nitrogen fixation ($kg\ N_2\ ha^{-1}\ a^{-1}$)	References
Alnus glutinosa	±30	100	8.4		Akkermans (1971)
	15–20	100	45.4	56	Akkermans (1971)
	±20	500	41.9		Akkermans (1971)
	4	50	7.9	200	Akkermans and Lock (unpublished)
Alnus rubra	7	—	11.7	140	Zavitkovski and Newton (1968)
	30	—	24.4	209	Zavitkovski and Newton (1968)
Ceanothus greggi	—	20	3[a]	0.1	Kummerov *et al.* (1978)
Ceanothus velutinus	—	—	25[b]	±60	Delwiche *et al.* (1965)
	—	—	14.7	32	McNabb *et al.* (1977)
Hippophaë rhamnoides	6–7	9	5.3	—	Oremus (1979)
	3	1	2.5	4	Stewart and Pearson (1967)
	11	1	39.0	42	Stewart and Pearson (1967)
	13	1	64.2	58	Stewart and Pearson (1967)
	16	1	2.7	2	Stewart and Pearson (1967)
	1–2	32	0.4	2	Akkermans (1971)
	1–3	0.8	3.1	—	Akkermans (1971)
	4	10.5	0.9	—	Akkermans (1971)
	4	5	0.1	—	Akkermans (1971)
	5–8	6	1.4	—	Akkermans (1971)
	5–8	0.6	1.9	—	Akkermans (1971)
	8–9	0.5	0.01	—	Akkermans (1971)
	9–12	1	1.1	—	Akkermans (1971)
	10–15	1.5	3.0	15	Akkermans (1971)
Purshia tridentata	—	8.1	0.05[c]	0.06	Dalton and Zobel (1977)

[a] Original value: 12.9 fresh weight m^{-2}.

[b] Original value: 100 g fresh weight m^{-2} (assumption).

[c] Original value: 1.91 kg fresh weight ha^{-1}.

1976), only few studies are based on quantitative assays of both nodule biomass and annual fixation rate. Some examples will be presented here. For *Ceanothus* an annual fixation rate of 32 kg N ha^{-1} was calculated by McNabb *et al.* (1977). In a pure stand of 20-year-old *A. glutinosa* in wet peat soil in The Netherlands, an average of 0.127 g N-fixed g nodule dry wt^{-1} a^{-1} was measured (Akkermans and van Dijk 1976). The nodule biomass in this particular grove amounted to 454 kg dry wt ha^{-1}. From these data an annual rate of fixation of 56 kg N ha^{-1} a^{-1} was calculated. Significantly higher values were found in young stands of *A. glutinosa*, with a higher mean nodule activity in spite of a lower nodule biomass. Estimates were made in a four-year-old stand with a density of 2.6 trees m^{-2}, a nodule biomass of 79 kg dry wt ha^{-1} and an average nitrogenase activity of 65 μmol C_2H_4 g nodule dry wt h^{-1}. The annual fixation rate was found to be about 200 kg N ha^{-1} a^{-1} (Akkermans and Lock, unpublished).

Ecological and practical importance

Many nodulated non-legumes are important colonizers on nitrogen-poor soils and can introduce significant amounts of nitrogen into the ecosystem (see above). The best known examples are some *Alnus* species such as *A. viridis*, *A. crispa*, and *A. incana* which are distributed in the Northern hemisphere. Since legumes and other nitrogen-fixing symbioses are almost absent in wide areas of Scandinavia and Canada, most of the soil nitrogen in these areas originates from actinorhizae. Paleontological studies of the pollen distribution of *Alnus*, *Hippophaë*, *Dryas*, and *Myrica* in northern temperate regions demonstrate the wide distribution of these species in the past (Tallantire 1974; see further in Silvester 1976) and illustrate their important role in the nitrogen cycle. Alder trees, especially *A. rubra* (Oregon), *A. glutinosa* (Europe) and *A. incana* (Europe) are used in forestry, either in pure stands or in mixed cultures with pine or poplar (van der Meiden 1961; Mikola 1966, 1974, 1975; Trappe *et al.* 1968). This has recently been summarized in the proceedings of three symposia (Briggs, Bell, and Atkinson 1978; Torrey and Tjepkema 1979; Gordon, Wheeler, and Perry 1979).

Alder is used for wind hedges around orchards (Fig. 3.4(a)), along ditches and as stabilizer of eroded areas and coal spoils (Kohnke 1941; Beunis 1956; Lowry, Brokaw, and Breeding 1962). Successful afforestation of bogs in Finland with *A. incana* after industrial exploitation of peat has been demonstrated by Mikola (1974, 1975). Favourable effects of alder on growth of other species such as pine growing in admixture with it have frequently been reported (Holmsgaard 1960; Mikola 1966, 1974, 1975; Tarrant and Trappe 1971). Other examples of nodulated non-legumes which may have potentials in forestry include *Ceanothus* (Fig. 3.4(b)) and *Purshia* (Fig. 3.4(c)), both native to North America.

Fig. 3.4 (a) Wind hedges of *Alnus glutinosa* around orchards in Voorne (The Netherlands). (b) *Ceanothus velutinus* as understory on open sites in pine forests in Oregon, USA. (c) *Purshia tridentata* as understory in open forest of *Pinus ponderosa* on dry soils in Oregon, USA.

3.4 *Parasponia*-type root nodules

History

A definite example of a *Rhizobium*–non-legume symbiosis was discovered in the Ulmaceae by Trinick (1973). Nodulated specimens were found as weeds in tea plantations in Papua-New Guinea and were initially identified as *Trema aspera* (Trinick 1973) or *T. cannabina* var. *scabra* (Trinick 1976; Trinick and Galbraith 1976; Coventry, Trinick, and Appleby 1976). Further studies have shown that these specimens were incorrectly identified and belong to *Parasponia rugosa* Bl. (Akkermans 1978*b*,*c*; Akkermans, Abdulkadir, and Trinick 1978*a*,*b*). *Trema* and *Parasponia* are closely related genera and have frequently been confused in the past, as mentioned in earlier literature (Koorders and Valeton 1910; de Wit 1949; van Steenis 1962; Backer and Bakhuizen van den Brink 1965). Examples of frequent mistakes and re-identification were also found in the Herbarium Bogoriensis (Bogor, Indonesia). This was especially the case with specimens of *T. cannabina* var. *scraba* and *P. rugosa* (Akkermans *et al.* 1978*a*).

Parasponia spp are small trees up to 15 m high which grow as pioneer plants in mountain areas of Malesia, Indonesia, and Papua-New Guinea (Fig. 3.5). *Trema* spp are generally larger and have a wider distribution pattern in the tropics (Fig. 3.5). *P. parviflora* Miq. is a

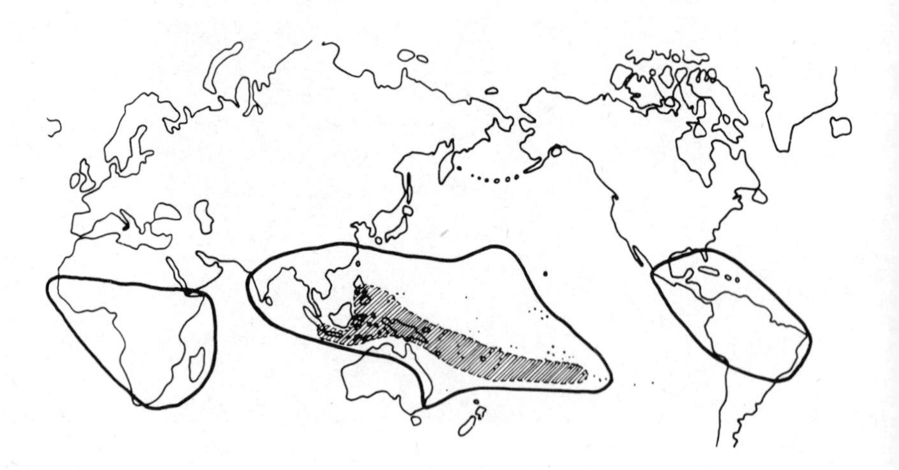

Fig. 3.5. Approximate range of *Trema* Lour. (solid line) and *Parasponia* Miq. (shaded area). Data according to Soemadmo (1977).)

prominent pioneer on volcanic soils in Java, and its ecological importance has been extensively described in botanical studies on the flora of Java (Koorders and Valeton 1910; Clason 1935; de Voogd 1940; Backer and Bakhuizen van den Brink 1965). First comments on the occurrence of root nodules on specimens of Ulmaceae was made by the Dutch forester Ham (1909) in a discussion on a symposium on 'Green manure in Indonesia'. He mentioned: '. . . that plants other than legumes have been used as green manure. . . . One of the tree species which occur everywhere in the higher areas in Java is "Anggroeng" (Sundanese for "Koerai"),† and these plants have root nodules. It is tempting to suggest that also this species can collect nitrogen.' [Translated from Dutch.] It is uncertain whether Ham had *Parasponia* or *Trema* spp in mind. The observation of Ham was confirmed by Backer and van Slooten in 1924. In their study of the role of weeds on Javanese tea plantations the following comment was made on *T. orientalis* [*loc. cit.*, translated from Dutch]: 'Referring to a comment of Ham in 1909 that "Koerai" bear root nodules, we tried to confirm these observations in 1923 in Buitenzorg, Bogor (± 245 m above sea level) and at Tjinjiroean (1500 m above sea level). In Buitenzorg none of the specimens of *Trema* sp. were nodulated. However, at Tjinjiroean, Spruit and van Slooten found one specimen that bore well-developed root nodules with bacteria in it. The assumption of Ham that this tree possibly is a nitrogen collector could therefore be correct.' Also in this case the identification of the '*Trema*' is questionable. The single nodulated specimen probably was a *Parasponia* sp.

Further comments on nitrogen fixation in Ulmaceae were made by Clason (1935) in his excellent study of the ecology of *P. parviflora*, a pioneer on volcanic ash of Mount Kelut (Java): '*Parasponia* possesses root nodules, nitrogenous food being possibly obtained in this way, so that *Parasponia* is thus adapted as a pioneer on virgin soils' [loc. cit., p. 512]. Nodulated specimens collected by Clason and Clason-Laarman are still present in the Herbarium at Bogor (Fig. 3.6). Unfortunately, foresters and botanists have given little attention to this phenomenon. Comments in Dutch literature on nodulation in Ulmaceae of the Indonesian flora have been neglected until recently (Akkermans 1978*b*, *c*, *d*; Akkermans *et al.* 1978*a*, *b*).

In a recent survey of root systems of the Ulmaceae and Urticaceae in Indonesia (Akkermans *et al.* 1978*b*), it was shown that nitrogen-fixing root nodules occurred only in *P. parviflora* Miq. (see Table 3.1). These results were confirmed by Becking (1979). Possibly, symbioses of *Rhizobium* spp with members of the Ulmaceae occur only in the Asiatic genus

† 'Anggroeng' is the Javanese name for *Trema orientalis*. This name has occasionally also been used for *P. parviflora*.

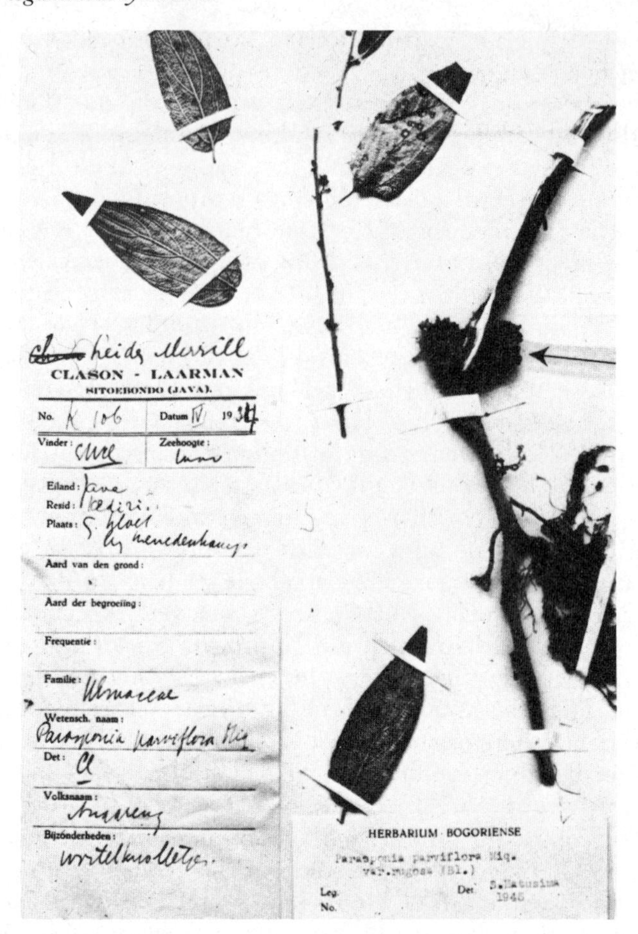

Fig. 3.6. Herbarium specimen of *Parasponia parviflora* Miq., collected by Clason-Laarman (1934) from the Herbarium Bogoriensis at Bogor, Indonesia. The arrow denotes the presence of root nodules [in Dutch: 'Wortelknolletjes'].

Parasponia. A third nodulated species, viz. *P. andersonii* Planch. has recently been described by Trinick (1979).

Nodules have not yet been observed in *T. floridana* from Florida (Lie, personal communication) and *T. orientalis* (L.), *T. cannabina* Lour var. *glabrescens* (Pl) de Wit, *T. cannabina* var. *scraba*, and *T. aspera* in Indonesia (Becking 1979), Australia, and Papua-New Guinea (Trinick 1976). Similar negative results have been reported for *T. orientalis* (syn. *T. quinensis*?) in Pretoria (Grobbelaar, personal communication). It should be noted that most of these field observations have not been confirmed by inoculation experiments with plants grown under controlled conditions.

Development and function of Parasponia *root nodules*

The microsymbiont of a nodulated specimen of the Ulmaceae was originally described as 'bacteria' (Backer and van Slooten 1924). Trinick (1973, 1976) proved that nodules of *P. rugosa* (described as *Trema* spp, see p. 93) are induced by a promiscuous, slow-growing strain of *Rhizobium* which cross-inoculates with certain tropical legumes, such as *Vigna* spp. Similar strains have been isolated from root nodules of *P. parviflora* by Akkermans and Roelofsen (unpublished). This microorganism was found to be a slow-growing *Rhizobium* strain which cross-inoculated with *Vigna unguiculata* (cow-pea). Inoculation of sterile seedlings of *P. parviflora* with the isolate induced the formation of nodules.

Root nodules of *Parasponia* fix nitrogen at a rate comparable to that of perennial nodules of other non-legumes. *In situ* activity of small nodules (2–4 mm) of field-grown *P. parasponia* in Indonesia was about 15 μm C_2H_4 g nodule fr. wt^{-1} h^{-1}, while activity of larger nodules (20–30 mm) varied from 0.8–3.1 μmol C_2H_4 g^{-1} h^{-1} (Akkermans 1978*b*; Akkermans *et al.* 1978*a*).

Parasponia nodules differ markedly from leguminous nodules in various ways. *Rhizobium* is mainly present in highly branched infection threads within the host cells (Trinick 1976, 1979; Trinick and Galbraith 1976). Only few bacteria are released from the infection threads and differentiate into bacteroids. In leguminous nodules, however, most rhizobia are transformed into bacteroids. Moreover, *Parasponia* nodules lack haemoglobin (Conventry *et al.* 1976) while this pigment is always present in leguminous nodules and may function in the oxygen transport to the bacteroids. These observations on *Parasponia* nodules clearly show that both the formation of bacteroids and leghaemoglobin are determined by the host and cannot be denoted as obligate prerequisites of an effective *Rhizobium*–plant symbiosis.

Ecological and practical importance

Parasponia, as well as *Casuarina* and leguminous trees, grow rapidly in virgin soils in South-East Asia (Clason 1935) and probably play an important role in soil fertility. These preliminary observations indicate that these species may be useful in reafforestation of eroded areas in the tropics. Application of nitrogen-fixing '*Trema orientalis*' in forestry in Africa was proposed by de Wildeman (1925). No use has yet been made of nodulated *Parasponia* however. Although it is unlikely that nodulation occurs in *Trema*, further study of nitrogen fixation in other members of the Ulmaceae is needed.

Acknowledgement

Thanks are due to Dr J. W. Woldendorp for his critical comments on the manuscript.

References

Akkermans, A. D. L. (1971). Nitrogen fixation and nodulation of *Alnus* and *Hippophaë* under natural conditions. Thesis, Leiden.

—— (1978*a*). In *Interactions between non-pathogenic soil microorganisms and plants* (ed. Y. R. Dommergues and S. V. Krupa) p. 335. Elsevier, Amsterdam.

—— (1978*b*). Symbiotic nitrogen fixation by non-legumes in Indonesia. Report, Netherlands Foundation for the Advancement of tropical Research WOTRO.

—— (1978*c*). *Vakbl. Biol.* **58**, 82.

—— (1978*d*). *Acta bot. neerl.* **27**, 240.

—— and Dijk, C. van (1976). In *Symbiotic nitrogen fixation in plants* (ed. P. S. Nutman) *International Biological Programme*, Vol. 7, p. 511. Cambridge University Press.

—— and Houwers, A. (1979). In *Symbiotic nitrogen fixation in the management of temperate forests* (ed. J. C. Gordon, C. T. Wheeler, and D. A. Perry) p. 23. Oregon State University Press.

—— Abdulkadir, S., and Trinick, M. J. (1978*a*). *Pl. Soil* **49**, 711.

—— —— —— (1978*b*). *Nature, Lond.* **274**, 190.

—— Roelofsen, W., and Blom, J. (1979). In *Symbiotic nitrogen fixation in the management of temperate forests* (ed. J. C. Gordon, C. T. Wheeler, and D. A. Perry) p. 160. Oregon State University Press.

—— Straten, J. van, and Roelofsen, W. (1977). In *Recent developments in nitrogen fixation. Proc. 2nd Int. Symp. on Nitrogen Fixation* (ed. W. Newton, J. R. Postgate, and C. Rodriguez-Barrueco) p. 591. Academic Press, London.

Allen, E. K. and Allen, O. N. (1950). *Proc. Soil Sci. Soc. Am.* **14**, 179

—— —— and Klebesadal, L. J. (1964). In *Science in Alaska. 14th Alaskan Science Conference Proc.* (ed. G. Dahlgren) p. 54. Anchorage, Alaska.

Allen, J. D., Silvester, W. B., and Kalin, M. (1966). *N.Z. Jl Bot.* **4**, 57.

Angulo Carmona, A. F. (1974). *Acta bot. neerl.* **23**, 257.

—— Dijk, C. van, and Quispel, A. (1976). In *Symbiotic nitrogen fixation in plants* (ed. P. S. Nutman) *International Biological Programme*, Vol. 7, p. 475. Cambridge University Press.

Athar, M. and Mahmood, A. (1972). *Pakistan J. Bot.* **4**, 209.

Backer, C. A. and Bakhuizen van den Brink Jr., R. C. (1965). *Flora van Java*, Vol. II. Noordhoff, Groningen, Netherlands.

—— and Slooten, D. F. van (1927). *Geïllustreerd Handboek der Javananse Theeonkruiden en hunne Beteekenis voor de Cultuur*. Batavia Drukkerijen Ruygrok.

Baker, D. and Torrey, J. G. (1979). In *Symbiotic nitrogen fixation in the management of temperate forests* (ed. J. C. Gordon, C. T. Wheeler, and D. A. Perry) p. 38. Oregon State University Press.

Becking, J. H. (1961). *Nature, Lond.* **192**, 1204.

—— (1966). *Annls Inst. Pasteur, Paris.* Suppl. **111**, 211.

—— (1970*a*). *Pl. Soil* **32**, 611.

—— (1970*b*). *Int. J. Syst. Bact.* **20**, 201.

—— (1974). In *Bergey's manual of determinative bacteriology* (ed. R. E. Buchanan and N. E. Ribbon) 8th edn., p. 701. Williams and Wilkins, Baltimore.

—— (1975). In *The development and function of roots* (ed. J. G. Torrey and D. T. Clarkson) p. 508. Academic Press, London.

—— (1976). In *Proc. First Int. Symp. on Nitrogen Fixation* (ed. W. E. Newton and C. J. Nyman) Vol. 2, p. 581. Washington State University Press.

—— (1977). In *A treatise on dinitrogen fixation.* Section III: *Biology* (ed. R. W. F. Hardy and W. S. Silver) p. 185. John Wiley, New York.
—— (1979*a*). *Pl. Soil* **51**, 289.
—— (1979*b*). *Pl. Soil* **53**, 541.
—— Boer, W. D. de, and Houwink, A. (1964). *Antoie van Leeuwenhoek* **30**, 343.
Benecke, U. (1969). *Pl. Soil* **30**, 145.
—— (1970). *Pl. Soil* **33**, 30.
Bermudez de Castro, F., Miguel, C., and Rodriguez-Barrueco, C. (1976). *Ann. Microbiol. Inst. Pasteur, Paris* **127A**, 307.
Berry, A. and Torrey, J. G. (1979). In *Symbiotic nitrogen fixation in the management of temperate forests* (ed. J. C. Gordon, C. T. Wheeler, and D. A. Perry) p. 69. Oregon State University Press
Beunis, P. C. (1956). *Tijdschr. ned. Heidemaatsch.* **67**, 179.
Bollen, W. B., Lu, K. C., Trappe, J. M., and Tarrant, R. F. (1969). Research Note PNW 103, p. 38. Pacific Northwest Forest and Range Experiment Station, Forest Service, USDA, Portland.
Bond, G. (1951). *Ann. Bot., N.S.* **15**, 447.
—— (1952). *Ann. Bot., N.S.* **16**, 467.
—— (1962). *Nature, Lond.* **193**, 1103.
—— (1963). In *Symbiotic associations. Symp. Soc. Gen. Microbiol.* (ed. P. S. Nutman and B. Mosse) p. 72. Cambridge University Press, London.
—— (1967). *Phyton* **24**, 57.
—— (1970). In *Nitrogen nutrition of the plant* (ed. E. A. Kirby) p. 1. University of Leeds.
—— (1971). In *Biological nitrogen fixation in natural and agricultural habitats. Pl. Soil* Special Volume (ed. T. A. Lie and E. G. Mulder) p. 317. Nijhoff, The Hague.
—— (1974). In *Biology of nitrogen fixation* (ed. A. Quispel) p. 342. North Holland, Amsterdam.
—— (1976*a*) In *Symbiotic nitrogen fixation in plants* (ed. P. S. Nutman) *International Biological Programme*, Vol. 7, p. 443. Cambridge University Press.
—— (1976*b*). *Proc. R. Soc.* **B193**, 127.
—— and Mackintosh, A. H. (1975). *Proc. R. Soc.* **B192**, 1.
—— Fletcher, W. W., and Ferguson, T. P. (1954). *Pl. Soil* **5**, 309.
—— MacConnel, J. T., and McCallum, A. H. (1956). *Ann. Bot., N.S.* **20**, 501.
Briggs, D. G., Bell, S. de, and Atkinson, W. A. (1978). *Utilization and management of alder.* Pacific Northwest Forest and Range Experiment Station, Forest Services. USDA, Portland.
Burgeff, H. (1932). *Saprophitismus und Symbiose.* Jena.
Callaham, D. and Torrey, J. G. (1977). *Can. J. Bot.* **55**, 2306.
—— del Tredici, P., and Torrey, J. G. (1978). *Science, N.Y.* **199**, 899.
Canizo, A., Miguel, C., and Rodriguez-Barrueco, C. (1978). *Pl. Soil* **49**, 195.
Chaudhary, A. H. (1978). In *International symposium on limitations and potentials for biological nitrogen fixation in the tropics.* (ed. J. Döbereiner, R. H. Burris, and A. Hollaender) p. 359. Plenum Press, New York.
—— (1979). *Pl. Soil* **51**, 163.
Clason, E. W. (1935). *Bull. Jard. bot. Buitenz.* Serie III, **13**, 509.
Conventry, D. R., Trinick, M. J., and Appleby, C. A. (1976). *Biochim. biophys. Acta* **420**, 105.
Crocker, R. L. and Major. J. (1955). *J. Ecol.* **43**, 427.
Dalton, D. A. and Naylor, A. W. (1975). *Am. J. Bot.* **62**, 76.
—— and Zobel, D. B. (1977). *Pl. Soil* **48**, 57.
Danilewicz, K. (1961). *Acta microbiol. pol.* **14**, 321.
Decker, H. F. (1966). In *Soil development and ecological succession in a deglaciated area of Muir Inlet, Southeast Asia* (ed. A. Mirsky) p. 73. Institute of Polar Studies, Report 20, Ohio.

DEITSCHMAN, G. H., JORGENSEN, K. R., and PLUMMER, A. P. (1974). In *Seeds of woody plants in the United States* (ed. C. S. Schopmeyer) p. 686. Agricultural Handbook No. 450. Forest Service, USDA, Washington DC.
DELWICHE, C. C., ZINKE, P. J., and JOHNSON, C. M. (1965). *Pl. Physiol., Lancester* **40**, 1045.
DIJK, C. VAN (1978). *New Phytol.* **81**, 601.
—— (1979). In *Symbiotic nitrogen fixation in the management of temperate forests* (ed. J. C. Gordon, C. T. Wheeler, and D. A. Perry) p. 84. Oregon State University Press.
—— and MERKUS, E. (1976). *New Phytol.* **77**, 73.
ENGLER, A. and PRANTL, K. (1894). *Die natürlichen Pflanzenfamilien.* III 6a, 246. Duncker & Humblot, Berlin.
FERGUSON, T. P. and BOND, G. (1953). *Ann. Bot., N.S.* **17**, 175.
FESSENDEN, R. J. (1979). In *Symbiotic nitrogen fixation in the management of temperate forests* (ed. J. C. Gordon, C. T. Wheeler, and D. A. Perry) p. 403. Oregon State University Press.
—— KNOWLES, R., and BROUZES, R. (1973). *Soil Sci. Soc. Am. Proc.* **37**, 893.
FLESCHNER, M. D., DELWICHE, C. C., and GOLDMAN, C. R. (1976). *Am. J. Bot.* **63**, 945.
GARDNER, I. C. (1965). *Archs Mikrobiol.* **51**, 365.
—— (1976). In *Symbiotic nitrogen fixation in plants* (ed. P. S. Nutman) *International Biological Programme* Vol. 7, p. 485. Cambridge University Press.
—— and BOND, G. (1957). *Can. J. Bot.* **35**, 305.
GATNER, E. M. S. and GARDNER, I. C. (1970). *Archs Mikrobiol.* **70**, 183.
GORDON, J. C. and WHEELER, C. T. (1978). *New Phytol.* **80**, 179.
—— —— and PERRY, D. A. (1979). *Symbiotic nitrogen fixation in the management of temperate forests. Proc. Symp. Corvallis 1979.* Oregon State University Press.
HAM, S. P. (1909). *Handel. 10de Congr. Ned. Indisch Landbouw Syndicaat* II^2, 26.
HOEPPEL, R. E. and WOLLUM, A. G. (1971). *Can. J. Bot.* **49**, 1315.
HOLMSGAARD, E. (1960). *Forst. ForsVæs. Danm.* **26**, 253.
HUTCHINSON, J. (1964, 1967). *The genera of flowering plants*, Vols. I and II. Oxford University Press.
KÄPPEL, M. and WARTENBERG, H. (1958). *Archs Mikrobiol.* **30**, 46.
KHAN, A. G. (1971). *Pakistan J. Bot.* **3**, 71.
KNOWLTON, S., BERRY, A., and TORREY, J. G. (1979). In *Symbiotic nitrogen fixation in the management of temperate forests* (ed. J. C. Gordon, C. T. Wheeler, and D. A. Perry) p. 479. Oregon State University Press.
KOHNKE, H. (1941). *J. For.* **39**, 333.
KOORDERS, S. S. H. and VALETON, Th. (1910). *Bijdrage no. 12 tot de Kennis der Boomsoorten op Java.* G. Kolff, Batavia.
KREBILL, R. G. and MUIR, J. M. (1974). *Nwest Sci.* **48**, 266.
KROCHMAL, A. and MCCRAIN, J. (1975). USDA Forest Service Research Note SE-215.
KUMMEROW, J., ALEXANDER, J. V., NEEL, J. W., and FISHBECK, K. (1978). *Am. J. Bot.* **65**, 63.
LALONDE, M. (1977). In *Recent developments in nitrogen fixation. Proc. 2nd. Int. Symp. on Nitrogen Fixation* (ed. W. Newton, J. R. Postgate, and C. Rodriguez-Barrueco) p. 569. Academic Press, London.
—— (1978). *Can. J. Bot.* **56**, 2621.
—— (1979). In *Bot. Gaz.* Special Volume (ed. J. G. Torrey and J. Tjepkema) **140**, 35.
—— and CALVERT, M. (1979). In *Symbiotic nitrogen fixation in the management of temperate forests* (ed. J. C. Gordon, C. T. Wheeler, and D. A. Perry) p. 25. Oregon State University Press.
—— and KNOWLES, R. (1975). *Can. J. Microbiol.* **21**, 1058.
—— and QUISPEL, A. (1977). *Can. J. Microbiol.* **23**, 1529.
—— KNOWLES, R., and FORTIN, J. A. (1975). *Can. J. Microbiol.* **21**, 1901.

LAWRENCE, D. B. (1958). *Am. Scient.* **46**, 89.
—— SCHOENIKE, R. E., QUISPEL, A., and BOND, G. (1967). *J. Ecol.* **55**, 793.
LECHEVALIER, H. A. and LECHEVALIER, M. P. (1970). In *The actinomycetales* (ed. H. Prauser) p. 393. Fischer, Jena.
—— —— (1979). In *Symbiotic nitrogen fixation in the management of temperate forests* (ed. J. C. Gordon, C. T. Wheeler, and D. A. Perry) p. 111. Oregon State University Press.
LLOYD, A. B. (1969). *J. gen. Microbiol.* **57**, 35.
LOWRY, G. L., BROKAW, F. C., and BREEDING, C. H. J. (1962). *J. For.* **60**, 196.
MACCONNELL, J. T. and BOND, G. (1957). *Ann. Bot., N.S.* **21**, 185.
MACKINTOSH, A. H. and BOND, G. (1970). *Phyton* **27**, 79.
MCNABB, D. H. (1979). *Poster Harvard Symposium 1978* (ed. J. G. Torrey and J. Tjepkema). *Bot. Gaz.* Special Volume.
MCNABB, D. H., GEIST, J. M., and YOUNGBERG, C. T. (1977). *Am. Soc. Agron.* Abstract.
MCNIEL, R. E. and CARPENTER, P. L. (1974). *Hort. Sci.* **9**, 381.
MEDAN, D. and TORTOSA, R. D. (1976). *Boln Soc. argent. Bot.* **17**, 323.
MEIDEN, H. A. VAN DER (1961). *Ned. Boschb Tijdschr.* **33**, 168.
MIAN, S. and BOND, G. (1978). *New Phytol.* **80**, 187.
—— —— and RODRIGUEZ-BARRUECO, C. (1976). *Proc. R. Soc.* **B194**, 285.
MIEHE, H. (1918). *Flora* **111–12**, 431.
MIGUEL, C. and RODRIGUEZ-BARRUECO, C. (1974). *Pl. Soil* **41**, 521.
—— CANZIO, A., COSTA, A., and RODRIGUEZ-BARRUECO, C. (1978). In *Limitations and potentials for biological nitrogen fixation in the tropics* (ed. J. Döbereiner, R. H. Burris, and A. Hollaender) p. 121. Plenum Press, New York.
MIKOLA, P. (1966). The value of alder in adding nitrogen in forest soils. Report, Research conducted under grant authorized by US Public Law 480. Grant No. EG-FI-131, Helsinki.
—— (1974). *Proc. Int. Symp. Forest Drainage*, Jyväskylä-Oulu, Finland, p. 329.
—— (1975). *Silva fenn.* **8**, 101.
MILLBANK, J. W. (1974). In *The biology of nitrogen fixation* (ed. A. Quispel) p. 238. North Holland, Amsterdam.
—— (1977). In *A treatise on dinitrogen fixation.* Section III *Biology* (ed. R. W. F. Hardy and W. S. Silver) p. 125. John Wiley, New York.
MITCHELL, W. W. (1968). In *Biology of alder* (ed. J. M. Trappe, J. F. Franklin, R. F. Tarrant, and G. M. Hansen) p. 45. Pacific Northwest Forest and Range Experiment Station Forest Service, USDA, Portland, Oregon.
MONTEMARTINI, L. (1905). *Annali Bot.* **3**, 101.
—— (1906). *Atti Accad. naz. Lincei Rc.* 15.
MORRIS, M., EVELEIGH, D. E., RIGGS, S. C., and TIFFNY Jr., W. N. (1974). *Am. J. Bot.* **61**, 867.
MORRISON, T. M. and HARRIS, G. P. (1958). *Nature, Lond.* **182**, 1746.
MOSTAFA, M. A. and MAHMOUD, M. Z. (1951). *Nature, Lond.* **167**, 446.
MOWRY, H. (1933). *Soil Sci.* **36**, 409.
OREMUS, P. A. I. (1975). *Verh. K. ned. Akad. wet. Afd. Natuurkunde II* **66**, 106.
—— (1979). *Pl. Soil* **52**, 59.
—— (1980). *Pl. Soil* **56**, 123.
PIZELLE, G. (1975). *C.r. hebd. Séanc. Acad. Sci., Paris* **281**, 1829.
—— and THIERY, G. (1977). *Physiol. Vég.* **15**, 333.
PLOTHO, O. VON (1941). *Archs Mikrobiol.* **12**, 1.
POMMER, E. H. (1956). *Flora* **143**, 603.
—— (1959). *Ber. dt. bot. Ges.* **72**, 138.
QUISPEL, A. (1954*a*). *Acta bot. neerl.* **3**, 495.
—— (1954*b*). *Acta bot. neerl.* **3**, 512.

—— (1955). *Acta bot. neerl.* **4**, 671.
—— (1960). *Acta bot. neerl.* **9**, 380.
—— (1974*a*). In *The biology of nitrogen fixation* (ed. A. Quispel) p. 499. North Holland, Amsterdam.
—— (1974*b*). In *The biology of nitrogen fixation* (ed. A. Quispel) p. 719. North Holland, Amsterdam.
—— and TAK, T. (1978). *New Phytol.* **81**, 587.
ROBERG, M. (1934). *Jb. wiss. Bot.* **79**, 472.
—— (1938). *Jb. wiss. Bot.* **86**, 344.
RODRIGUEZ-BARRUECO, C. (1966). *Phyton* **23**, 103.
—— (1968). *J. gen. Microbiol.* **52**, 189.
—— and BOND, G. (1968). *J. gen. Microbiol.* **52**, 189.
ROGERS, R. D. and WOLLUM II, A. G. (1974). *Soil Sci. Soc. Am. Proc.* **38**, 756.
ROSSI, S. (1964). *Annls Inst. Pasteur, Paris* **106**, 505.
RUDDICK, S. M. and WILLIAMS, S. T. (1972). *Soil Biol. Biochem.* **4**, 93.
RUNDEL, P. W. and NEEL, J. W. (1978). *Flora* **167**, 127.
RUSSELL, S. A. and EVANS, H. J. (1970). Report to the Pacific Northwest Forest and Range Experiment Station.
SABET, Y. S. (1946). *Nature, Lond.* **157**, 656.
SCHAEDE, R. (1933). *Planta* **19**, 389.
—— (1962). *Die pflanzlichen Symbiosen*. G. Fischer Verlag, Stuttgart.
SCHUBERT, K. R. and EVANS, H. J. (1976). *Proc. natn. Acad. Sci. U.S.A.* **73**, 1207.
SEVERINI, G. (1922). *Annali Bot.* **15**, 29.
SHIBATA, K. and TAHARA, M. (1917). *Bot. Mag., Tokyo* **31**, 157.
SILVER, W. S. (1969). *Proc. R. Soc.* **B172**, 389.
—— and MAGUE, T. (1970). *Nature, Lond.* **227**, 378.
SILVESTER, W. B. (1976). In *A treatise on dinitrogen fixation.* Section IV: *Agronomy and ecology* (ed. R. W. F. Hardy and A. H. Gibson) p. 141. John Wiley, New York.
SIMONOV, I. N., ZHIZNEVSKAYA, G. Y., KHAILOVA, G. F., ILYASOVA, V. B., KUDRYATSEVA, N. N., and TIBILOV, A. A. (1978). *Jzvest. Timir. Selskok. Akad.* **1**, 142.
SMALL, J. G. C. and LEONARD, O. A. (1969). *Am. J. Bot.* **56**, 187.
SOEPADMO, E. (1977). In *Flora Malesiana* (ed. C. G. G. J. van Steenis) Ser. 1, Vol. 8, p. 31.
STEENIS, C. G. G. J. VAN (1962). *Philipp. J. Sci.* **91**, 507.
STEWART, W. D. P. (1962). *J. exp. Bot.* **13**, 250.
—— and PEARSON, M. C. (1967). *Pl. Soil* **24**, 348.
—— FITZGERALD, G. P., and BURRIS, R. H. (1967). *Proc. natn. Acad. Sci. U.S.A.* **58**, 2071.
STRATEN, J. VAN, AKKERMANS, A. D. L., and ROELOFSEN, W. (1977). *Nature, Lond.* **266**, 257.
STUTZ, H. C. (1972). In *Wildland shrubs—their biology and utilization* (ed. C. M. McKell, J. P. Blaisdell, and J. R. Goodin). USDA Forest Service General Technical Report INT-1. Utah, Ogden.
SUESSENGUTH, K. (1953). In *Die natürlichen Pflanzenfamilien* (ed. A. Engler and K. Prantl) 20d, 1 Duncker & Humblot, Berlin.
TALLANTIRE, P. A. (1974). *New Phytol.* **73**, 529.
TARRANT, R. F. and TRAPPE, J. M. (1971). In *Biological nitrogen fixation in natural and agricultural habitats. Plant and Soil* Special Volume (ed. T. A. Lie and E. G. Mulder) p. 335. M. Nijhoff, The Hague.
TJEPKEMA, J. (1978). *Can. J. Bot.* **56**, 1365.
TORREY, J. G. and CALLAHAM, D. (1978). *Can. J. Bot.* **56**, 1357.
—— and TJEPKEMA, J. (eds.) (1979). *Bot. Gaz.* Special Volume **140**.
TRAPPE, J. M., FRANKLIN, J. F., TARRANT, R. F., and HANSEN, G. M. (1968). *The biology of alder.* Pacific Northwest Forest and Range Experiment Station, Forest Service, USDA, Portland.

TRINICK, M. J. (1973). *Nature, Lond.* **244**, 459.
—— (1976). In *Proc. 1st Int. Symp. Nitrogen Fixation* (ed. W. E. Newton and C. J. Nyman) Vol. 2, p. 507. Washington State University Press.
—— (1979). *Can. J. Microbiol.* **25**, 565.
—— and GALBRAITH, J. (1976). *Archs. Microbiol.* **108**, 159.
TROTTER, A. (1902). *Bull. Soc. Bot. It.* 50.
—— (1906). *Bull. Soc. bot. It.* 53.
TYSON, J. H. and SILVER, W. S. (1979). In *Bot. Gaz.* Special Volume (ed. J. G. Torrey and J. Tjepkema).
UEMURA, S. (1952). *Bull. Govt Forest Exp. stn Meguro* **52**, 1.
—— (1971). In *Biological nitrogen fixation in natural and agricultural habitats. Plant and Soil* Special Volume (ed. T. A. Lie and E. G. Mulder) p. 349. M. Nijhoff, The Hague.
UGOLINI, F. C. (1968). In *Biology of alder* (ed. J. M. Trappe, J. F. Franklin, R. F. Tarrant, and G. M. Hansen), p. 115. Pacific Northwest Forest Range Experiment Station Forest Service, USDA, Portland.
VLAMIS, J., SCHULTZ, A. M., and BISWELL, H. H. (1964). *J. Range Mgmt* **7**, 73.
VOOGD, C. N. A. DE (1940). *Trop. Natuur* **29**, 37.
WAGLE, R. F. and VLAMIS, J. (1961). *Ecology* **42**, 745.
WAUGHMAN, G. J. (1972). *Pl. Soil* **37**, 521.
—— (1977). *J. exp. Bot.* **28**, 949.
WEBSTER, S. R., YOUNGBERG, C. T., and WOLLUM II, A. G. (1967). *Nature, Lond.* **216**, 392.
WHEELER, C. T. (1969). *New Phytol.* **68**, 675.
—— (1971). *New Phytol.* **70**, 487.
—— and BOWES, B. G. (1974). *Pflanzenphysiol.* **71**, 71.
—— and LAWRIE, A. (1976). In *Symbiotic nitrogen fixation in plants* (ed. P. S. Nutman) *International Biological Programme*, Vol. 7, p. 497. Cambridge University Press.
—— GORDON, J. C., and CHING, TE MAY. (1979). *New Phytol.* **82**, 449.
WILDEMAN, E. DE (1925). *Revue Bot. appl. Agric. trop.* **5**, 864.
WILLIS, J. C. (1966). *A dictionary of flowering plants and ferns*, 7th edn. Cambridge University Press.
WIT, H. C. D. DE (1949). *Bull. Jard. Bot Buitenz.* Serie 111, **18**, 181.
WOLLUM, II, A. G., YOUNGBERG, C. T., and CHICHESTER, F. W. (1968). *For. Sci.* **14**, 114.
YOUNGBERG, C. T. and HU, L. (1972). *For. Sci.* **18**, 211.
ZAK, B. (1973). In *Ectomycorrhizae* (ed. G. C. Marks and T. T. Kozlowski) p. 43. Academic Press, London.
ZAVITKOVSKI, I. and NEWTON, M. (1968). In *Biology of alder* (ed. J. M. Trappe, J. F. Franklin, R. F. Tarrant, and G. M. Hansen) p. 209. Pacific Northwest Forest and Range Experiment Station, Forest Service USDA, Portland.

4 Environmental physiology of the legume–*Rhizobium* symbiosis

T. A. LIE

4.1 Introduction

Without the intervention of man and/or grazing animals, natural environments with a dominant population of legumes seldom occur. This is quite surprising, since legumes, by virtue of the capacity to fix nitrogen in association with bacteria of the genus *Rhizobium*, are independent of soil nitrogen, which is often the limiting factor for plant growth (cf. Döbereiner and Campelo 1977; Mulder, Lie, and Houwers 1977). One of the reasons is that the symbiosis is less tolerant to extreme conditions, and growth of the symbiotic system is already hampered under conditions where growth of the same plant, provided with combined nitrogen, is apparently unaffected (cf. Lie 1974). Therefore, the range of environments for the symbiotic system is narrower than that of the nitrogen-fertilized plant.

Plants in the field are continuously exposed to environmental changes, either short-term or more prolonged, often cyclic, fluctuations. Basically, the growth potential is dictated by the genetic constitution of the plant, but the environment can modify or even limit the ultimate growth. However, in response to stress, the plant often displays a remarkable adaptation, obviously in an attempt to minimize any harmful effect of the environment. This adaptation, essential for the plant to survive, can be due to a phenotypic plasticity of the individuals, enabling them to make full use of the prevalent conditions. Another possibility is based on the genetic variability, always present in a plant population, resulting in the selection of ecotypes, genetically more adapted to the locality (Bennett 1970; Cooper 1970).

Environmental factors mainly operate through the plant component of the symbiotic system. Also, the symbiotic properties of rhizobial strains are expressed in the plant, i.e. when fully integrated in metabolic processes of the host. Therefore, attention is focused on the contribution of the plant, which is essential for the understanding of the limits and potentialities of a symbiotic association in a particular environment.

The purpose of this chapter is to consider two aspects of the environment on the symbiosis. At first the limits of the symbiosis, imposed by some environmental stress, will be presented. This includes various items, which have been reviewed in recent papers (cf. Vincent 1965; Dart 1974; Lie 1974; Gibson 1976*a*,*b*; Pate 1976; Munns 1977), and to avoid repetition the choice of subjects is selective rather than exhaustive. It mainly deals with recent developments and is therefore complementary to earlier papers. In the second part, the flexibility of the symbiotic system will be considered. In particular, the use of genetic variability to overcome some environmental stresses will be demonstrated.

4.2 Establishment of the symbiosis

Although there are many gaps in our knowledge, recent findings give more insight in the mode of infection of the roots by *Rhizobium*. As several reviews are available (Dart 1974; Fåhraeus and Sahlman 1977) only a short description will be presented here.

The main point of entry for *Rhizobium* is the root hair which is infected by way of an infection thread, and most of the studies are concerned with this type of symbiosis. There are some exceptions like the peanut, *Arachis hypogaea*, where no infection threads have been found. Here, the root hairs are present at the sides of emerging lateral roots, but although *Rhizobium* causes curling and deformation it does not enter the root hair. Instead, it penetrates the root intercellularly as zooglea, at the junction of the root hair and the epidermal and cortical cells (Chandler 1978).

Root exudates

Before the bacteria enter the roots, they have to multiply in the rhizosphere at the expense of root exudates (cf. van Egeraat 1975*a*). This is presumably a non-specific reaction, since many other bacteria will multiply in the rhizosphere of legumes. Conversely, *Rhizobium* can also multiply, although less readily, in the rhizosphere of non-leguminous plants (Vincent 1974). A selective stimulation of *R. leguminosarum* strains by homoserine, a major component of root exudate, was observed by van Egeraat (1975*b*). No growth was obtained with *R. trifolii* and *R. phaseoli* when homoserine was used as a source of nitrogen, whereas an inhibitory effect on *R. meliloti* was found.

Among the amino acids exuded by the roots, tryprophan has received particular attention, because it is easily converted by *Rhizobium* to the plant hormone indole-acetic acid (IAA). It is assumed to play a role in the infection mechanism, but it is more likely that IAA stimulates the formation and elongation of the root hairs (Fåhraeus and Ljunggren 1967). IAA alone will not cause deformation of root hairs (Fåhraeus and Ljunggren 1967) as often stated. However, since infection is closely linked with active growth of the root hair, infection is presumably favoured by IAA.

Recent work by Currier and Strobel (1976, 1977) showed that rhizobia are attracted to the roots. From root exudates of *Lotus corniculatus* (birdsfoot trefoil), a glycoprotein, chemotactin, was isolated, which strongly attracts a strain of trefoil *Rhizobium*. However, it is not always correlated with nodulation, and even non-leguminous plants may attract *Rhizobium* strains. Therefore, chemotaxis is not very specific although, in general, rhizobia are more attracted by the host plants which they do nodulate.

Attachment

Before infection takes place a close contact between rhizobia and the root hair must be established. Sahlman and Fåhraeus (1963) showed that nodule bacteria became firmly attached, perpendicular to the root hairs. More recently, the perpendicular attachment was attributed to a marked polarity of rhizobial cells (Bohlool and Schmidt 1976). However, the polar attachment is not specific for legumes, but *Rhizobium* was also found to settle on the roots of non-leguminous plants (Menzel, Uhlig, and Weichsel 1972) and even at oil–water interfaces (Marshall, Cruickshank, and Bushby 1975).

The hypothesis that plant lectins, specific proteins with sugar-binding properties, are involved in the recognition of the specific rhizobial symbionts by legumes, has received considerable attention (Hamblin and Kent 1973; Bohlool and Schmidt 1974; Dazzo and Hubbell 1975*a*, *b*). Lectins are assumed to be responsible for the specific attachment of compatible rhizobial cells on the root hairs, by acting as a molecular bridge between the common or cross-reactive antigens of the roots and *Rhizobium* cells (Dazzo and Hubbell 1975*b*). Recently, a lectin, trifolin, was detected in the root hairs of clover. It can be eluted by 2-deoxyglucose (Dazzo and Brill 1977), the probable haptenic determinant of the rhizobial antigen (Dazzo and Hubbell 1975*b*). Host specificity can be partly explained, therefore, by a preferential binding of its own (homologous) *Rhizobium* strains by specific lectins. Although in many instances there is a good correlation between lectin binding and host-specific infectivity, there are still many controversial observations. The most serious objection is that many *Rhizobium* strains capable of nodulating a host plant, do not always bind with the plant-specific lectin, and that binding is obtained with non-specific *Rhizobium* strains (Bohlool and Schmidt 1974; Law and Strijdom 1977; Brethauer and Paxton 1977).

Also, only a small percentage of the compatible rhizobial population was found to have binding properties (Law and Strijdom 1977). Recent experiments (Bhuvaneswari and Bauer 1978) suggest that the ability of rhizobial cells to develop lectin receptors depends on the growing conditions of the bacteria. Of eleven *Rhizobium japonicum* strains, only five showed binding properties with *Glycine max* lectin, when the bacteria were cultivated in a synthetic medium. When cultured in root exudates of *G.max*, or in association with roots of *G.max* seedlings, all the *Rhizobium* strains develop specific receptors for the lectins. *Rhizobium* strains from other cross-inoculation groups fail to bind with *G.max* lectins, except for two strains belonging to the 'cowpea' group.

Infection

Deformation of root hairs is a prelude to infection if a compatible *Rhizobium* strain is in contact with the legume roots. To some extent deformation can be obtained by using culture filtrates or a crude polysaccharide of *Rhizobium* (Yao and Vincent 1969; Hubbell 1970; Solheim and Raa 1973). In this case, only branching and moderate curling occurred. Marked curling, however, or the formation of a shepherd's crook—a deformation at the root hair tip or a side branch—was only obtained in the presence of living, compatible rhizobial cells (Yao and Vincent 1976).

Infection of the root hairs take place, almost without exception, in the markedly curled root hairs (shepherd's crook). Presumably, the rhizobial cells become entrapped in the folds and pockets enabling the accumulation of relatively high concentrations of metabolites, essential for infection (Fåhraeus and Sahlman 1977). An occasional infection, arising from a flat surface of the root hair, is due to a close contact of two adjacent root hairs and the bacteria became entrapped between these two hair surfaces (Fåhraeus and Sahlman 1977). In the absence of wall-degrading enzymes (cf. Dart 1974) infection and the formation of infection threads must be regarded as an invagination process (Nutman 1956) owing to a reorientation of growth of the root hair wall. The participation of pectolytic enzymes, proposed by Ljunggren and Fåhraeus (1961), is still a matter of controversy (Lillich and Elkan 1968; Solheim and Raa 1971). A possible source of error is the presence of substances in the seeds, phenolic compounds, potent inhibitors of pectolytic enzymes (Fåhraeus and Sahlman 1977). Recently, a weak pectolytic activity of rhizobial cells, growing on an agar plate, has been reported (Hubbell, Morales, and Umali-Garcia 1978), but this work awaits confirmation.

Growth of the infection thread is directed by the host-cell nucleus which precedes the tip of the infection thread (Fåhraeus 1957). This is particularly shown in those cases where the infection thread initiated in a side branch, instead of growing to the base of the root hair, grows towards the tip of the root hairs, where the nucleus is (Fahraeus and Sahlman 1977). Normally, the infection thread grows centripetally towards the stele, traversing the cortical cells. Abortion of the infection thread has often been observed, e.g. in *P. sativum* cv. Afghanistan, which is resistant to certain *Rhizobium* strains (Lie 1971, 1978*a*; Degenhardt, Larue, and Paul 1976)). Disorientation of the infection thread was observed in roots treated with the chelate Fe-EDTA, a specific inhibitor of nodulation in legumes (Lie and Brotonegoro 1969). Instead of growing into the inner cortical cells, the infection threads remain in a few, sub-epidermal cells, and by branching and curling, a cluster of threads was formed, ultimately filling the whole cell (Lie 1974).

Nodule development

Prazmowski (1870) was the first to show that division of plant cells starts some distance from the advancing infection thread. This was later confirmed by Libbenga and Bogers (1974). Auxins (Thiman 1936) and cytokinins (Philips and Torrey 1970, 1972) are known to be produced by *Rhizobium* in pure cultures. Although difficult to prove, it is attractive to assume that these phytohormones are also produced inside the infection threads and act as triggers for cell division (Torrey 1961). Another unknown growth factor, presumably diffusing from the xylem, is also needed for cell division in the cortical cells (Libbenga and Bogers 1974).

Further steps in the symbiosis include the release of bacteria from the infection thread into the disomatic plant cells, i.e. cells having the double number of chromosomes (Wipf and Cooper 1940), the transformation of bacterial cells into bacteroids (cf. Bergersen 1974) and the formation of leghaemoglobin and the enzyme nitrogenase. Our knowledge of these steps is meagre.

Much can be learned from symbiotic systems with a genetic defect. The study of *Trifolium* mutants (cf. Nutman 1969) has provided some data pertaining to host genes controlling, for example the transformation to bacteroids or the release of bacteria from the infection threads. Bergersen (1957) assumed that cells of certain bacterial strains, which produce swellings instead of normal nodules, remain in the infection thread owing to the absence of disomatic cells.

Ineffectiveness in *Vicia faba* by a *Rhizobium* strain, normally effective on *Pisum*, was found to be due to the lack of transformation in bacteroids in *V. faba* but not in peas (van de Berg 1978). Several mutants of rhizobial strains were also obtained with a lesion in genes specifying either a step in nodulation or nitrogen fixation. An interesting mutant strain failed to fix nitrogen owing to the inability to produce component 2 of the enzyme complex nitrogenase, but the nodules produced still contained leghaemoglobin (Maier and Brill 1976). In conclusion we must assume that the unravelling of this complex process depends on the availability of symbiotic mutants, lacking only in a single step of the symbiosis. Another possibility is the study of the failure of the symbiosis caused by inhibitors or some environmental stress.

4.3 Some environmental factors limiting the symbiosis

The capacity of a certain plant–*Rhizobium* association to fix nitrogen under optimal conditions can be defined as 'potential nitrogen fixation' as opposed to 'actual nitrogen fixation' (Lie 1971). The latter may be used to describe the capacity to fix nitrogen under less favourable conditions of growth, usually encountered in the field where some limiting factor may prevent the full expression of the symbiosis.

There are many factors limiting the symbiosis, but under field conditions soil moisture, temperature, and light are presumably the most important. Reference is made to earlier reviews on the effect of soil pH and nutritional factors (Lie 1974; Munns 1977; Lie 1980).

Soil moisture

Legumes are intolerant to shortage (water stress) and excess (waterlogging) of water and this is primarily due to the ultrasensitivity of the symbiosis. In the field, it is common to obtain waterlogged conditions after a heavy rainfall, but a few days of sunny weather, combined with wind, will induce temporary wilting of the plants. In general the symbiotic system will recover from short exposures of water stress or waterlogging, but prolonged exposures may lead to permanent damage and shedding of the nodules (Wilson 1931*a*).

Water stress

Infection is restricted in dry soils owing to the absence of normal root hairs. Instead, short, stubby root hairs appear, which are inadequate for infection by *Rhizobium*. On watering, these abnormal root hairs may resume normal growth, resulting in a slender outgrowth, which eventually may become infected. Nodules initiated under conditions of adequate water supply, are retarded in growth if exposed to dry conditions, resulting in partially developed organs, embedded in the root cortex (Worrall and Roughley 1976).

The functioning of the nodule is severely restricted by water stress. Loss of water by evaporation from the nodule surface and especially by the export of nitrogenous material from nodules into the xylem, must be compensated for if the nodule is to function well. For *Pisum* a balance sheet was constructed of the water budget by Minchin and Pate (1973). It appears that the nodule acquires only 13 per cent of the water by surface uptake from the medium, 20 per cent from the phloem during the transport of carbon from the shoot, but the major part (67 per cent) is derived from lateral movement from the adjacent roots. This may explain the presence of active nodules in the upper dry strata of the soil, provided that the roots have access to water in deeper layers of the soil.

Shortage of water induces a rapid inactivation of nitrogen fixation, but the activity can be restored upon watering, if the moisture loss does not exceed 20 per cent of the fresh weight of the nodule (Sprent 1971). In the latter case, more permanent damage of the nodule structure results, often followed by shedding of the nodule (Wilson 1931*a*). Plants with restricted meristematic growth of the nodule (round nodules) are more sensitive than plants with elongated nodules, which can resume growth if water is supplied, owing to the meristimatic tissue at the nodule apex (Engin and Sprent 1973).

Water stress applied to detached nodules or nodulated roots, reduces both nitrogen fixation and respiration, and the reduction is, within certain limits, proportional to the degree of water loss of the nodules (Sprent 1971). During periods of drought, osmotic damage to fixation may occur owing to the high salt concentrations near or on the nodule (Sprent 1972). This may also occur in water culture, if the nodule zone is above the liquid, allowing the salts to concentrate on the nodule surface (Minchin and Pate 1975).

Interestingly, water stress also induces the formation of ethanol (Sprent and Gallacher 1976) a phenomenum commonly associated with oxygen deficiency under waterlogged conditions (van Straten and Schmidt 1974*a*, *b*). Pankhurst and Sprent (1975) suggest that water stress creates a barrier for oxygen and other gasses. Indeed part of the inhibitory effect can be alleviated by increasing the oxygen concentration.

It was observed that irrigation of field soil, after a period of drying increased nitrogen fixation more than ten-fold (Sprent 1976). In *Vicia faba* it was observed that nitrogen fixation decreased, once wilting of the lower leaves occurred. It is likely that photosynthesis in these leaves is arrested, and since the lower leaves are presumably the main providers of photosynthates to the nodules (Pate 1968), it is possible that the initial effect of drought is caused by a limited supply of the carbon source (Sprent 1976).

Waterlogging

A complex of factors, adversely affecting plant growth, is attributed to waterlogging but the major factor detrimental to root growth and function is presumably a low availability of oxygen. At the same time a build-up of carbon dioxide may occur, which at high concentrations may inhibit nodule formation (Grobbelaar, Clarke, and Hough 1971*a*).

Another gas known to be produced in anaerobic soils is ethylene (Lynch 1976), which, at very low concentrations, restricts nodulation (Grobbelaar *et al.* 1971*a*; Drennan and Norton 1972; Dart 1977; Lie, unpublished). Many products derived from microbial fermentation are phytotoxic, and contribute to the waterlogging syndrome (Lynch 1976).

As far as the author is aware, there is no detailed study of the effect of anaerobiosis on infection by *Rhizobium*. However, it is well known that root-hair formation is reduced at low oxygen concentrations, and therefore infection may be restricted. Nodule development is retarded and the nodules remain very small at low oxygen concentrations (Gallacher and Sprent 1978).

Excess water is particularly detrimental to nitrogen fixation. Even a thin layer of water on the nodule surface reduces fixation to almost zero (Sprent 1969; Schwinghamer, Evans, and Dawson 1970). Detached nodules immersed in water can be used for studies of nitrogen fixation if the oxygen concentration is increased to 80–90 per cent (Sprent 1969; Houwaard 1978), supporting the hypothesis that oxygen supply is the limiting factor.

Using detached nodules incubated in a small vessel creates anaerobiosis, and three volatile substances are produced: ethanol, acetaldehyde, and acetone (van Straten and Schmidt 1974*a*, *b*). The reduction of nitrogen fixation is inversely related to the accumulation of ethanol in the vessel, and added ethanol was found to be inhibitory to nitrogen fixation. However, controversy still exists as to the role of ethanol in intact nodulated plants (van Straten and Schmidt 1975).

In the field, a temporary layer of water around the nodule must occur regularly and this can be an important regulator of nitrogen fixation. With prolonged exposure to waterlogging, the plant may devise special structural changes on the nodule surface (ruptures, lenticells, protuberances) obviously to acquire more oxygen for the functioning of the nodule (Gallacher and Sprent 1978). There is a variation between the susceptibility amongst plants, *Pisum* being very susceptible whereas *V. faba* is rather tolerant excess water (Minchin and Pate 1975; Gallacher and Sprent 1978). A rapid adaptation of the nodulated root to low oxygen tensions was noted in intact *G. max* plants, indicating the flexibility of the intact system (Hardy, Criswell, and Havelka 1977).

Temperature

Temperature has a marked effect on the symbiosis, and it appears that nearly all stages of the development and functioning are more or less affected by temperature. It operates mainly in a non-specific way through plant metabolic processes such as respiration, photosynthesis, transport, and transpiration. The temperature range for the symbiotic system is narrower than that of the plant supplied with fertilizer nitrogen, and the symbiosis collapses when it is exposed to extreme temperatures.

All leguminous plants so far investigated have a normal Calvin photosynthetic cycle, with an optimum temperature of 15–25 °C. At higher temperatures photosynthesis is severely reduced. No tropical legumes have been found with C_4 photosynthesis, which functions optimally at 30–40 °C (Black 1973; Black, Brown, and Moore 1978).

The root system can be regarded as a heterotrophic organ, depending on the shoot for the supply of carbohydrates from photosynthesis. The amount used by the root is considerable, almost 50 per cent of the daily acquired photosynthates is used for respiration (Minchin and Pate 1973). It is well known that respiration is increased with higher temperatures, and this may imply that less carbon will be available for the symbiosis. Lowering the night temperature (Roponen, Valle, and Ettala 1970; Minchin and Pate 1974) increases the amount of nitrogen fixed, presumably by saving carbon from respiration. At high temperatures the amount of roots produced is low, and the roots produced are thin, often unbranching, and with very few laterals and root hairs (Lie 1974; Frings 1976).

Inhibitory effects of extreme temperatures

Nodulation of temperature legumes may take place, albeit slowly at temperatures as low as 7 °C (Stalder 1952; Roughley 1970), but tropical legumes are already adversely affected by temperatures around 20 °C (Mes 1959; Dart and Mercer 1965; Gibson 1971). Infection is delayed at low temperatures, but once infected, nodule development is very rapid (Roughley 1970). However, nodules formed at low temperatures contain low amounts of bacteroid-containing tissue; almost none at 7 °C (Roughley 1970). Nitrogen fixation itself is not cold-sensitive. Nodules formed at a higher temperature will even fix nitrogen at tempertures as low as 2 °C (Dart and Day 1971*a*). Of course there are differences in sensitivity to low temperatures among the plant species: tropical plants being more sensitive than temperate plants.

High temperatures reduce the numbers of lateral roots and root hairs. With *P. sativum* grown in water culture, very few root hairs were formed at 30 °C, a temperature known to be inhibitory to nodulation (Diener 1950; Lie 1974; Frings 1976). The few root hairs formed have an abnormal form, are bulbous and very short (Lie 1974; Frings 1976), and are obviously inadequate for infection by *Rhizobium*.

The inhibitory effect of high temperatures on nodulation is localized (Frings 1976), just as the effect of EDTA on nodulation is (Lie and Brotonegoro 1969). If the upper part of the root system is exposed to a temperature of 30 °C, nodules will only develop on the lower part, which is kept at 20 °C, the optimum temperature for *P. sativum*. The reverse is also true (Frings 1976).

At high temperatures a rapid degeneration of nodules takes place, resulting in a shortening of the period for nitrogen fixation. There are differences amongst the bacterial strains in the reponse to high temperatures (Pankhurst and Gibson 1973; Lie 1974), and selection of rhizobial strains for high temperatures is feasible.

Rhizobium strains from specific environments

Plants bearing effective nodules can be found under extreme environmental conditions, indicating that symbiotic systems adapted to these environments are available. In the arctic regions, legumes have been found with red nodules a few centimetres above the frozen soil layers (Allen, Allen, and Klebesadel 1964). On the other hand, nodulated legumes are present in tropical regions and in semi-arid regions where temperatures may reach high levels (Tadmor, Shannon, and Evenari 1971).

A systematic study was made of *Rhizobium* strains occurring naturally in an arctic environment, northern Scandinavia, and southern regions of Sweden. When the northern and southern *Rhizobium* strains were tested on clover at 20 °C, no or only small differences in nodulation and nitrogen fixation capacity were detected. When the comparison was made at 10 °C, however, the northern strains were superior to the southern strains, as a result of an earlier appearance of the nodules and a higher nitrogen-fixing activity (Ek-Jander and Fåhraeus 1971).

We recently tested a large number of *Rhizobium* strains isolated from *P. sativum* grown in Turkey and Israel (Middle East strains). Many of these strains performed poorly when tested on *P. sativum* growing at 20 °C. This was due to slow and poor nodulation, resulting in plants with few, small nodules. Nodulation and nitrogen fixation were improved when tested at 25 °C. This is in contrast with the European *Rhizobium* strains, which prove better at 20 than at 25 °C. In particular, a strain presumably coming from Scandinavia was severely inhibited at 25 °C, owing to a rapid degeneration of root nodules (Lie 1974; Table 4.1).

Temperature-sensitive (t_s) nodulation

Temperature-sensitive nodulation is a very useful tool to study nodulation since by simply varying the temperature, nodulation can be easily controlled. We obtained a *P. sativum* cv. Iran which, in symbiosis with certain *Rhizobium* strains, fails to form nodules at 18–20 °C, but does so when exposed for a short period to 25 °C. The time needed at the higher temperature is very short, 24 hours, and confined to the second and/or third day after inoculation (Lie 1971). During this period the nodulation process is cold-sensitive, i.e. when the plants were exposed to a low temperature during the second or third day, and otherwise growing at

TABLE 4.1

Nitrogen fixation of Rhizobium *strains from western Europe (PF_2), Scandinavia (310a), or the Middle East (Ng 1) on pea cv. Rondo growing at different temperatures (as percentage from PF_2 at 20 °C)*

Rhizobium strain	Temperature (°C)		
	20	25	30
PF_2	100	80	0
310a	90	25	0
Ng 1	7	52	0

the conducive temperature (25 °C), nodulation is also inhibited. Infection of the root hairs takes place and, when the critical period is passed, growth and fixation, proceed normally at 20 °C. The temperature-sensitive nodulation is host-controlled, and determined by a single dominant gene (Lie, HilleRisLambers, and Houwers 1976).

An effect of *Rhizobium* strains was also observed. Formerly, when only *Rhizobium* strains from cultivated *P. sativum* in Europe were available, only three classes were distinguished (Lie 1971). Recently, a new group of *Rhizobium* strains was isolated from soils of the Middle East, which nodulates pea cv. Iran abundantly at both temperatures (Lie 1978*a*; Table 4.2).

Light

Light affects the symbiosis mainly through photosyntheis, by controlling the supply of carbohydrates for the development and functioning of the nodule. Photosynthesis is independent of the wavelength of light, provided that the same number of quanta is absorbed by the chlorophyll of the plant. Rather high light intensities are required to obtain measurable effects, and within a certain range, a linear relationship exists between light intensity, nodulation, and nitrogen fixation (Lie 1964).

An additional effect of light on nodule morphogenesis was demonstrated and evidence has been provided for the role of phytochrome in nodule formation (Lie 1964, 1969*a*). This process is highly dependent on the wavelength of light and dramatic affects, e.g. stem elongation, and seed germination, are already obtained at very low light intensities.

Non-photosynthetic effects of light on nodulation

Control of root-nodule formation by phytochrome was first suspected

TABLE 4.2

Classification of Rhizobium *strains from Europe and the Middle East according to the capacity to nodulate pea cv. Iran and to fix nitrogen at 20 or 26 °C*

	European strains						Middle East strains	
Type	'Temperature-sensitive' type		'Intermediate' type		'Temperature-insensitive' type		'Temperature-insensitive' type	
Temperature (°C)	20	26	20	26	20	26	20	26
Nodule number per plant	0–1	5–10	1–5	10–30	10–20	10–30	>50	>50
Nitrogen fixed per plant	very low (I)		moderate (I/E)		moderate (I/E)		high (E)	
Rhizobium strains	PRE		PF_2, 313 etc. (18 strains)		310a		Tom. Ng etc. (50 strains)	

when leguminous plants, inoculated with *Rhizobium*, were grown in light of different wavelengths (Lie 1964, 1971). These results demonstrate that nodulation is poor in blue and maximal in red light. To eliminate errors due to differences in photosynthesis, light intensities of equal number of quanta were used so that approximately the same amount of dry matter was produced in all treatments. To prevent differences in stem elongation, symbiotic systems with limited shoot growth were used, i.e. decapitated plants and rooted leaves (Lie 1964, 1969*a*).

Further support was obtained by the demonstration that nodulation is under the red and far-red light control—classical evidence for the implication of the photoreversible phytochrome in the nodulation process (Lie 1964, 1969*a*, 1971*a*). Essentially, the plant shoots were exposed to a few minutes of red (660 nm) or far-red (730 nm) light at the end of the photoperiod, five days after inoculation with *Rhizobium*. Far-red light inhibited nodulation markedly, and this inhibitory effect is partially alleviated by subsequent irradiation with red light.

Similar effects were obtained when the roots only were exposed for a few minutes to red and far-red light (Lie 1969). This is in accordance with the fact that roots of etiolated pea plants contain phytochrome (Furuya and Hillman 1964). The formation of lateral roots is also under the red/far-red light control (Furuya and Torrey 1964). Earlier Rudin (1956) found that roots exposed for long periods (days) to white or blue light, formed few nodules.

An interesting observation was made that the interaction of red and far-red light on nodulation only took place if the same plant organ (either the shoot or the roots) was exposed to both light treatments, but not if each was given to different part of the plant. This implies that the effects of light are not transmissible. Further support was obtained from the fact that stem elongation only occurs if the shoot was treated with far-red light, but not if the roots were so treated (Lie 1969*a*),

We are still ignorant about the precise action of far-red light on nodulation, but by treating the plants at different days after inoculation with far-red light, the most sensitive period appears to be the second and third day after inoculation. Therefore, nodule initiation is presumably the target rather than infection. Antoniw and Sprent (1978) concluded from their experiments with plants growing in light cabinets with different ratios of red and far-red light, that the development of nodules is reduced by far-red light.

Under natural conditions, an excess of far-red light was found under a leaf canopy in forests (Vezina and Boulter 1966) and under *Medicago sativa* (Robertson 1966). This is due to a preferential filtering of red light by chlorophyll. Therefore, shading is not only detrimental by reducing photosynthesis, but presumably also by inhibiting root-nodule formation.

Photosynthesis and the energy supply

Nodulation and nitrogen fixation can be accomplished in the complete absence of light, provided that enough carbohydrates are available. This was shown for etiolated seedlings derived from big seeds (Wilson 1931*b*; McGonagle 1949; Lie 1969*a*) and root cultures (Raggio, Raggio, and Torrey 1965; Grobbelaar *et al.* 1971*a*). However, in general plants obtain their carbon and energy from photosynthesis. The importance of light on the symbiosis was established already more than fifty years ago (cf. Wilson 1940), but recently interest in this subject was revived by the demonstration that photosynthate is presumably the key limiting factor in nitrogen fixation under field conditions (Hardy and Havelka 1976).

Increasing the light intensity and/or the concentration of carbon dioxide, to improve photosynthesis, also favours nodulation and nitrogen fixation (cf. Wilson 1940; Hardy and Havelka 1976). Similar effects are obtained by directly spraying the leaves with sugars (van Schreven 1959). On the other hand, the application of combined nitrogen reduces nodulation and nitrogen fixation. It is now well established that the inhibitory effect of combined nitrogen is due to the diversion of photosynthates to the roots and hence the deprivation of the nodules of carbohydrates (Small and Leonard 1969). Bacteroids from plants, treated with combined nitrogen, still have nitrogenase activity, although the intact plant is severely inhibited in nitrogen fixation (Houwaard 1978). These findings support the hypothesis that the inhibition of nitrogen fixation in intact plants is due to a reduced supply of photosynthates to the bacteroids.

In the field, self-shading and mutual-shading reduce the light available to plants, especially to the lower leaves. Since the lower leaves are responsible for feeding of the nodules (Pate 1968), nitrogen fixation is severely reduced. The inhibitory effect is more intense when a closed canopy is reached within a crop (Hardy and Havelka 1976; Sprent 1976). Lodging of the crop also reduces nitrogen fixation. In the case of *Glycine max* inhibition of nitrogen fixation is observed within four days of lodging. The reduction can be so severe that the amount of nitrogen fixed by the lodged plants is only one-third of that of the unlodged control (Hardy and Havelka 1976). Planting density was also shown to affect nitrogen fixation and part of this effect is ascribed to differences in light availability within the crop.

Defoliation, either artificially or by grazing, reduces nitrogen fixation, and recovery is dependent on the rate of defoliation (Butler, Greenwood, and Soper 1959). On the other hand, increasing the size of the shoot by grafting on a second top, increases nitrogen fixation almost twofold (Streeter 1974).

In recent years much attention has been paid to the architecture of the plants to improve the light-capturing ability within a crop. In legumes this may well be important for the prolongation of nitrogen fixation. Yield would be considerably improved if the nitrogen-fixing period can be extended long enough to provide sufficient nitrogen during the period of high demand, the pod-filling stage.

Carbon requirements of the symbiotic system. The importance of photosynthates for the symbiosis is clearly shown by the detailed studies pertaining to the distribution of photosynthates and assimilated nitrogen to different part of the nodulated legume (Minchin and Pate 1973, 1974; Pate 1976, 1977; Pate and Herridge 1978). During the early vegetatitive period in *P. sativum*, about 74 per cent of the daily acquired photosynthate is transported downwards, 42 per cent being diverted to the roots, and 32 per cent to the nodules. Roughly half (15 per cent) of the carbon input of the nodules is returned to the shoot as fixation products and the rest is mainly used for respiration (12 per cent) and growth of the nodule (5 per cent). Of the carbon required by the roots, approximately 35 per cent is used for respiration, so that nodulated roots use approximately 50 per cent of the daily photosynthates for respiration (Minchin and Pate 1973; Pate 1976).

Other studies with *Vigna unguiculata* (Herridge and Pate 1977) and *Lupinus* sp (Pate and Herridge 1978) also show that roughly half of the photosynthates is used for respiration of the roots.

The amount of photosynthates cycled through the nodules in young nodulated *Lupinus*, *Pisum*, and *Vigna* plants is very high, 24–30 per cent of the net photosynthates, and it is reduced to approximately 10 per cent in older plants. About half of the carbon is returned to the shoot as nitrogenous compounds, and the remainder is largely used in respiration. The amount used for nodule growth is rather small, 5–8 per cent of the photosynthates in the early stages of plant growth (Minchin and Pate 1973; Herridge and Pate 1977; Pate and Herridge 1978). In Table 4.3 an estimate of the amount of photosynthates used for nitrogen fixation is given. There is a good agreement between the different legumes, 4–7 mg carbon is needed for each mg N_2 fixed. Interestingly, almost similar results were found for plants growing on nitrate. In peas 6.2 mg C is respired for each mg N-NO_3 and 5.9 mg C per mg N_2 fixed (Minchin and Pate 1973).

In plants like *Pisum* and *Vigna* the amount of downward transport decreases with age, particularly during the generative phase when the pods and seeds compete successfully for carbon and nitrogen with other plant organs. The demand for nitrogen during the seed filling stage can be so high that protein-nitrogen from other plant organs is required, inducing a rapid death of the plant tissue (the 'self-destruction' theory of Sinclair and de Wit (1975)). It seems of interest to maintain a steady supply of photosynthates to the roots, so that nitrogen fixation can be

TABLE 4.3

The requirement of carbon for nitrogen fixation in some legumes and the non-legume Alnus glutinosa

Plant species		mg C/mg N	mg CH_2O/mg N	References
Alnus glutinosa		4.8	12	Akkermans and van Dijk (1980)
Glycine max		7.6	19	Bond (1941)
Lupinus albus		4.0–6.5	10–16.3	Pate and Herridge (1978)
Pisum sativum		4.1	10.3	Minchin and Pate (1973)
P. sativum	day (18 °C)	5.4	13.6	Minchin and Pate (1974)
	night (12 °C)	2.1	5.3	Minchin and Pate (1974)
P. sativum		6.8	17	Mahon (1977)
Trifolium subterraneum		1.2–4	3–10	Gibson (1966)
Vigna unguiculata		6.8	17	Herridge and Pate (1977)

maintained during the critical stage of heavy demand of nitrogen. This is the case with *Lupinus albus* where a constant high level (50 per cent) of photosynthates is transported to the roots, also during the seed-filling stage (Pate and Herridge 1978).

Carbon dioxide and photorespiration

Wilson (1940) showed that carbon dioxide enrichment increased nodulation and nitrogen fixation by improving photosynthesis of plants growing in enclosed jars. Mulder and van Veen (1960) observed that carbon dioxide supplied to the roots, stimulated nitrogen fixation, presumably by dark assimilation of carbon dioxide by the roots and nodules. Indeed high levels of phosphoenolpyruvate carboxylase were detected in root nodules, but the net contribution of carbon dioxide assimilation is difficult to estimate, due to the high rate of decarboxylation reactions in the nodules (Lawrie and Wheeler 1975; Christeller, Laing, and Sutton 1977).

In field-grown *G. max* a dramatic increase in nitrogen fixation was observed by carbon dioxide enrichment. The enrichment of carbon dioxide in the air from 0.03 to 0.12 per cent resulted in a four to five-fold increase in nitrogen fixed. This was due to more root nodules formed, a higher efficiency of the nodules and a prolongation of the active period of nitrogen fixation (Hardy and Havelka 1976). Similar effects were obtained with *Arachis hypogea* and *P. sativum* (Hardy *et al.* 1977). Moreover, the contribution of nitrogen fixation increased from 25 per cent (control plants) to 80 per cent (carbon-dioxide-enriched plants) so that less nitrogen was taken up from the soil by the carbon-dioxide-treated plants.

In later experiments it was found that lowering the oxygen concentration in the plant atmosphere had almost the same effect on nitrogen fixation as increasing carbon dioxide (Quebedeaux, Havelka, Livak, and Hardy 1975). The conclusion was drawn that both increasing pCO_2 or lowering pO_2 reduced photorespiration and thus more photosynthates were available for nitrogen fixation. Lowering the pO_2 below ambient concentrations also prevents the reproductive growth of plants (Quebedeaux *et al.* 1975; Quebedeaux and Hardy 1976).

The observation that a reduction of photorespiration results in an increase of nitrogen fixation is of considerable interest. In legumes photorespiration may result in a loss of 30–50 per cent of the carbon dioxide taken up (Zelitch 1975). It is now well established that glycolate is the substrate for photorespiration. The production of glycolate is catalysed by the enzyme ribulose diphosphate (RuDP) carboxylase-oxygenase, which has a dual function depending on the CO_2/O_2 ratio: (i) at high CO_2/O_2 ratios, carboxylation of ribulose 1,5-diphosphate (RuDP) to produce two molecules of phospoglycerate and (ii) at low

CO_2/O_2 ratios, oxygenation of RuDP to form one molecule of glycerate and one molecule of glycolate. The latter is used for photorespiration (Jensen and Bahr 1977).

Several attempts have been made to reduce photorespiration in C_3 plants using chemicals to inhibit either the synthesis or the metabolism of glycolate (Zelitch 1975; Servaites and Ogren 1977), so far with little success (Servaites and Ogren 1977). Another possibility is the selection or modification of plants to obtain mutants with an altered RuDP carboxylase-oxygenase, which has a higher carboxylation and a lower oxygenation activity in air (Ogren 1976).

Recently, the hypothesis was advanced that C_4 plants utilize combined nitrogen more efficiently than the C_3 plants (Black *et al.* 1978). As a result, the yield of the C_4 plants can be almost twice as that of C_3 plants. The study of C_4 plants is of interest since it may help to increase net photosynthesis, and, therefore, nitrogen fixation in plants.

Energy waste by hydrogen evolution

Hydrogen evolution by root nodules was demonstrated in *G. max* and *V. unguiculata* (Hoch, Little, and Burris 1957; Bergersen 1963; Dart and Day 1971*a*). Negative results were obtained with *P. sativum* (Dixon 1967) and the non-legumes *Alnus glutinosa*, *Hippophaë rhamnoides*, and *Casuarina cunninghamiana* (Dart and Day 1971*b*). Hoch, Schneider, and Burris (1960) provided evidence that this type of hydrogen evolution requires oxygen and is catalysed by nitrogenase. The need for oxygen is obviously associated with the role of oxygen in the production of ATP in oxydative phosphorylation (Bergersen and Turner 1968).

The metabolic energy to reduce protons rather than nitrogen, can be regarded as a waste of energy otherwise allocated to nitrogenase. The quantity of hydrogen produced by root nodules represents a considerable amount of energy, and evolution of hydrogen therefore may signify an inefficient use of photosynthates by the nodule. Dixon (1967) found at first that *P. sativum* nodules do not produce hydrogen, but that gas was taken up if provided. This was due to hydrogenase present in the bacteroids (Dixon 1968, 1972). Earlier Phelps and Wilson (1941) reported the presence of hydrogenase in root nodules, but Dixon (1968) showed that the uptake of hydrogen is coupled with ATP production. The failure to observe net hydrogen production in *P. sativum* nodules (Dixon 1967) is due to an active hydrogenase, enabling the nodule to recover part of the energy otherwise lost as hydrogen. Using different *Rhizobium* strains on peas it was shown that the formation of hydrogenase depends on the *Rhizobium* strain used. Conversely by using the same *Rhizobium* strain on three host plants (*P. sativum*, *Vicia faba*, and *V. bengalensis*) the control of the host on hydrogenase formation was demonstrated (Dixon 1972).

Evans and coworkers developed methods to estimate the loss of energy by hydrogen evolution. In a series of papers, they showed that in the majority of agronomically important legumes, including *G. max*, *M. sativa*, and *Trifolium* spp, 30–60 per cent of the energy flow through nitrogenase is lost by hydrogen evolution. In cases where energy supply is limited, this means that if the hydrogen evolved can be utilized, nitrogen fixation must be considerably increased (Schubert and Evans 1976, 1977; Ruiz-Argüeso, Hanus, and Evans 1978; Evans, Ruiz-Argüeso, and Russell 1978). Interestingly, *V. unguiculata* and most of the non-leguminous plants like *Alnus* spp are very efficient losing almost no hydrogen.

It may be assumed that *Rhizobium* strains with an active hydrogenase may produce nodules with a higher nitrogen-fixing efficiency, owing to a better utilization of the photosynthates, assumed to be the limiting factor under field conditions. Using *Rhizobium* strains differing in the amount of hydrogen evolved, it was indeed found in *G. max* and *V. unguiculata* that the highest yield was obtained with *Rhizobium* strains, which produce no or only low amounts of hydrogen in air (Schubert, Jennings, and Evans 1978). However, more data are required to prove that the ability to recycle hydrogen is always correlated with high efficiency of nitrogen fixation and high yield.

4.4 Flexibility of the symbiotic system

Certain environmental stresses can be alleviated by better management procedures such as irrigation and drainage, liming, and application of fertilizers. Climatic factors such as light intensity, daylength, temperature, and seasonal changes, however, cannot be easily modified and we have to rely on the selection of symbiotic systems, adapted to the prevaling conditions. An important step is to select systems with appropriate life cycles, which are synchronized with the seasonal climatic changes. Such plants can be obtained in localities with a similar climate, as shown by the success of introducing plants from the Mediterranean into Australia (Morley and Katznelson 1965).

It is now well demonstrated that selection for *Rhizobium* strains adapted to some environmental stress is feasible (cf. Lie 1974). For the successful isolation of these specific *Rhizobium* strains it is imperative that a trap plant should be used which is not restricted in growth in the particular environment. We propose the use of wild or primitive legumes, related to the crop plants (Mulder, Lie, Dilz, and Houwers 1966; Lie 1978*a*) in view of their adaptability to the local environment and greater genetic variability (Bennett 1970).

Phenotypic variability

Adaptation may be due to phenotypic plasticity of the individual plant, and some environmental stresses can be partly relieved by compensatory mechanisms. This compensation is important to protect the plant against short-term environmental changes, often encountered in the field. The most general reaction is the regulation of the amount of nodules formed per plant. As shown by Nutman (1958) an inverse relationship exists between the number of nodules and the nodule size, so that the total mass of nodules formed remains more or less constant. In those cases where infection is restricted, e.g. in acid soils, the individual nodule can be very large (Jensen 1943; Lie 1969*b* and unpublished). This is also the case when the number of nodules is reduced by using Fe-EDTA, an inhibitor of nodule formation (Lie and Brotonegoro 1969).

Plants growing at low temperatures have nodules with a low specific activity and this is compensated by a greater amount of nodules formed (Gibson 1969; Roughley 1970) or a higher concentration of leghaemoglobin (Davidson, Gibson, and Birth 1970). Similarly, *P. sativum* inoculated with a *Rhizobium* strain with a low nitrogenase activity at 25 °C, produce almost twice the amount of nodules at this temperature in comparison with plants growing at the optimal temperature, 20 °C. This was achieved by a continuous production of new nodules at 25 °C, whereas at 20 °C nodulation was halted (Lie, unpublished).

Plants may also compensate by increasing the specific nitrogenase activity, for example when transferred to higher light intensities (Gibson 1976*a*). Increasing the supply of photosynthates by grafting a second shoot on a plant also doubled the nitrogenase activity within two days (Streeter 1974). However, in all cases the plants will produce more nodules within one week, so that the specific nitrogenase activity returns to the old level. An interesting observation was made in nodules of *Phaseolus* plants transferred from a low to a high light intensity (Antoniw and Sprent 1978). The increase in nodule size, after transfer to a higher light intensity, is mainly due to an increase in size of the plant cells, containing bacteroids. These authors also showed that the effect of light intensity is expressed in the amount of nodules formed, and that the specific activity of the nodules is the same at high and at low intensities. All these results suggest that an increase of photosynthate causes a transient increase of specific nitrogenase activity. After a few days, however, more nodules are formed and although the total nitrogenase activity is increased, the specific activity returns to the same old level.

A rapid adaptation to changes in pO_2 was also observed in root nodules when attached to the intact plants. Decreasing the pO_2 from 0.21 atm to 0.06 atm reduced nitrogen fixation to 55–58 per cent during the first hours. However, within 24 hours the nitrogen-fixing activity was returned to the old level, indicating that adaptation to low oxygen concentration had taken place. Nodules adapted to low oxygen levels, when returned to a pO_2 of 0.21 atm, had to readapt again to the new situation (Hardy *et al.* 1977). Such an adaptation to pO_2 was not observed with detached nodules and the inhibitory effect of low oxygen levels on fixation was also more pronounced than in the intact plants. Therefore, some caution must be exercised to relate the data obtained from detached nodules and from intact plants.

Genotypic variability

Because of their agronomic importance our knowledge of symbiotic nitrogen fixation is mainly based on data obtained from advanced or cultivated legumes. In contrast to the primitive and wild genotypes, the cultivated plants are extremely uniform owing to a limited genetic base. One of the reasons for this is that many crop plants are now cultivated in regions far from their native habitat (cf. Simmonds 1976). The number of introductions is usually limited and the number of successful genotypes in a certain region is even more restricted. It is also understandable that only the successful introductions are further used for breeding. Therefore, the longer the plant is under cultivation in a certain area, the greater the uniformity will be (Bennett 1970). High-yielding cultivated genotypes are now widely used, and by replacing the primitive genotypes they endanger our source of genetic variability.

How dangerously narrow the genetic base of many important crop plants can be was shown in a survey in the United States. Although many varieties of each crop are available, only a few major varieties dominate, covering almost 90 per cent of the planted area of a crop (cf. Day 1974). Currently used *G. max* cultivars in the United States, for example are related to ten or eleven varieties; even these lines have many common genetic backgrounds, nearly all coming from Manchuria. In the most productive varieties the genetic material Lincoln is repeatedly being introduced (Hartwig 1973) and the cultivar Mandarin can be traced back in 84 per cent of the cultivars grown in the northern-central states of USA (Hymowitz 1976).

Symbiotic uniformity of cultivated legumes

The capacity to fix nitrogen is obviously an important factor for a plant to survive under natural conditions and in primitive agriculture, where soil nitrogen is generally scarce. The well-nodulated plant, yielding better than the others, is presumably selected by the farmer and used for further breeding. We surmised that the homogeneous response of our modern cultivated legumes to *Rhizobium* is caused by a common genetic trait introduced long ago in the common ancestors of our cultivated crops, and perpetuated in modern genotypes.

In the breeding programme of *P. sativum* in Sweden, the pea cv. Monopol has been repeatedly used and many high-yielding cultivars can be traced back to this genotype (Gelin and Blixt 1964). Interestingly, this genotype is characterized by an extraordinarily high nodulating capacity, and it is tempting to assume that this property has contributed to the success of the high-yielding varieties. In Wageningen we have embarked in a programme to study the genetics of nodulation of pea plants. In our earlier experiments we observed that nodulation of the cultivated genotypes, although coming from different parts of the world, shows very little variation (Lie *et al.* 1976). This is also true for cultivated peas growing in tropical countries at high altitudes (Lie, unpublished). Further experiments showed that large variations in the symbiosis can only be detected in wild or primitive pea genotypes (Lie 1978*a*).

It was also assumed that the clover cross-inoculation group, which consists of a single genus *Trifolium*, is very homogeneous. This was indeed the case with *Trifolium* spp from temperate regions but not when clovers from the Caucasus (*T. ambiguum*) and from the highlands in East and Central Africa (*T. tembense*, *T. semipilosum*, etc.) were included (Norris and 't Mannetje 1964). A strong symbiotic specialization had taken place in the African and Caucasean clovers, presumably due to the development for a long period in isolated regions. The specialization can be so intense that even different introductions, collected from the same species from nearby places, will only form an effective symbiosis with *Rhizobium* strains isolated from their own growth area (Norris and 't Mannetje 1964).

A rapid change due to selection and breeding is illustrated in the case of the successful breeding of *T. ambiguum* for Australian alpine regions. This species, consists of a mixture of diploid and polyploid plants. It requires specific *Rhizobium* strains and nodulation is particularly slow, even with the best *Rhizobium* strain available. Within the plant population a small percentage of plants nodulate promptly and by selection and crossing within these early-nodulating plants over four generations, a suitable cultivar was obtained which fixed nitrogen satisfactory (Hely 1972). However, the genetic variability in nitrogen fixation was probably reduced and the selection was carried out with only one *Rhizobium* strain.

The performance of a symbiosis is equally dependent on the host plant and the *Rhizobium* strains present in the soil. There is reason to believe that selection and breeding will also result in plant genotypes, not only adapted to local conditions but also with the highest affinity to the local dominant *Rhizobium* strains. This type of specialization can be achieved within a short period, if the number of *Rhizobium* strains is rather limited, which is presumably the case if plants are introduced in a new area with some casual *Rhizobium* strains on the seeds. This may well explain the small variability in nitrogen fixation of *T. subterraneum* in Australia (Nutman 1961). In Canada the trait 'creeping-rootedness' was introduced in *M. sativa* by using *M. falcata* (creeping alfalfa) in the crossing programme. When introduced in Australia, these Canadian cultivars were found to be poor in nitrogen fixation with local *Rhizobium* strains, and this may well explain their lack of success under Australian conditions. The best *Rhizobium* strains for these plants were obtained from nodules of plants growing in Canada, demonstrating that symbiotic specialization with local *Rhizobium* strains had taken place during the process of breeding (Gibson 1962).

In *G. max*, failure to form root nodules or to fix nitrogen in some important cultivars can be traced back to some introductions from Manchuria. These results show how dangerous it can be if the symbiotic properties of the legume are neglected in the breeding programme.

Gene centres as a source for genetic variability

Vavilov (1951) demonstrated that certain regions of the earth are particularly rich in genetic variants of plants, related to or even ancestral forms of our cultivated crop plants. These regions, originally called 'centres of origin' by Vavilov are now known as 'centres of diversity' in view of the abundance of wild or primitive genotypes in a restricted area (Zohari 1970).

The value of these centres as a source for disease resistance in plants is well established (Leppik 1970; Watson 1970; Day 1974). We demonstrated that these centres are also an important source for genetic variation of host plants and *Rhizobium* for symbiotic nitrogen fixation (Lie 1978*a*). As mentioned in the previous section all the cultivated pea genotypes show little variation in nodulation and nitrogen fixation when in symbiosis with *Rhizobium* strains isolated from cultivated peas from Europe (European strains). The two exceptional, non-nodulating lines belong to primitive pea genotypes from Iran and Afghanistan (Lie 1971; Lie *et al.* 1976), one of the centres of diversity of peas (Govorov 1928). When more primitive and wild pea genotypes were analysed, it appears that symbiotic variability is very large in the primitive genotypes and that the trait 'non-nodulation' is widespread in peas of Afghanistan (Table 4.4). There is another noteworthy observation: peas belonging to the ecotypes abyssinicum and fulvum, occurring respectively in Ethiopia and Israel, form nodules with no or low nitrogenase activity (Lie 1978; Table 4.5). Further experiments showed that these symbiotic defects are due to the use of non-compatible European *Rhizobium* strains. These primitive and wild genotypes require specific *Rhizobium* strains occurring only in their centres of diversity as shown for the pea cv. Afghanistan (Lie 1978*a*; Table 4.6). Interestingly, the distribution of *Rhizobium* for the cultivated pea is much more widespread, indicating its wide disemination by man.

Indigenous *Rhizobium* strains, a valuable gene pool

When a leguminous crop is introduced in a new area, one of the problems agronomists have to face, is the presence of *Rhizobium* strains from related, wild or primitive legumes. In general these *Rhizobium* strains are not very effective in nitrogen fixation when in symbiosis with the cultivated crop. This is also the case with the majority of *Rhizobium* strains, isolated in Turkey and Israel, when in symbiosis with *P. sativum* cv. Rondo (Lie, unpublished results). These bacteria are a nuisance

TABLE 4.4

Symbiotic classification of primitive Pisum *genotypes from different geographical regions*[a]

Class[b]	Ethiopia	Israel	Turkey	Iran	Afghanistan	India
Resistant	0	4	0	1	19	0
Intermediate	11	20	1	2	10	4
Susceptible	7	0	8	1	7	2

[a] Lie (1978*a*).

[b] No nodulation with all *Rhizobium* strains used (restraint), nodulation with only a few strains (intermediate), and nodulation with all strains (susceptible). The *Rhizobium* strains were isolated from cultivated peas in Europe.

TABLE 4.5

Symbiotic properties of Rhizobium *strains isolated from cultivated peas in Europe (PF_2) and those isolated from wild or primitive peas in Turkey (Tom) and Israel (F13) on different pea ecotypes*

Pisum sativum	*Rhizobium* strain		
	PF_2	Tom	F13
cv. Rondo (ecotype sativum, cultivated pea)	+++ (E)	+++ (E)	+ (E/I)
cv. Afghanistan (ecotype asiaticum)	–	+++ (E)	+ (E/I)
cv. H 18 (ecotype humile)	–	+++ (E)	+++ (E)
cv. WP 7370 (ecotype abyssinicum)	++ (I)	+++ (E)	+++ (E)
cv. Fu 3 (ecotype fulvum)	++ (I)	++ (I)	+++ (E)

–, no nodules; +, a few; ++, moderate number and +++, abundant nodules.
I, no nitrogen fixation; E/I, moderate; and E, high levels of nitrogen fixation.

TABLE 4.6

Distribution of Rhizobium *strains capable of nodulating pea cvs. Rondo and Afghanistan in soils of different countries*

Country	No. of soil samples	No. soils containing *Rhizobium*	
		cv. Rondo	cv. Afghanistan
Afghanistan	1	1	1
Egypt	8	7	0
Ethiopia	1	1	0
Europe	55	12	0
Greece	8	7	0
Iran	6	6	6
Israel	55	52	52
Kenya	3	0	0
Mali	3	0	0
South Africa	4	4	0
Tunisia	1	1	0
Turkey	120	115	115

since they compete with the introduced *Rhizobium* strains for sites on the roots of the plant.

On the other hand it can be demonstrated that these indigenous bacteria have useful properties not found in *Rhizobium* strains from cultivated crops. This is shown in our search for acid-tolerant *Rhizobium* strains for example. Many well-nodulated wild legumes were found in a small, acid plot in the Apenines. A large number of *Rhizobium* strains were isolated from indigenous clovers, including *Trifolium medium*. One of these *Rhizobium* strains from *T. medium* proved to be acid tolerant in symbiosis with *T. pratense* (Mulder *et al.* 1966).

Another example can be found in our current research on *Rhizobium* strains from primitive peas in the Middle East. The results shown in Tables 4.5 and 4.6 demonstrate that the European strains are highly specialized and almost restricted to the cultivated pea, whereas the Middle East strains are very promiscuous, capable of nodulating nearly all ecotypes. The promiscuous character of the Middle East *Rhizobium* strains may be of interest if we need *Rhizobium* strains with a wide host range. Another interesting feature is that some of these Middle East strains have a higher temperature optimum for symbiosis, even when used with the cultivated pea (Lie, Soe-Agnie, Muller, and Göktan 1980; Table 4.1).

There is a growing concern that advanced, high-yielding cultivars tend to replace the related primitive and wild plants in large areas.

This is particularly dangerous when the plants are introduced in the centres of diversity, e.g. the introduction of high-yielding Mexican wheat in Turkey (Bennett 1970). Valuable plant material has been lost rapidly in the last decades, and the establishment of a gene pool to conserve wild and plant material is well recognized.

We suspect that it might also happen with *Rhizobium*. Although this bacterium can live as a common saprophyte in the soil, its distribution and multiplication is closely related to the presence of the host plant (Nutman 1963). Normally, sown legumes are inoculated with an effective *Rhizobium* strain. Besides being effective, the inoculant strain should compete strongly with other bacteria to nodulate the host plant and to survive in large numbers in the soil (Parker, Trinick, and Chatel 1977). In the long run, therefore, repeated use of *Rhizobium* strains as an inoculant may lead to the replacement of the multifarious microbial flora by a few dominant *Rhizobium* strains. The situation is aggravated by the disappearance of indigenous leguminous plants, the usual hosts of native *Rhizobium* strains.

Evaluation of *Rhizobium* strains from wild and primitive legumes has hardly started, but from the few data available the impression is that these bacteria, similar to wild plants, have a greater genetic variability than *Rhizobium* strains from cultivated legumes. Perhaps they deserve more attention, and selection within this group is more rewarding than in the specialized *Rhizobium* strains from cultivated crops.

References

Akkermans, A. D. L. and Dijk, C. van (1980). In *Nitrogen fixation*, Vol. 1: *Ecology* (ed. W. J. Broughton) pp. 56–103. Oxford University Press.

Allen, E. K., Allen, O. N., and Klebesadel, L. J. (1964). In *Science in Alaska, Proc. 14th Alaska Science Conference* (ed. G. Dahlgren) p. 54. Anchorage, Alaska.

Antoniw, L. D. and Sprent, J. I. (1978). *Ann. Bot.* **42**, 399.

Bennett, E. (1970). In *Genetic resources in plants* (ed. O. H. Frankel and E. Bennett) p. 115. Blackwell, Oxford.

Berg, E. H. R. van de (1978). *Pl. Soil* **48**, 629.

Bergersen, F. J. (1957). *Aust. J. biol. Sci.* **10**, 233.

—— (1963). *Aust. J. biol. Sci.* **16**, 669.

—— (1974). In *The biology of nitrogen fixation* (ed. A. Quispel) p. 473. North Holland, Amsterdam.

—— and Turner, G. L. (1968). *J. gen. Microbiol.* **53**, 205.

Bhuvaneswari, T. V. and Bauer, W. D. (1978). *Pl. Physiol., Lancaster* **62**, 71.

Black, C. C. (1973). *A. Rev. Pl. Physiol.* **24**, 253.

—— Brown, R. H., and Moore, R. C. (1978). In *Limitations and potentials for biological nitrogen fixation in the tropics* (ed. J. Döbereiner, R. H. Burris, and A. Hollaender) p. 95. Plenum Press, New York.

Bohlool, B. B. and Schmidt, E. L. (1974). *Science, N.Y.* **185**, 269.

—— —— (1976). *J. Bact.* **125**, 1188.

Bond, G. (1941). *Ann. Bot.* **5**, 313.

BRETHAUER, T. S. and PAXTON, J. D. (1977). In *Cell wall biochemistry related to specificity in host–plant pathogen interactions* (ed. B. Solheim and J. Raa) p. 381. Universiteitsforlaget, Oslo.

BUTLER, G. W., GREENWOOD, R. M., and SOPER, K. (1959). N.Z. *Jl agric. Res.* **2**, 415.

CHANDLER, M. (1978). *J. exp. bot.* **29**, 749.

CHRISTELLER, J., LAING, W. A., and SUTTON, W. D. (1977). *Pl. Physiol., Lancaster* **60**, 47.

COOPER, J. P. (1970). In *Genetic recourses in plants* (ed. O. H. Frankel and E. Bennett) p. 131. Blackwell, Oxford.

CURRIER, W. W. and STROBEL, G. A. (1976). *Pl. Physiol., Lancaster* **57**, 820.

—— —— (1977). *FEMS Microbiol. Lett.* **1**, 243.

DART, P. J. (1974). In *The biology of nitrogen fixation* (ed. A. Quispel) p. 381. North Holland, Amsterdam.

—— (1977). In *A treatise on dinitrogen fixation III* (ed. R. W. F. Hardy and W. S. Silver) p. 381. John Wiley, New York.

—— and DAY, J. (1971*a*). In *Biological nitrogen fixation in natural and agricultural habitats* (ed. T. A. Lie and E. G. Mulder) *Pl. Soil*, Special Volume, p. 167. Martinus Nijhoff, The Hague.

—— —— (1971*b*). Rothamsted Experimental Station Report for 1970, part 1, p. 86.

—— and MERCER, F. V. (1965). *Aust. J. agric. Res.* **16**, 321.

DAVIDSON, J. L., GIBSON, A. H., and BIRCH, J. W. (1970). In *Proc. 11th Int. Grassl. Cong.*, p. 542. Surfer's Paradise, Queensland.

DAY, P. R. (1974). *Genetics of host–parasite interaction*, p. 190. W. H. Freeman, San Francisco.

DAZZO, F. B. and BRILL, W. J. (1977). *Appl. Environ. Microbiol.* **33**, 132.

—— and HUBBELL, D. H. (1975*a*). *Appl. Microbiol.* **30**, 172.

—— —— (1975*b*). *Appl. Microbiol.* **30**, 1017.

DEGENHARDT, T. L., LARUE, T. A., and PAUL, E. A. (1976). *Can. J. Bot.* **54**, 1633.

DIENER, T. (1950). *Phytopath. Z.* **16**, 129.

DIXON, R. O. D. (1967). *Ann. bot.* **31**, 179.

—— (1968). *Archs Mikrobiol.* **62**, 272.

—— (1972). *Archs Mikrobiol.* **85**, 193.

DÖBEREINER, J. and CAMPELO, A. B. (1977). In *A treatise on dinitrogen fixation* (ed. R. W. F. Hardy and A. H. Gibson) Section IV, p. 191. John Wiley, New York.

DRENNAN, D. S. H. and NORTON, C. (1972). *Pl. Soil* **35**, 53.

EGERAAT, A. W. S. M. VAN (1975*a*). *Pl. Soil* **42**, 367.

—— (1975*b*). *Pl. Soil* **42**, 381.

EK-JANDER, J. and FÅHRAEUS, G. (1971). In *Biological nitrogen fixation in natural and agricultural habitats* (ed. T. A. Lie and E. G. Mulder) *Pl. Soil*, Special Volume, p. 129. Martinus Nijhoff, The Hague.

ENGIN, M. and SPRENT, J. I. (1973). *New Phytol.* **72**, 117.

EVANS, H. J., RUIZ-ARGÜESO, T., and RUSSELL, S. A. (1978). In *Limitations and potentials for biological nitrogen fixation in the tropics* (ed. J. Döbereiner, R. H. Burris, and A. Hollaender) p. 209. Plenum Press, New York.

FÅHRAEUS, G. (1957). *J. gen. Microbiol.* **16**, 374.

—— and LJUNGGREN, H. (1967). *Physiologia Pl.* **12**, 145.

—— and SAHLMAN, R. (1977). *Annls Acad. Reg. Sci. Upsaliensis* (Kungl. Vetenskaps samhällets i Uppsala årsbok) **20**, 107.

FRINGS, J. F. J. (1976). The *Rhizobium*–pea symbiosis as affected by high temperatures. Thesis Wageningen. H. Veenman en Zonen B.V., Wageningen.

FURUYA, M. and HILLMAN, W. S. (1964). *Planta* **63**, 31.

—— and TORREY, J. G. (1964). *Pl. Physiol., Lancaster* **39**, 987.

GALLACHER, A. E. and SPRENT, J. I. (1978). *J. exp. Bot.* **29**. 413.

GELIN, O. and BLIXT, S. (1964). *Agric. Hort. Genet.* **22**, 149.
GIBSON, A. H. (1962) *Aust. J. agric. Res.* **13**, 388.
—— (1966). *Aust. J. biol. Sci.* **19**, 499.
—— (1969). *Aust. J. biol. Sci.* **22**, 829.
—— (1971). In *Biological nitrogen fixation in natural and agricultural habitats* (ed. T. A. Lie and E. G. Mulder) *Pl. Soil*, Special Volume, p. 139. Martinus Nijhoff, The Hague.
—— (1976*a*). In *Symbiotic nitrogen fixation in plants* (ed. P. S. Nutman) p. 385. Cambridge University Press.
—— (1976*b*). In *Proc. 1st Int. Symp. Nitrogen fixation* (ed. W. E. Newton and C. J. Nyman) Vol. 2, p. 401. Washington State University Press, Pullman.
—— (1977). In *A treatise on dinitrogen fixation* (ed. R. W. F. Hardy and A. H. Gibson) Section IV, p. 400. John Wiley, New York.
GOVOROV, L. J. (1928). *Trudy prikl. Bot. Genet. Selek.* **19**, 497.
GROBBELAAR, N., CLARKE, B., and HOUGH, M. C. (1971*a*). In *Biological nitrogen fixation in natural and agricultural habitats* (ed. T. A. Lie and E. G. Mulder) *Pl. Soil*, Special Volume, p. 203. Martinus Nijhoff, The Hague.
—— —— —— (1971*b*). In *Biological nitrogen fixation in natural and agricultural habitats* (ed. T. A. Lie and E. G. Mulder) *Pl. Soil*, Special Volume, p. 215. Martinus Nijhoff, The Hague.
HAMBLIN, J. and KENT, S. P. (1973). *Nature, Lond.* **245**, 28.
HARDY, R. W. F. and HAVELKA, U. D. (1976). In *Symbiotic nitrogen fixation in plants* (ed. P. S. Nutman) p. 421. Cambridge University Press.
—— CRISWELL, J. G., and HAVELKA, U. D. (1977). In *Recent developments in nitrogen fixation* (ed. W. Newton, J. R. Postgate, and C. Rodriguez-Barrueco) p. 451. Academic Press, London.
HARTWIG, E. G. (1973). In *Soybeans: improvement, production and uses* (ed. B. E. Caldwell) p. 187. American Society of Agronomy, Madison.
HELY, F. W. (1972). *Aust. J. agric. Res.* **23**, 437.
HERRIDGE, D. F. and PATE, J. S. (1977). *Pl. Physiol., Lancaster* **60**, 759.
HOCH, G. E., LITTLE, H. N., and BURRIS, R. H. (1957). *Nature, Lond.* **179**, 430.
—— SCHNEIDER, K. C., and BURRIS, R. H. (1960). *Biochim. biophys. Acta* **37**, 273.
HOUWAARD, F. (1978). *Appl. Environ. Microbiol.* **35**, 1061.
HUBBELL, D. H. (1970). *Bot. Gaz.* **131**, 337.
—— MORALES, V. M. and UMALI-GARCIA, M. (1978). *Appl. environ. Microbiol.* **35**, 210.
HYMOWITZ, T. (1976). In *Evolution of crop plants* (ed. N. S. Simmonds) p. 159. Longman, London.
JENSEN, H. L. (1943). *Proc. Linn. Soc. N.S.W.* **68**, 207.
JENSEN, R. G. and BAHR, J. T. (1977). *A. Rev. Pl. Physiol.* **28**, 379.
LAW, I. J. and STRIJDOM, B. W. (1977). *Soil Biol. Biochem.* **9**, 79.
LAWRIE, A. C. and WHEELER, C. T. (1975). *New Phytol.* **74**, 437.
LEPPIK, E. E. (1970). *A. Rev. Phytopath.* **8**, 323.
LIBBENGA, K. R. and BOGERS, R. J. (1974). In *The biology of nitrogen fixation* (ed. A. Quispel) p. 430. North Holland, Amsterdam.
LIE, T. A. (1964). Nodulation of leguminous plants as affected by root secretions and red light. Thesis, Wageningen. Veenman en Zonen, N.V., Wageningen.
—— (1969*a*). *Pl. Soil* **30**, 391.
—— (1969*b*). *Pl. Soil*, **31**, 391.
—— (1971). In *Biological nitrogen fixation in natural and agricultural habitats* (ed. T. A. Lie and E. G. Mulder) *Pl. Soil*, Special Volume, p. 117. Martinus Nijhoff, The Hague.
—— (1974). In *The biology of nitrogen fixation* (ed. A. Quispel) p. 555. North Holland, Amsterdam.
—— (1978*a*). *Ann. Appl. Biol.* **88**, 445.

—— (1980). In *Handbook of nutrition and food* (ed. M. Rechigl Jr) C.R.C. Press, Cleveland, Ohio.
—— and BROTONEGORO, S. (1969). *Pl. Soil* **30**, 339.
—— HILLERISLAMBERS, D., and HOUWERS, A. (1976). In *Symbiotic nitrogen fixation in plants* (ed. P. S. Nutman) p. 319. Cambridge University Press.
—— SOE-AGNIE, I. E., MULLER, G. J. L., and GÖKTAN, D. (1980). In *Soil microbiology and plant nutrition* (ed. W. J. Broughton, C. K. Sohn, J. C. Rajarao, and B. Lim) p. 194. University of Malaya Press, Kuala Lumpur.
LILLICH, T. T. and ELKAN, G. H. (1968). *Can. J. Microbiol.* **14**, 617.
LJUNGGREN, H. and FAHRAEUS, G. (1961). *J. gen. Microbiol.* **26**, 521.
LYNCH, J. M. (1976). *Crit. Rev. Microbiol.* **5**, 67.
MCGONAGLE, M. P. (1949). *Proc. R. Soc. Edinb.* **B63**, 219.
MAHON, J. D. (1977). *Pl. Physiol., Lancaster* **60**, 817.
MAIER, R. J. and BRILL, W. J. (1976). *J. Bact.* **127**, 763.
MARSHALL, K. C., CRUICKSHANK, R. H., and BUSHBY, H. V. A. (1975). *J. gen. Microbiol.* **91**, 198.
MENZEL, G., UHLIG, H., and WEICHSEL, G. (1972). *Zentbl. Bakt. ParasitKde. Abt. II* **127**, 348.
MES, M. G. (1959). *Nature, Lond.* **184**, 2032.
MINCHIN, F. R. and PATE, J. S. (1973). *J. exp. Bot.* **24**, 259.
—— —— (1974) *J. exp. Bot.* **25**, 295.
—— —— (1975). *J. exp. Bot.* **26**, 60.
MORLEY, F. W. and KATZNELSON, J. (1965). In *The genetics of colonizing species* (ed. H. G. Baker and G. L. Stebbins) p. 269. Academic Press, New York.
MULDER, E. G. and VEEN, W. L. VAN (1960). *Pl. Soil* **13**, 265.
—— LIE, T. A., and HOUWERS, A. (1977). In *A treatise on dinitrogen fixation* (ed. R. W. F. Hardy and H. Gibson) Section IV, p. 221. John Wiley, New York.
—— —— DILZ, K., and HOUWERS, A. (1966). In *Proc. 9th Int. Congr. Microbiol.*, Moscow, p. 133.
MUNNS, D. N. (1977). In *A treatise on dinitrogen fixation* (ed. R. W. F. Hardy and A. H. Gibson) Section IV, p. 353. John Wiley, New York.
NORRIS, D. O. and 'T MANNETJE, L. (1964). *E. Afr. agric. For. J.* **29**, 214.
NUTMAN, P. S. (1956). *Biol. Rev.* **31**, 109.
—— (1958). In *Nutrition of the legumes* (ed. E. G. Hallsworth) p. 87. Butterworth, London.
—— (1961). *Aust. J. agric. Res.* **12**, 212.
—— (1963). *Symbiotic associations* (ed. P. S. Nutman and B. Mosse) p. 51. Cambridge University Press.
—— (1969). *Proc. R. Soc.* **B172**, 417.
OGREN, W. L. (1976). In *CO_2 metabolism and plant productivity* (ed. R. H. Burris and C. C. Black) p. 119. University Park Press, Baltimore, Maryland.
PANKHURST, C. E. and GIBSON, A. H. (1973). *J. gen. Microbiol.* **74**, 219.
—— and SPRENT, J. I. (1975). *J. exp. Bot.* **26**, 287.
PATE, J. S. (1968). In *Recent aspects of nitrogen metabolism in plants* (ed. J. Hewitt and C. V. Cutting) p. 219. Academic Press, London.
—— (1976). In *Symbiotic nitrogen fixation in plants* (ed. P. S. Nutman) p. 335. Cambridge University Press.
—— (1977). In *A treatise on dinitrogen fixation* (ed. R. W. F. Hardy and W. Silver) Section III, p. 473. John Wiley, New York.
—— and HERRIDGE, D. F. (1978). *J. exp. Bot.* **29**, 401.
PARKER, C. A., TRINICK, M. J., and CHATEL, D. L. (1977). In *A treatise on dinitrogen*

fixation (ed. R. W. F. Hardy and A. H. Gibson) Section IV, p. 331. John Wiley, New York.
Phelps, A. S. and Wilson, P. W. (1941). *Proc. Soc. exp. Biol.* **47**, 473.
Philips, D. A. and Torrey, J. G. (1970). *Physiologia Pl.* **23**, 1057.
—— —— (1972). *Pl. Physiol., Lancaster* **49**, 11.
Prazmowski, A. (1870). *Landwn VersStnen* **37**, 161.
Quebedeaux, B. and Hardy, R. W. F. (1976). In *CO_2 metabolism and plant productivity* (ed. R. H. Burris and C. C. Black) p. 177. University Park Press, Baltimore.
—— Havelka, U. D., Livak, K. L., and Hardy, R. W. F. (1975). *Pl. Physiol., Lancaster* **56**, 761.
Raggio, M., Raggio, N., and Torrey, J. G. (1965). *Pl. Physiol., Lancaster* **40**, 601.
Robertson, G. H. (1966). *Ecology* **47**, 640.
Roponen, I. E., Valle, E., and Ettala, T. (1970). *Physiologia Pl.* **23**, 1198.
Roughley, R. J. (1970). *Ann. Bot.* **34**, 631.
Rudin, P. E. (1956). *Phytopath. Z.* **26**, 57.
Ruiz-Argüeso, T., Hanus, J., and Evans, H. J. (1978). *Archs Microbiol.* **166**, 113.
Sahlman, K. and Fahraeus, G. (1963). *J. gen. Microbiol.* **33**, 425.
Schreven, D. A. van (1959). *Pl. Soil* **11**, 93.
Schwinghamer, E. A., Evans, H. J., and Dawson, M. D. (1970). *Pl. Soil* **33**, 192.
Schubert, K. R. and Evans, H. J. (1976). *Proc. natn. Acad. Sci. U.S.A.* **73**, 1207.
—— —— (1977). In *Recent developments in nitrogen fixation* (ed. W. Newton, J. R. Postgate, and C. Rodriguez-Barrueco) p. 469. Academic Press, London.
—— Jennings, N. T., and Evans, H. J. (1978). *Pl. Physiol., Lancaster* **61**, 398.
Servaites, J. C. and Ogren, W. L. (1977). *Pl. Physiol., Lancaster* **60**, 461.
Simmonds, N. W. (1976). *Evolution of crop plants.* Longman, London.
Sinclair, T. R. and Wit, C. T. de (1975). *Science, N.Y.* **189**, 565.
Small, J. G. C. and Leonard, O. A. (1969). *Am. J. Bot.* **56**, 187.
Solheim, B. and Raa, J. (1971). *Pl. Soil* **35**, 275.
—— —— (1973). *J. gen. Microbiol.* **77**, 241.
Sprent, J. I. (1969). *Planta* **88**, 372.
—— (1971). *New Phytol.* **70**, 9.
—— (1972). *New Phytol.* **71**, 603.
—— (1976). In *Symbiotic nitrogen fixation in plants* (ed. P. S. Nutman) p. 405. Cambridge University Press.
—— and Gallacher (1976). *Soil Biol. Biochem.* **8**, 317.
Stalder, L. (1952). *Phytopath. Z.* **18**, 376.
Straten, J. van and Schmidt, E. L. (1974*a*). *Soil Biol. Biochem.* **6**, 231.
—— —— (1974*b*). *Soil Biol. Biochem.* **6**, 347.
—— —— (1975). *Appl. Microbiol.* **29**, 432.
Streeter, J. G. (1974). *J. exp. bot.* **25**, 189.
Tadmor, N. H., Shannon, L., and Evenari, M. (1971). *Agron. J.* **63**, 91.
Thiman, K. V. (1936). *Proc. natn. Acad. Sci. U.S.A.* **22**, 511.
Torrey, J. G. (1961). *Expl Cell Res.* **23**, 281.
Vavilov, N. I. (1951). *Chronica bot.* **13**, 1.
Vezina, P. E. and Boulter, D. K. W. (1966). *Can. J. Bot.* **44**, 1267.
Vincent, J. M. (1965). In *Soil nitrogen* (ed. W. Bartholomew and F. C. Clark) p. 385. American Society of Agronomy, Madison.
—— (1974). In *The biology of nitrogen fixation* (ed. A. Quispel) p. 265. North Holland, Amsterdam.
Watson, I. A. (1970). In *Genetic resources in plants* (ed. O. H. Frankel and E. Bennett) p. 441. Blackwell, Oxford.

WILSON, J. K. (1931*a*). *J. Am. Soc. Agron.* **23**, 670.
—— (1931*b*). *Phytopathology* **21**, 1083.
WILSON, P. W. (1940). *The biochemistry of symbiotic nitrogen fixation*. University of Wisconsin Press, Madison.
WIPF, L. and COOPER, D. C. (1940). *Am. J. Bot.* **27**, 821.
WORRALL, V. S. and ROUGHLEY, R. J. (1976). *J. exp. Bot.* **27**, 1233.
YAO, P. Y. and VINCENT, J. M. (1969). *Aust. J. biol. Sci.* **22**, 413.
—— —— (1976). *Pl. Soil* **45**, 1.
ZELITCH, D. (1975). *A. Rev. Biochem.* **44**, 123.
ZOHARI, D. (1970). In *Genetic resources in plants* (ed. O. H. Frankel and E. Bennett) p. 33. Blackwell, Oxford.

5 Nitrogen fixation in some terrestrial environments

G. J. WAUGHMAN, J. R. J. FRENCH, and K. JONES

5.1 Introduction

Progress in understanding the ecology of nitrogen fixation in natural habitats benefited greatly from the development of the acetylene reduction technique. In many situations the rate of nitrogen fixation is too low to be even detected, much less quantified, by earlier methods. However, in spite of this difficulty much progress had been made prior to 1968, and relevant details of this earlier work are included here.

This chapter is organized by what might be loosely referred to as ecosystems, with additional sections dealing with animals and with ecological succession. At some future date it may be possible to fully integrate the role of nitrogen fixation associated with animals into discussions of specific ecosystems, but in view of the fact that the study of nitrogen fixation in animals is a comparatively recent development of this subject, we feel that the information is currently more usefully presented as a separate section.

In each section we have attempted to place nitrogen fixation in an ecological perspective, but due to the different amounts of information available, the extent to which this has been possible varies from system to system.

5.2 Forests

The nitrogen status of forests

Most of the nitrogen of a forest is in large storage components, and is turned over only very slowly. For example, in a mixed deciduous forest in the south-eastern United States, 80 per cent of the forest nitrogen is in the soil organic matter, 11 per cent in the vegetation, 3 per cent in the litter, 4 per cent in micro-organisms, and 2 per cent in 'free soil pools' (Mitchell, Waide, and Todd 1975). Similar data are provided for Douglas fir by Gessel, Cole, and Steinberger (1973) with the suggestion that in older forests a higher proportion of the forest nitrogen is in the vegetation and forest floor.

There is no shortage of nitrogen in forests. Table 5.1 shows the amounts of nitrogen in soils below larch and Douglas fir in the English Lake District. These values are high when compared with the nitrogen contents of most habitats, for example, salt marshes (Jones 1974). However, the turnover of nitrogen in forests, especially coniferous forests, is very slow (Williams 1972) and the nitrogen released is subject

TABLE 5.1
Total nitrogen of soil layers of Douglas fir and larch (mg Ng^{-1} dry soil)

Layer	Douglas fir	Larch
Litter	13.3	18.4
Humus	19.8	26.7
Soil	18.2	13.8
Sub-soil	9.0	4.3

Compare with total nitrogen of 0.5 mg N g^{-1} dry soil for the lower zones of a salt marsh and 2.5 mg N for the upper grazed zone.

TABLE 5.2
Estimates of nitrogen fixation in the forest floor

Location	Tree-type	Nitrogen fixed ($kg\ ha^{-1}\ a^{-1}$)	Reference
NW USA	Mixed conifers	4.9	Denison *et al.* (1977)
NW USA	Douglas fir	1.0	Gessel *et al.* (1973)
Sweden	Scots pine (15 years old)	0.5	Granhall and Lindberg (1978)
Sweden	Scots pine (120 years old)	0.3	Granhall and Lindberg (1978)
Sweden	Scots pine/ Norway spruce	1.2–37.9	Granhall and Lindberg (1978)
UK	Douglas fir	3.1–5.1	Jones *et al.* (1974)
Canada	Maple/aspen	0.5–1.0	Knowles and O'Toole (1975)
France	Beech	1.7–6.7	Remacle (1977)
Australia	Conifers	50.0	Richards (1973)
New Zealand	Mixed conifers	12.5	Silvester and Bennett (1973)
SE USA	Mixed hardwoods	10.8	Todd *et al.* (1978)

to a variety of fates which may include leaching and denitrification. Therefore, although there is no overall shortage of nitrogen, there is a shortage of nitrogen available for the growth of trees, and indeed, this lack of combined nitrogen is a major factor in limiting forest productivity. It is in this context, the provision of combined nitrogen for the growth of trees, that nitrogen fixation in forests should be examined.

Distribution of heterotrophic nitrogen-fixing bacteria in forest soils

Some estimates of nitrogen fixation are presented in Table 5.2. In most cases the values are low and can be explained either by the lack of nitrogen-fixing organisms, or through conditions being unfavourable for fixation.

Free-living heterotrophic nitrogen-fixing bacteria are found in all forest soils, with the anaerobic and facultatively anaerobic genera, *Clostridium* and *Bacillus*, predominating. However, as is shown in Table 5.3, several workers have isolated aerobic nitrogen-fixing bacteria from soils, even acidic ones, and Granhall and Lindberg (1978) suggest that they play a part in nitrogen fixation in some Swedish forest soils.

Nitrogen-fixing bacteria are found throughout the soil profile (Table 5.4) but are usually more numerous and more active in humus of the fermenting layer (Granhall and Lindberg 1978; Jones 1978; Silvester and Bennett 1973) where conditions might be expected to be suitable for nitrogen-fixing bacteria.

Nitrogen-fixing bacteria are also found above the soil and litter surface, with significant rates of fixation occurring on the boles of living trees (Todd, Meyer, and Wide 1978) and on woody litter (Granhall and Lindberg 1978; Todd, Waide, and Cornaby 1975; Todd *et al.* 1978). Several workers have also shown an association between nitrogen-fixing bacteria and fungi involved in the decay of timber (Aho, Seidler, Evans, and Aju 1974; Baines and Millbank 1978; Cornaby and Waide 1973; Seidler, Aho, Raju, and Evans 1972; Sharp 1975*a*, *b*) and suggest that the breakdown products of the wood are available for the growth of nitrogen-fixing bacteria. The bacteria concerned are species of *Klebsiella* and *Enterobacter* which fix nitrogen in low oxygen tensions and are found where the wood is moist. No transfer of combined nitrogen to the fungi has been demonstrated.

It would appear from the ubiquity of nitrogen-fixing bacteria in forest soils that their absence is not the reason for the low rates of nitrogen fixation which have been measured.

Factors effecting nitrogen fixation by heterotrophic bacteria in forest soils

The major factor limiting nitrogen fixation by bacteria in forest soils is shortage of energy. The addition of a carbohydrate source greatly

TABLE 5.3

Forest soils from which nitrogen-fixing bacteria have been isolated

Location	Tree-type	Aerobes	Anaerobes	Facultative anaerobes	References
Canada	Mixed conifers		+		Brouzes *et al.* (1969)
Sweden	Scots pine	+	+	+	Granhall and Lindberg (1978)
	Norwegian spruce	+	+	+	
Norway	Birch	+	+		Hanssen, Lid-Tovsuik, and Tovsuik (1973)
UK	Douglas fir	−	+	+	Jones (1976, 1978)
	Hemlock	−	+	+	
	Larch	−	+	+	
	Lawson cyprus	−	+	+	
	Scots pine	−	+	+	
	Sitka spruce	−	+	+	
	Alder	−	+	+	
	Beech	−	+	+	
	Oak	−	+	+	
	Sycamore	−	+	+	
Finland	Birch	+			Kallio and Kallio (1975)
	Pine	+			
France	Oak	+			Remacle (1970)
	Hazel	+			
	Hornbeam	+			
	Beech	−	+		Remacle (1977)
	Spruce	−	+		
New Zealand	Slash pine	+	+		Richards (1973)
	Hoop pine	+			
Japan	*Cryptomeria* sp	+	+		Susaki, Tanabe, Matsuguchi, and Araragi (1975)
	Chaemaecyparis sp	+	+		

TABLE 5.4
Most probable number of nitrogen-fixing bacteria in larch soil layers

Layer	Number of bacteria (g^{-1} dry wt soil)
Litter	130
Humus	8000
Soil	1400

stimulates bacteria already active and initiates nitrogen fixation by dormant ones (Brouzes, Lasik, and Knowles 1969; Jones, King and Eastlick 1974; Knowles and O'Toole 1975; Remacle 1977). The positive correlation between nitrogen-fixing activity on the one hand, and the location of soil organic matter, and soil respiration and fermentation on the other (Granhall and Lindberg 1978; Silvester and Bennett 1973), can be explained by the provision of available carbohydrate at low ambient oxygen concentrations.

The low pH values found in most forest soils, depress nitrogen fixation and restrict the distribution of aerobic nitrogen-fixing bacteria. Whilst increasing the pH of forest soils does not necessarily increase the rate of nitrogen fixation, it will boost the levels of carbohydrate-stimulated nitrogen fixation (Jones *et al.* 1974; Jones 1978). There is some evidence that increasing the pH aids mineralization of nitrogen in forest soils (Jones 1978) and may be responsible for the increase growth of Douglas fir on limed soils observed by Heilman and Ekuan (1973).

Addition of various minerals such as phosphate, sulphate, iron, and molybdenum with or without carbohydrate have no stimulatory effect on nitrogen fixation in larch or Douglas fir soils (Jones 1978) and are presumed not to limit the process in forest soils.

Apart from the availability of an energy source the two main factors controlling nitrogen fixation in the forest are moisture and temperature, and these factors together can explain the seasonal variation in the rate of nitrogen fixation recorded in forests (Granhall and Lindberg 1978; Jones *et al.* 1974; Todd *et al.* 1978). For example, nitrogen fixation in Douglas fir soils studied by Jones *et al.* (1974) was highest in the spring, autumn, and early winter when temperature and moisture were adequate, but absent in winter due to low temperatures and in summer due to lack of moisture.

It is perhaps surprising that higher rates of nitrogen fixation have not been found in deciduous forests where there is an autumnal leaf fall coupled with a higher soil pH and a higher rate of nutrient turnover than in coniferous forest soils.

Nitrogen fixation in the rhizosphere of forest trees

Stimulation of nitrogen-fixing bacteria in the rhizosphere of plants is a well documented phenomenum (see review of Döbereiner (1974) and there is evidence that it occurs in forest soils. The high values for nitrogen fixation in Australian coniferous soils (50 kg ha^{-1} a^{-1}) have been attributed to fixation in the rhizosphere (Richards 1973), and Silvester and Bennett (1973) showed that bacteria in the rhizosphere of *Podocarpus* where responsible for relatively high rates of fixation and not 'nodules' as had previously thought to have been the case. Granhall and Lindberg (1978) detected rhizosphere nitrogen fixation in Scots pine and Norway spruce in Swedish soils, and Remacle (1977) attributes all the soil nitrogen fixation in a beech forest in France to rhizosphere activity. The rhizosphere of the forest ground flora has also been shown to contribute fixed nitrogen to forests (Granhall and Lindberg 1978; Remacle 1977). Repeated efforts have failed to detect nitrogen fixation in the rhizosphere of larch and Douglas fir soils (Jones, unpublished data), and Jain and Vlassak (1975) were unable to isolate either aerobic or anaerobic nitrogen-fixing bacteria from the rhizosphere of oak and *Pinus* trees.

Nitrogen fixation in forest soils by organisms other than bacteria

The distribution of blue–green algae is restricted in forests due to their intolerance of acid conditions. However, Granhall and Lindberg (1978) have detected relatively high rates of nitrogen fixation by crusts of blue–green algae in Swedish coniferous forests. High rates of fixation have also been demonstrated for moss/blue–green algal associations found in some forests (Billington and Alexander 1978; Granhall and Lindberg 1978).

Mixed forest of alder and Douglas fir have been shown to fix nitrogen at rates as high as 38 kg $ha^{-1}a^{-1}$ (Tarrent and Miller 1978) and although the two species tend to compete it has been suggested that by appropriate management the beneficial aspects of alder could be exploited (Denison, Caldwell, Bormann, Eldred, Swanberg, and Anderson 1977; Gessel *et al.* 1973; Newton, El Hassan, and Zavitkoyski 1968).

Haines, Haynes, and White (1978) showed that growing sycamore seedlings with *Trifolium* spp and vetches was more beneficial than could be explained solely by the nitrogen provided by the legumes. The legumes did not compete as they grew in the cooler season. They provided cover for several years and helped in the control of weeds.

Nitrogen fixation in the canopy

Nitrogen-fixing bacteria are readily isolated from the leaves of temperate coniferous and deciduous trees (Jones 1970; Jones *et al.* 1974;

Jones 1976), but it would appear that the bacteria are accidental inhabitants of the leaves and are not adapted to conditions on the leaf surface. Ruinen (1974) suggests that this is the normal situation with phyllosphere/phylloplane nitrogen-fixing bacteria.

The activity of nitrogen-fixing bacteria on leaves (at least in temperate forests) is low, this is due to a shortage of carbohydrate and a requirement for low oxygen concentrations. Nitrogen-fixing bacteria isolated from larch and Douglas fir leaves in the English Lake District include anaerobes, facultative anaerobes, or mixed cultures which initially fix nitrogen in the presence of oxygen but loose that ability in culture (Jones *et al.* 1974; Jones 1976). Denison *et al.* (1977) were able to demonstrate small amounts of acetylene reduction by 'scuzz' bacteria on the surface of conifer needles in aerobic conditions and isolated *Azotobacter*-type bacteria from the leaves. Data for nitrogen-fixation in the phyllosphere of forest trees is presented in Table 5.5. The values are low and reflect the relatively inhospitable conditions on the leaf surface. The estimate for Douglas fir in the English Lake District has been revised by taking into account the nitrogen content of the leaves (after Gessel *et al.* 1973) compared with that of the whole canopy (after Ovington 1957). There is the potential for foliar uptake of the products of phyllosphere nitrogen fixation. Jones (1976) has demonstrated direct uptake by larch shoots of ^{15}N-labelled glycine (one of the amino acids liberated by nitrogen-fixing bacteria isolated from larch leaves) and the translocation of the ^{15}N to actively growing parts of the plant.

Lichens have been shown to fix nitrogen in the forest environment (Becker, Reeder, and Stetler 1977; Billington and Alexander 1978; Denison *et al.* 1977; Denison 1973; Hallgren and Nyman 1977; Huss-Danell 1977; Kallio 1978; Kallio and Kallio 1978). Some of the available data for lichen nitrogen fixation in the canopy are shown in Table 5.6. Lichens are also found on the boles of trees, on woody litter, and in clearings. In certain environments such as subarctic forests (Kallio 1978) and in tropical rain forests (Edmisten 1970) lichens appear to be major contributors of fixed nitrogen. Millbank (1978) has shown that

TABLE 5.5
Estimates of nitrogen fixation in the canopy

Tree type	Nitrogen fixed (kg N ha^{-1} a^{-1})	Reference
Scots pine	0.001–0.022	Granhall and Lindberg (1978)
Norway spruce	0.047	Granhall and Lindberg (1978)
Douglas fir	2.1	Jones *et al.* (1974; see text)
Beech	8.3	Remacle (1977)
Mixed deciduous	0.22	Todd *et al.* (1978)

the alternating drying and wetting, to which most lichens are subject, results in a leakage of fixed nitrogen to the surrounding environment.

The significance of nitrogen fixation in forest soils

Various values have been given for the annual requirements of trees for nitrogen. Ovington (1954, 1957) calculated an annual requirement of 17 and 38 kg N ha^{-1} for two Douglas fir forests in the United Kingdom, Gessel *et al.* (1973) set a figure of 50 to 70 kg N ha^{-1} for Douglas fir in the north-western United States and Mitchell *et al.* (1975) a figure of 142 kg N ha^{-1} for a mixed deciduous forest in the south-east United States. Most of this nitrogen comes from the turnover of nitrogen already in the forest system. New nitrogen comes from three sources: that already in the soil, in precipitation, and nitrogen fixation. These inputs are roughly balanced, in the long-term at least, by output due to leaching and denitrification (Gessel *et al.* 1973; Mitchell *et al.* 1975; Todd *et al.* 1974; Todd *et al.* 1978). There is evidence of a gradual increase in the nitrogen content of the forest as a whole (Richards 1964). Todd *et al.* (1975) suggests that gaseous inputs (nitrogen fixation) and outputs (denitrification) of nitrogen represent major components of the forest nitrogen cycle and appear to dominate total gains and losses. In other words only a small amount of nitrogen is being transformed at any one time, and a significant amount of this could come from nitrogen fixation. This appears to be supported by the fact that seedlings of larch and Douglas fir actively select ^{15}N-labelled bacterial fixation products from soil (Jones *et al.* 1974; Jones 1978), and translocate the fixed nitrogen to the actively growing parts of the seedlings.

Thus although nitrogen fixation appears to be occurring at a slow rate in forests, it is important because of the nature of the nitrogen compounds which are made available to the growing trees.

5.3 Polar and alpine tundra

The tundra environment

The term tundra refers to a landscape occurring between the tree line and the area of permanent snow at high latitudes or high altitudes. The regions are characterized by a short growing season, freeze and thaw cycles, permafrost, intense solar radiation in alpine tundra, and a dryness which prevents forest development. Small areas of both circum-polar and alpine tundra occur in the southern hemisphere, although much larger developments of both types exist in the northern hemisphere (Billings 1974). Vegetation cover dominated by mosses, lichens, and sedges becomes increasingly sparce at higher elevations and higher

TABLE 5.6
Nitrogen fixation by lichens mainly in the canopy of forest trees

Tree type	Nitrogen fixed (kg ha^{-1} a^{-1})	Reference
Douglas fir	2–11	Denison (1973)
Sitka spruce	12	Denison (1977)
Tropical rain forest	88	Edmisten (1970)
Columbian rain forest	1.5–8	Forman (1975)

latitudes, giving way to patterned ground and fell fields. In climatically less severe parts of the tundra, a cover of low shrubs may develop in which members of the Ericaceae, *Betula* spp (dwarf birch), and *Salix* spp (dwarf willow) are abundant. Many of the vascular species are adapted to this environment, and possess metabolic features which prevent them growing in warmer regions (Billings 1974).

Net primary production is low, ranging from about 50 g m^{-2} a^{-1} for lichens and cushion vegetation, to 250 g m^{-2} a^{-1} for heath shrub and carex communities, with an average of approximately 90 g m^{-2} a^{-1} for all vegetation types with 100 per cent ground cover (Bliss 1962, 1966; Bliss and Wein 1972; Haag 1973). The average uptake of nitrogen into tundra vegetation has been estimated at 2–4 g N m^{-2} a^{-1} (Rodin and Bazilevich 1967; Stutz and Bliss 1975) of which less than 0.2 g m^{-2} a^{-1} is supplied by precipitation (Eriksson 1952; Barsdate and Alexander 1975).

Nitrogen fixation

The most northern arctic tundra consists of patterned ground. In the Barrow region of Alaska, a number of studies on patterned ground have demonstrated that nitrogen is fixed predominantly by *Nostoc cummune* in the peaty troughs, and by various lichens in the polygon tops (Alexander, Billington, and Schell 1974); lichens recorded at Barrow include *Peltigera* spp and *Stereocaulon tomentosum* (Table 5.7). Similar results have been obtained in the North West Territories where rates of nitrogenase activity for blue–green algae in the mesic meadows are comparable with those in the polygon troughs (Stutz and Bliss 1975; Jordan, McNicol, and Marshall 1978). Amounts of nitrogen fixed annually range from less than 2 mg m^{-2} in the polar deserts and polygon tops to over 200 mg m^{-2} in the troughs and meadows (Table 5.8). The higher rates in the troughs and meadows possibly results from the more favourable water and nutrient supply in this situation. Water is considered to be the main limiting factor during the summer, with a slower rate of activity also observed in conditions of low light and low tem-

TABLE 5.7
Specific nitrogenase activity in some tundra lichens measured by acetylene reduction

Species	Location	nmol C_2H_4 g dry wt^{-1} h^{-1}	Temperature (°C)	Month	Reference
Peltigera aphthosa	Iceland	354–2326	13–19	July	Crittenden (1975
P. aphthosa	N. Sweden	40, 560	14–19	July, Aug.	Granhall and Selander (1973)
P. aphthosa	Kevo, Finland	40–1540	−1–12	April	Alexander and Kallio (1976)
P. aphthosa	Barrow, Alaska	1360, 1000	15	July, Aug.	Alexander *et al.* (1974)
P. scabrosa	Barrow, Alaska	980–940	15	July, Aug.	Alexander *et al.* (1974)
P. scabrosa	N. Sweden	3500–5500	19	Aug.	Granhall and Selander (1973)
P. canina	Iceland	1029–3952	13–17	July	Crittenden (1975)
P. canina	Barrow, Alaska	320, 2770	15	July, Aug.	Alexander *et al.* (1974)
Stereocaulon paschale	N. Sweden	200, 1300	8–19	July, Aug.	Granhall and Selander (1973)
S. paschale	Kevo, Finland	0–1200	0–20	[Laboratory]	Kallio (1973)
S. paschale	Kevo, Finland	0–470	−1–16	April	Alexander and Kallio (1976)
S. alpinum	Iceland	0–276	11–17	July	Crittenden (1975)
S. vesuvianum	Iceland	19–69	10–16	July	Crittenden (1975)
S. tomentosum	Barrow, Alaska	480	15	July, Aug.	Alexander *et al.* (1974)
Nephroma arcticum	N. Sweden	1600, 4400	19	Aug.	Granhall and Selander (1973)
N. arcticum	Kevo, Finland	0–6410	0–20	[Laboratory]	Kallio *et al.* (1972)
N. expallidum	N. Sweden	420	19	Aug.	Granhall and Selander (1973)
Placopsis gelida	Iceland	18(cm^2 h^{-1})	14	July	Crittenden (1975)
Solorina crocea	Kevo, Finland	20–3010	−5–15	[Laboratory]	Kallio *et al.* (1972)
Lobaria linita	Barrow, Alaska	500, 1670	15	July, Aug.	Alexander *et al.* (1974)

Estimates of the amount of nitrogen fixed annually at some circumpolar and alpine sites

Site	Location	Main organisms involved	Estimates of nitrogen fixed (mg m^{-2} a^{-1})	Reference
Patterned ground	Barrow, Alaska	Lichens and algae	69	Alexander *et al.* (1974)
Patterned ground	Barrow, Alaska	Lichens and algae	15	Alexander and Schell (1973)
Patterned ground	Barrow, Alaska	Lichens and algae	94	Barsdate and Alexander (1975)
Lichen heath	Norway	Lichens and bacteria	56	Lid-Torsvik (1971)
Birch wood	Norway	Lichens, legumes, and bacteria	165	Lid-Torsvik (1971)
Wet meadow	Norway	Algae	255	Lid-Torsvik (1971)
Gravel ridge	Devon Island, Canada	Algae	7–30	Stutz and Bliss (1975)
Mesic meadow	Devon Island, Canada	Algae and bacteria	120–380	Stutz and Bliss (1975)
Wet solifluction slopes	S. Orkney Islands	*Nostoc* and *Collema*	0–1.8	Horne (1972)
Mixed mire ombrotrophic peat	N. Sweden	Bacteria	29–150	Granhall and Selander (1973)
Mixed mire wet depressions	N. Sweden	Blue–green algae	9400–12 000	Granhall and Selander (1973)
Glacial till	Glacier Bay	*Dryas drummondii*	1200	Crocker and Major (1955)
Subalpine (alt. 2240 m)	Colorado	Blue–green algae	4000	Porter and Grable (1969)
Subalpine meadow	Colorado	Bacteria	1000	Porter and Grable (1969)

perature (Schell and Alexander 1973; Alexander and Schell 1973; Alexander *et al.* 1974). No intrinsic diurnal pattern of nitrogenase activity was found to occur in either *Nostoc* or *P. aphthosa*, therefore the marked daily rhythm observed in *P. aphthosa* is thought to be a response to light and temperature variations (Alexander *et al.* 1974).

Nostoc is probably the most widespread and generally abundant blue–green alga in polar regions, and as such it may be the single most important source of nitrogen in tundra ecosystems (see, for example Tichomirov (1957), Holm-Hansen (1964), and other references quoted here). Other nitrogen-fixing genera of blue–green algae recorded include *Anabaena*, *Scytonema*, *Gloeocapsa*, *Calothrix*, and *Tolypothrix*, although reports suggest that these species make a contribution to the nitrogen balance mainly in pools and wet depressions (Tichomirov 1957; Alexander 1974; Granhall and Selander 1973; Jordan *et al.* 1978). The importance of moisture was emphasized by Cameron (1972) in a study of Antarctic dry valleys where the duration of available water was found to be a prime factor influencing the distribution of blue–green algae, and by Horne (1972) who concluded that dessication and low temperature limited ^{15}N-fixation by *Nostoc* and *Collema pulposum* on Signy Island in Antarctica. In adidtion, Kallio (1973) found a very strong positive correlation between moisture and acetylene reduction in the lichen *Stereocaulon paschale*. Thus, in latitudinal tundra the supply of water appears to be one of, if not the most important factor limiting nitrogen fixation.

Low temperatures and aridity are not suitable for high rates of nitrogen fixation; however, Henriksson and Simu (1971) found that the lichens *Collema tuniforme* and *Peltigera rufescens* recovered in less than one day to nearly maximum rates of nitrogenase activity after 30 weeks of dessication, and activity has been reported at temperatures well below 0 °C for the lichens *Solorina crocea*, *Peltigera aphthosa*, and *P. canina* (Kallio, Suhonen, and Kallio 1972; Englund and Meyerson 1974; Alexander and Kallio 1976). Therefore it appears that the symbiotic association in lichens may give the blue–green alga some protection against these conditions, which may partially account for the abundance of lichens in tundra environments. Other tundra lichens for which nitrogen fixation has been reported are included in Table 5.7 together with some specific activities.

Azotobacter spp have been reported from tundra soils (Russell 1940; Boyd and Boyd 1962; Jordan *et al.* 1978), but Jurgensen and Davey (1971) failed to isolate any aerobic bacteria from samples of Alaskan tundra soils, however they did find both *Bacillus polymyxa* and *Clostridium pasteurianum*. Cameron, King, and David (1970) found that heterotrophic nitrogen-fixing bacteria occurred in dry valleys of South Victoria Land (Antarctica) only when blue–green algae were present.

Fogg and Stewart (1968) concluded that the contribution by heterotrophic bacteria to the pool of biologically fixed nitrogen in sites on Signy Island was insigificant compared with the amount fixed by autotrophic organisms. Putative nitrogen-fixing species of *Clostridium*, *Azotobacter*, *Bacillus*, and *Klebsiella*, have been isolated from tundra soil in the North West Territories by Jordan *et al.* (1978) who found the in-situ bacterial nitrogenase activity to be very low; however, activity in these samples was stimulated by the addition of glucose. Slow rates of mineralization probably restrict nitrogen fixation by bacteria, although some acetylene reduction tests have suggested that heterotrophic nitrogen fixation may be locally important in tundra situations (Granhall and Selander 1973; Waughman and Bellamy 1980; Baker 1974).

Nodulated vascular plants are of little general importance in the high latitude tundra, but at lower latitudes their role in the nitrogen balance may be locally significant. On the arctic slope of Alaska five genera and twenty-three species of legumes have been recorded (Wiggins and Thomas 1962) and many, such as *Lupinus arcticus*, are widespread, with important pioneer roles on rocky slopes and sandy areas (e.g. Bliss and Cantlon 1957; Corns 1974). In the sub-arctic alpine tundra at Eagle Summit and in the Alaska Range, Alexander and Schell (1973) concluded, on the basis of acetylene-reduction tests, that legumes and other higher plants made an important contribution to the pool of biologically fixed nitrogen. Both *Alnus crispa* and *Dryas drumondii* bear nitrogen-fixing root nodules, and these plants may be locally abundant in the low arctic scrub together with *Betula* spp and *Salix* spp (Crocker and Major 1955). No other species of *Dryas* fix nitrogen, but most species of *Alnus*, many of which occur in sub-arctic regions, are reported to possess root nodules (Johnson 1968; Bond 1976).

The role of nitrogen fixation in Arctic and alpine environments

When compared to many other ecosystems, the tundra is seen to possess a relative abundance of nitrogen-fixing organisms, suggesting a shortage of available nitrogen: low concentrations of soluble nitrogen in soil have in fact been observed by numerous investigators (e.g. Russell 1940; Warren-Wilson 1966; Bliss and Wein 1972; Haag 1973). These small amounts must be due, in part at least, to low soil temperatures slowing the rate of organic decomposition, ammonification, and nitrification (Boyd 1958; Douglas and Tedrow 1959). The application of nitrogen fertilizers to tundra soils has been shown to stimulate vegetation growth in the absence of added phosphorus and potassium (Bliss 1966; Bliss and Wein 1972; Haag 1973), thus providing further evidence of nitrogen shortage. In alpine tundra, particularly at lower latitudes, the high summer temperatures stimulate mineralization of nitrogen to a greater extent than in polar tundra. For example Rehder

(1976*a*, *b*) found that the amount of nitrogen mineralized in the soil beneath some alpine tundra communities in Europe was greater than the amount incorporated into the above ground production. This result suggests that the supply of nitrogen in at least some alpine tundra ecosystems may be less of a problem than in the arctic tundra.

Biological nitrogen fixation is an important, if not essential source of nitrogen throughout much of the latitudinal tundra, contributing more than 75 per cent of the annual supply in some areas (Barsdate and Alexander 1975). Although it would be interesting to obtain an estimate of the amount of nitrogen fixed annually, any estimates are complicated by the heterogeneity of terrain, and by the fact that the relationship between nitrogenase activity and temperature, in at least some lichens and blue–green algae, varies according to the light intensity (Kallio *et al.* 1972; Waughman 1977). Some estimates are shown in Table 5.8, which includes alpine and sub-arctic sites. Most figures for the arctic tundra locations fall within the range 20–250 mg N m^{-2} a^{-1} with an average of about 100 mg N m^{-2} a^{-1}. This represents about 5 per cent of the estimated annual uptake of 2 g m^{-2}. Many of these estimates of nitrogen fixed in the tundra ecosystem are comparable with estimates for several other ecosystems with much higher net productivities (see other tables in this chapter), this again may be a manifestation of the slow mineralization, hence slow turnover rates of nitrogen within the tundra. Thus, more atmospheric nitrogen may need to be fixed to support any given net productivity in the tundra than is required for similar productivity in ecosystems situated in warmer regions. In this respect it is interesting to note that Barsdate and Alexander (1975) concluded that, as leaching losses in the permafrost are insignificant, more than 80 per cent of the nitrogen influx to the tundra ecosystem is retained. This view is supported by the fact that Stutz and Bliss (1975) were unable to detect any nitrogen in tundra drainage water.

5.4 Hot arid and semi-arid regions

The desert environment

Hot, dry regions of the earth, which occur predominantly between 15° and 45° North and South of the equator, receive intense solar radiation, giving rise to soil temperatures which may be as much as 15 °C higher than the temperature of the air: for example Beadle and Tchan (1955) recorded a temperature of 58 °C in an Australian semi-desert soil with the air temperature at 43 °C. The high temperatures also cause evaporation and upward movement of moisture in the soil. In some parts of Death Valley, California, where ground surface temperatures can reach 88 °C, evaporation of 380 cm of moisture per year is countered by an average annual rainfall of only 5 cm (Durrell 1962).

The upward movement of moisture frequently results in high salt concentrations near the soil surface. Amounts of ammonium-nitrogen are usually low (Beadle and Tchan 1955; Englund 1975), but the aerobic alkaline conditions coupled with the upward movement of moisture does permit moderately high concentrations of nitrate-nitrogen to accumulate locally (Beadle and Tchan 1955; Harmsen and Van Schrevan 1955; Englund 1975). The carbon–nitrogen ratio in semi-arid soils is usually about 10, but figures as low as 2 have been reported (Fletcher and Martin 1948; Beadle and Tchan 1955; Abd-el-Malek 1971). In moist conditions a loss of nitrogen by denitrification in virgin desert soils was noted by Macgregor (1972). This loss was enhanced when the soils were amended with carbon compounds.

Nitrogen fixation by bacteria, lichens, and algae

Very high temperatures, high salt concentrations, and the low organic content of these soils are not conducive to the growth of nitrogen-fixing bacteria, and although reports indicate that both *Azotobacter* and *Clostridium* are widespread in semi-arid soils, their numbers are generally very low (Tchan and Beadle 1955; Loub 1963; Abd-el-Malek 1971). Tchan and Beadle (op. cit.) estimated that the amount of nitrogen fixed by bacteria in their soil samples from South Australia was less than 0.1 kg ha^{-1} a^{-1} compared with up to 3.3 kg ha^{-1} a^{-1} by blue–green algae. In semi-desert areas these blue–green algae commonly occur on the underside of light quartz pebbles, whereas they are absent from beneath dark coloured pebbles (Tchan and Beadle 1955; Durrell 1962). Conditions under these quartz pebbles are suitable for the growth of algae because light can penetrate the pebble, and moisture condenses on the undersurface at night. Peat up to 6 mm thick is formed beneath similar pebbles in semi-arid parts of the Great Plains in North America (Williams 1943). In virgin soils from Tunisia, Englund (1975) estimated that fixation ranged between 12 and 75 ng N g^{-1} h^{-1} for blue–green algae, including *Nostoc punctiforme*, *N. ellipsosporum*, *N. piscinale*, *Anabaena aphanizomenoides*, *A. oryzae*, and *Nodularia harveyana*. In Death Valley, Durrell (1962) recorded 19 species of blue–green algae, many of which are known to fix atmospheric nitrogen. The success of blue-green algae in semi-desert situations is due, partially at least, to the colloidal sheath possessed by most members of the group; these sheaths help to prevent dessication caused by dry conditions and/or high salt concentrations. The sheaths also absorb water rapidly thus permitting the algae to become active quickly after periods of dehydration: for example air dried *Nostoc* can absorb water to 12 times its own weight in six minutes (Durrell 1962).

Acetylene reduction and ^{15}N methods have been used to detect nitrogenase activity in several desert lichens including *Collema cocco-*

phorus, *Lecidea crystalifera*, and *Parmelia conspersa* (Rogers, Lane, and Nicholas 1966), and *Peltigera rufescens* (Snyder and Wullstein 1973*a*). *Lecidea* and *Parmelia* possess green algal symbionts, therefore nitrogen fixation by these species may be due to associated bacteria. The possibility of polysymbiosis in lichens has been discussed by several workers (e.g. Cengia Sambo 1923; Krasilnikov 1949; Scott 1956).

In some dry regions blue–green algae grow in association with bacteria, fungi, and lichens in the thin rain crust which is a feature of many semi-arid soils. Such crusts have been described from both the United States (Booth 1941) and Russia (Bolyshov and Evdokinov 1944). Fletcher and Martin (1948) found the nitrogen content of an Arizona desert crust to be four times greater than in the soil below, and noted the presence of *Nostoc*. *Microcoleus*, *Scytonema*, and *Nostoc* were amongst the most frequent genera of blue–green algae in rain crusts of range and semi-desert soils of New Mexico studied by Shields, Mitchell, and Drouet (1957), and although *Azotobacter* was recorded they considered that the amount of nitrogen fixed by bacteria to be insignificant compared with the amount fixed by free-living blue–green algae and lichens. Shields *et al.* (op. cit.) found that up to 6000 p.p.m. of crust nitrogen was in the amino form, and less than 5 p.p.m. as simple inorganic nitrogen. These data were supported by the later work of Mayland and McIntosh (1966) who found that 50 per cent of the ^{15}N fixed biologically into wetted rain crusts was in the amino fraction, and only traces in the water-soluble nitrogen fraction. Any leaching that does occur in these crusts will do so only in the wet season which is when the nitrogen fixing process is active, thus with most of the nitrogen in the amino form losses during the wet period are reduced. Mayland has also reported that ^{15}N fixed by crustal micro-organisms can be later detected in grass seedlings (Mayland, McIntosh, and Fuller 1966). The fixed nitrogen is made available by direct excretion or via nitrifying bacteria which may be widespread in these soils (Shields *et al.* 1957; Mayland and McIntosh 1966; Mayland *et al.* 1966).

Using ^{15}N, Mayland *et al.* (1966) obtained estimates for nitrogen fixed by wetted crusts of between 0.12 and 0.18 kg nitrogen per hectare of crust per day; ^{15}N was detected in the crust three days after wetting, and fixation continued for 520 days: no change occurred in the nitrogen content of the control crusts. The more sensitive acetylene reduction method used by Macgregor and Johnson (1971) showed that activity can in fact be detected as soon as three hours after wetting. Based upon acetylene reduction, they estimated fixation to be about 3–4 g N ha^{-1} h^{-1} of crust. Some specific rates of nitrogenase activity in desert crusts are listed in Table 5.9. Rychert and Skujins (1974), also using the acetylene reduction method, found that crustal nitrogenase activity declined rapidly outside of the temperature range 15–25 °C, and was

TABLE 5.9

Rates of nitrogen fixation by soils from arid and semi-arid warm regions

Location	Soil type	Main groups of organisms involved	Rate	Reference
Great Basin Desert	Desert crust under *Atriplex* spp and *Eurotia lanata*	Algae, lichen	75–84 g N ha^{-1} h^{-1} (C_2H_4)	Rychert and Skujins (1974)
Great Basin Desert	Desert crust under *Atriplex* spp and *Eurotia lanata*	Algae, lichen	10–14 g N ha^{-1} h^{-1} (C_2H_4)	Rychert and Skujins (1974)
Nr. Tucson	Desert crust	Algae	6–9 g N ha^{-1} h^{-1} (^{15}N)	Mayland *et al.* (1966)
Sonoran Desert	Desert crust under grass and mesquite	Algae	3–4 g N ha^{-1} h^{-1} (C_2H_4)	Macgregor and Johnson (1971)
Tunisia	Dry sandy soil under incomplete grass cover	Algae	12–75 ng N g^{-1} h^{-1} (C_2H_4)	Englund (1955)
Tunisia	Damp loam	Algae	18–133 ng N h^{-1} h^{-1} (C_2H_4)	Englund (1955)
Oklahoma	Prairie	Bacteria	15–83 pmol C_2H_4 g dry wt^{-1} h^{-1}	Kapustka and Rice (1976)
Saskatchewan	Clay under native grassland	All groups	2.3 g N ha^{-1} h^{-1} (^{15}N)	Paul *et al.* (1971)
Tunisia	Bare saline alluvium	Bacteria	0 (+ve C_2H_4 in waterlogged rhizosphere)	Hauke-Pacewiczowa, Balandreau, and Dommergues (1970)
Ukrain	Alkali soil	Bacteria	4–19 g N ha^{-1} h^{-1} (^{15}N)	Maltseva (1973)
New Zealand	Soil under tussock grassland	Bacteria	0–433 pmol C_2H_4 g dry $wt.^{-1}$ h^{-1}	Line and Loutit (1973)
Saskatchewan	Clay under native grassland	All groups	25–150 g N m^{-2} h^{-1} (C_2H_4)	Vlassak, Paul, and Harris (1973)

Method of estimation indicated by ^{15}N or C_2H_4.

undetectable both below 5 °C and above 45 °C; acetylene reduction was also greatly reduced in low light intensities, and when the water potential of the crust was lower than −0.3 bars; nitrogenase activity was undetectable when the water potential was lower than −13 bars. From these characteristics Rychert and Skujins concluded that crustal fixation is likely to occur in the Spring and Autumn for about 120 days.

Nitrogen fixation associated with higher plants in deserts

Annual and ephemeral species of legumes grow in arid regions only when moisture is present. This, together with the need for an available endophyte and appropriate temperatures, suggests that this group may be of limited importance in the desert nitrogen economy. Leguminous perennial shrubs may be locally important in semi-arid regions, however such shrubs may be uncommon or absent from large areas of range and semi-desert soil (Farnsworth, Romney, and Wallace 1976), and although shrubs are the focus of nitrogen accretion in desert soils, there is evidence that the nitrogen content of leguminous shrubs may not be significantly higher than in non-legumes (Garcia-Moya and McKell 1970; Farnsworth *et al.* 1976). Some nodulated non-legumes with established nitrogen-fixing properties occur in semi-arid areas, such as *Casuarina* spp in Australia (Sprecht and Rayson 1957), and *Ceanothus* spp, *Elaegnus* spp, and *Shepherdia* spp in North America (Jepson 1936; Paul, Myers, and Rice 1971; Farnsworth 1975). In addition, many other non-legumes from arid zones have root nodules, including *Zygophyllum* spp, *Fagonia arabica*, and *Tribulus alatus* in the Zygophyllacea (Sabet 1946; Mostafa and Mahmoud 1946); and various species within the genera *Artemisia*, *Chrysothamnus*, *Opuntia*, *Mertensia*, *Kramaria*, *Lomatium*, and *Viola* (Farnsworth *et al.* 1976). The endophyte has been isolated from *Artemisia ludoviciana* (Farnsworth and Hammond 1968), and acetylene reduction by nodules in several of the above mentioned species suggests the presence of nitrogenase activity (Farnsworth and Clawson 1972; Wallace and Romney 1972). Nitrogenase activity has also been reported for root systems of several non-nodulated desert species (Wallace and Romney 1972), for the sand grain root sheaths of the grass *Oryzopis hymenoides* which contain a bacterium (Wullstein, Bruening, and Snyder 1972), and in the rhizosphere of corn (Raju, Evans, and Seidler 1972).

During the colonization phase of succession in semi-arid regions, nitrogen fixation by crustal organisms appears to be of great importance. Later, however, in the shrub stages, fixation by nodulated non-legumes and rhizosphere bacteria may also be of considerable significance to the nitrogen economy. In spite of the widespread occurrence in semi-arid areas of *Azotobacter* and *Clostridium*, no significant contribution to the pool of fixed nitrogen has been claimed for these free-living bacteria. One explanation for this may be that nitrogen-fixing bacteria are 'opportunistic' with respect to nitrogen fixation, performing this activity mostly in loose association with autotrophic species, rather than in the completely free-living situation.

Generalizations concerning the amount of nitrogen fixed biologically in environments discussed in this section are impossible because of the great variety of climates and types of vegetation cover which may be encompassed by the title. This is reflected by the estimates shown in Table 5.10. There is some evidence that the rate of nitrogen fixation and succession may be strongly related in some semi-arid regions, and that the rate of nitrogen fixation may increase as the prairie climax is approached (Kapustka and Rice 1976). Woodmansee, Dodd, Bowmann, Clark, and Dickinson (1978) have calculated that the annual nitrogen uptake of 29 kg ha^{-1} by vegetation in a short grass prairie ecosystem can be accounted for by the quantity of soil nitrogen mineralized each year, plus the quantity that is cycled internally, and concluded that the nitrogen cycle in the prairie climax ecosystem was very 'tight'. Therefore, the nitrogen fixed annually in semi-arid grassland climaxes (up to 3 kg ha^{-1}, Table 5.10) may create a short-term imbalance, but will offset occasional perturbations and thus keep the nitrogen budget in a long-term steady state. The topic of succession and nitrogen fixation in semi-arid grassland is discussed further on page 164.

5.5 Peatlands

Characteristics and distribution of peatlands

Peatlands cover about 150 million hectares of the earth's land surface, occurring mostly in either tropical countries, or at high latitudes where precipitation is in excess of evaporation. Eighty per cent of the world's peat resources occur within Finland, Canada, and the USSR with several other countries such as Scotland, Ireland, Sweden, Poland, and the United States having deposits of considerable size (Tibbetts 1969). Waterlogging is a pre-requisite of peat development; therefore the amount of available oxygen is small. Once formed, however, the upper layers of a peat deposit are occasionally aerated because of the fluctuating watertable. Several ecological gradients can be identified in peatlands of which the most important is the gradient from alkaline rich fens, with a regular supply of groundwater containing mineral nutrients, through fens and poor fens, to acid bogs fed only by precipitation. (See Sjors (1950) and Bellamy (1969) for details of this chemical gradient.)

TABLE 5.10

Estimates of the amount of nitrogen fixed annually at some sites located in arid and semi-arid warm regions

Site	Location	Main organisms involved	Nitrogen fixed ($kg\ ha^{-1}\ a^{-1}$)	References
Desert crust	Utah	Lichen and blue–green algae	10–100	Rychert and Skujins (1974)
Desert crust	Arizona	Blue–green algae	10.5	Mayland *et al.* (1966)
Range soil	S. Canada	Several groups	1.0	Paul *et al.* (1971)
Range soil	Oklahoma	Blue–green algae and *Azotobacter*	0.7–3.5	Kapustka and Rice (1976)
Tall grass prairie	Oklahoma	Legumes	0.3–0.7	Kapustka and Rice (1978)
Tall grass prairie	Oklahoma	Bacteria	3.5	Kapustka and Rice (1978)
Natural grassland	Oregon	Rhizosphere bacteria	0.1–3.0	Nelson *et al.* (1976)
Saskatchewan	Grasslands	Several groups	2.0	Vlassak *et al.* (1973)
Saskatchewan	Grasslands	*Elaeagnus commutata*	37.0 ($mg\ N\ m^{-2}\ day^{-1}$)	Vlassak *et al.* (1973)
Saskatchewan	Grasslands	*Shepherdia argentea*	136.0 ($mg\ N\ mg^{-2}\ day^{-1}$)	Vlassak *et al.* (1973)
Mulga scrub	New South Wales	*Azotobacter*	0.1	Tchan and Beadle (1955)
Mulga scrub	New South Wales	Blue–green algae	3.3	Tchan and Beadle (1955)
Tussock grassland	Otago, New Zealand	Bacteria	0–3.1	Line and Loutit (1973)
Alkali soil	Ukrain soil	Bacteria	2.8–14.0 ($kg\ ha^{-1}\ 28\ days^{-1}$)	Maltseva (1973)
Upland plantation of *Pinus radiata*	Nelson, New Zealand	Not stated	35.0	Stevenson (1959)

In most peats, as a result of the low pH and low redox potential, the ammonium ion is the predominant form of soluble inorganic nitrogen (Waughman 1980); nitrate-nitrogen occurs in rich fen peat, but in these situations it is either leached or rapidly absorbed by the vegetation (Cyplenkin and Schilin 1936; Persson 1962; Fibras 1952). Mineralization of nitrogen is very slow in bog peats from which nitrifying bacteria may be completely absent, however, as with most other types of microbial activity, mineralization is much faster in the more nutrient rich fen peats (Cyplenkin and Schilin 1936; Kaila, Soini, and Kivinen 1954; Latter, Cragg, and Heal 1967; Vavulo 1958; Christenen and Cook 1970).

Bacterial nitrogen fixation

During the past 50 years there have been numerous reports of nitrogen-fixing bacteria in peats from several countries including the United States (Snyder and Wyant 1932; Wilson and Wilson 1933), the USSR (Vavulo 1958; Popova 1961), the Arctic (Boyd and Boyd 1962), Sweden (Granhall and Selander 1973), Canada (Blasco and Jordan 1976), and England (Collins, D'Sylva, and Latter 1978). Fixation of gaseous nitrogen by heterotrophic micro-organisms in peat has also been detected using ^{15}N with a seven-day incubation period (Waughman and Bellamy 1972*a*), and by an increase in the total nitrogen content of enriched peat samples after 21 days (Kaila 1954); however, by contrast with the acetylene-reduction technique, neither of these methods is sufficiently sensitive to measure absolute or relative fixation rates by different types of peat. Although peat is an organic substrate the stimulation of nitrogenase activity by the addition of an energy source suggests that, as with most other soils, available carbohydrate is in short supply (Kaila 1954; Waughman and Bellamy 1972*b*; Murphy 1975; Blasco and Jordan 1976). Neither Murphy nor Blasco and Jordan could detect any acetylene reduction by the peat microflora using aerobic incubation; by contrast higher rates were recorded for assays started aerobically as compared with anaerobic assays by Granhall and Selander (1973) and Waughman (1976). Waughman (op. cit.) found that aerobic conditions initially suppressed acetylene reduction, but gave rise to higher rates later during time-course experiments, presumably when the oxygen had been depleted to levels suitable for nitrogen fixation. Other workers have found that a mixture of anaerobic and aerobic conditions stimulates nitrogen fixation in soil, or considered associations of anaerobic and aerobic bacteria to be an important aspect of soil nitrogen fixation (Pringsheim 1909; Shklyar 1956; Kalininskaya 1967; Rice, Paul, and Wetter 1967; Magdoff and Bouldin 1970). Specific reference to microbial associations in peat was made by Makrinov and Stepanova (1930), who proposed a relationship between cellulytic and nitrogen-fixing species, also by Kaila (1954) who failed to isolate *Azotobacter* from aerobically incubated peat samples which made considerable nitrogen gains. One bacterial species which may be involved in any such association in peat is *Bacillus polymyxa* which fixes nitrogen both aerobically and anaerobically (Moore and Becking 1963), and is considered by Kalininskaya (1967) to be an important partner in nitrogen-fixing microbial associations. *B. polymyxa* has been isolated from peat by Popova (1961), and by Blasco and Jordan (1976) who consider that bacterial nitrogenase activity is restricted to anaerobic or microaerophilic sites in the peat. This aspect of nitrogen fixation is particularly relevant to peatlands, because in this ecosystem regular watertable fluctuations create periods of both aerobic and anaerobic conditions in the upper (rooting) layer of the peat.

Bacterial nitrogenase activity is frequently undetectable in bog peats, even with long assay times, and when it is observed the rate is usually less than 0.5 nmol C_2H_4 g^{-1} h^{-1}. Blasco and Jordan (1976) isolated nitrogen-fixing bacteria from bog peat although bacterial nitrogenase activity was undetectable in the field, thus the lack of energy source in the poorly decomposed bog peat may restrict activity. Blasco and Jordan also found that 100 μg ammonium-nitrogen completely inhibited acetylene reduction by bog bacteria in laboratory experiments; as the concentration of ammonium-nitrogen is relatively high in many bog peats (Table 5.11) this may also contribute to the low levels of bacterial nitrogenase activity. By contrast with bogs, fen peats give rates which range from about 2–10 nmol C_2H_4 g^{-1} h^{-1} in poor fens, to over 50 nmol g^{-1} h^{-1} in the richer fens of England and Germany (see Table 5.12). Although these rates for bacterial activity in peat are low compared with nodulated species and blue–green algae, the bulk of the peat mass is such that the total amount of nitrogen added to the system is considerable, and may be an essential factor in the working of the ecosystem as a whole. Some estimates of the amounts of nitrogen fixed annually in peatland are included in Table 5.12.

TABLE 5.11

Acetylene reduction by heterotrophic micro-organisms, and the concentration of soluble nitrogen in different peat types from Southern Germany (Waughman and Bellamy 1980)

Peat type	Ph	nmol C_2H_4 ml peat^{-1} day^{-1}	μg ammonium-nitrogen g dry wt peat^{-1}
Fen	6.3	126	45
Poor fen	4.5	24	90
Bog	3.6	0.8	143

TABLE 5.12

Nitrogen fixation in peatland sites

Peatland type	Location	Main group of organisms involved	Average rate of measured activity	Estimate of nitrogen fixed ($kg\ ha^{-1}\ a^{-1}$)	Reference
Alder carr	Montreal Island	*Alnus rougosa*		165[a]	Daly (1966)
Subarctic mixed mire	N. Sweden	Bacteria	7.75 nmol C_2H_4 g dry wt^{-1} h^{-1}	1.5	Granhall and Selander (1973)
Subarctic mixed mire	N. Sweden	Algae associated with mosses	516 nmol C_2H_4 g dry wt^{-1} h^{-1}	94	Granhall and Selander (1973)
Poor fens and bogs	W. Scotland	*Myrica gale*		9	Bond (1951)
Treeless fen peat	S. Germany	Bacteria	127 nmol C_2H_4 ml $peat^{-1}$ day^{-1}	21	Waughman and Bellamy (1980)
Treeless bog peat	S. Germany	Bacteria	0.8 nmol C_2H_4 ml $peat^{-1}$ day^{-1}	0.7	Waughman and Bellamy (1980)
Poor fen peat	Scotland	Bacteria	12 nmol C_2H_4 ml $peat^{-1}$ day^{-1}	8.2	Waughman and Bellamy (1972*b*)
Low-level blanket peat	Co. Mayo, Ireland	Bacteria	247 nmol C_2H_4 m^{-2} h^{-1}		Dooley and Houghton (1973)
Drained and fertilized beat pasin	Ireland		0.02 nmol C_2H_4 g dry wt^{-1} h^{-1}		Murphy (1975)
Bog peat (Muskeg) with *Picea* and *Larix*	Ontario	Blue–green algae	55 nmol C_2H_4 g dry wt^{-1} h^{-1}		Blasco and Jordan (1976)
Treeless fen peat	E. Anglia, England	Bacteria	143 nmol C_2H_4 ml $peat^{-1}$ day^{-1}		Waughman and Bellamy (1972*a*)
Arctic slope peaty trough	Axel Heiberg Is. Canada	Bacteria	7 nmol C_2H_4 ml $peat^{-1}$ day^{-1}		Waughman and Bellamy (1980)

[a] Net accumulation in forest floor and soil

Nitrogen fixation by blue–green algae

Literature on peatland contains few references to blue–green algae, and two extensive surveys of peat bog microbiology report bacteria, fungi, and actinomycetes, but make no mention of blue–green algae (Vavulo 1958; Burgeff 1961). However, blue–green algae are important in many aquatic situations, and they may be significant during early stages of the hydrosere. Certainly blue–green algae do occur in mires, this fact has been demonstrated by workers whose main objective was the examination of algae, in contrast to the more general microbiological surveys quoted above (e.g. Flensburg 1967; Flensburg and Malmer 1970; Dooley and Houghton 1973; Granhall and Selander 1973; Flensburg and Sparling 1973). Dooley and Houghton concluded that nitrogen fixation by blue–green algae in low-level Irish blanket bog was insignificant compared with bacterial fixation, and reported the complete absence of blue–green algae below pH 5.4. Waughman (1976) found that nitrogenase activity in fen peat was not stimulated by light. However, Granhall and Selander (1973) who investigated a sub-arctic 'mixed mire', considered both heterotrophic nitrogenase activity in the drier parts of the mire, as well as nitrogen fixation by blue–green algae in the wet mineral rich depressions to be important. As might be expected, the photo-autotrophic activity decreased rapidly below the top 2–3 cm of the substratum. (A mixed mire refers to the skeletal peatland consisting of a mosaic of ombrotrophic hummocks and mineral-rich depressions.) Blasco and Jordan (1976) could detect bacterial nitrogenase activity in bog peat only in laboratory conditions, and concluded that the activity which they recorded in the field was due to blue–green algae. The light requirement of blue–green algae probably restricts the importance of this group in the many parts of the peatland ecosystem where tall sedges predominate; however they may be locally important in open pools and in ombrotrophic parts of the mire (bogs). Nitrogen fixation has been confirmed in many lichen species, and the studies on a sub-arctic mire by Granhall and Selander (1973) indicate that they make a significant contribution to the nitrogen economy of peatlands in tundra regions. Blue–green algal symbionts also occur in some bryophytes, and of these, nitrogenase activity has been detected in *Sphagnum* (Stewart 1966; Granhall and Hofsten 1976), but the ecological significance of this association to the nitrogen balance of bogs remains to be examined.

Nitrogen fixation associated with higher plants

Mycorrhizal relationships are widespread amongst the mire ericads, and the possibility that these associations might have a nitrogen-fixing function was investigated by Marthaler (1939), but his results were negative, as were those of Burgeff (1961). Several bog sedges possess swollen roots (Davies, Briarty, and Rieley 1973) but these appear to have no nitrogen-fixing ability (Rieley and Waughman, unpublished data). The most important nitrogen-fixing associations involving higher plants are those with *Alnus* and *Myrica gale*. *Alnus* ranges throughout temperate Eurasia and North America with many species occurring on peatlands. Nodulation and nitrogen fixation are reduced in *Alnus glutinosa* in the presence of high levels of combined nitrogen (MacConnell and Bond 1957; Stewart and Bond 1961), but more important factors regarding the distribution of this species in peatland are that a significant reduction in both seed germination and nodulation occurs below about pH 6 (Quispel 1958). *Alnus* grows only on fen peat, where it is often an important shrub or small tree in the hydrosere (McVean 1956*a*). It has an optimum pO_2 for nitrogen fixation of about 12 per cent (Bond 1961; Waughman 1972*a*), thus concentration of oxygen in fen peat would permit only a very low rate of activity. The concentration of oxygen in air in the vicinity of roots has, however, been measured at 17 per cent (McVean 1956*b*) and it is therefore possible that the nitrogen-fixing root nodules receive oxygen from the root's internal aerating system. *Myrica gale* also ranges throughout temperature Eurasia and North America, and like *Alnus* its nitrogen-fixing potential is reduced in high levels of combined nitrogen (Stewart and Bond 1961; Stewart 1963). However, the optimum pH for nitrogen fixation in *Myrica* is lower than in *Alnus*, with substantial nodulation occurring in the pH range 4.2–6.3 (Bond 1951) which is consistent with the fact that *Myrica* is essentially a plant of poor fens and bogs (Bond 1951). The absence of *Myrica* from very acid bogs may be due to: (i) a restriction of nodulation by the low pH; (ii) the high levels of ammonium-nitrogen, which may be greater than 100 p.p.m. in such sites (Waughman 1980); or (iii) the very anerobic bog peat providing a level of oxygen which it is too low for nitrogenase activity in this species (Bond 1961). Regarding the last point, it has been suggested that the negatively geotropic rootlets, which grow on *Myrica* nodules, may have an aerating function (Bond 1952; Fletcher 1955). Bond (1951) combined the information obtained from culture solution experiments with an estimate of the density of plants growing on poor fens and bogs in Western Scotland, and concluded that *Myrica* might fix up to 0.9 gN m^{-2} a^{-1} in field sites.

The significance of nitrogen fixation to the nitrogen economy of peatlands

From the results of over a thousand acetylene-reduction tests in peat from South German mires, it is clear that the pattern of rates of bacterial nitrogen fixation is similar to that of general microbial activity, being greatest in fens and lowest in bogs (Waughman and Bellamy 1980);

however, the concentration of soluble nitrogen in these peats has the reverse trend (Table 5.11). In bogs the nitrogen contributed by fixation is so low that the supply of nitrogen from precipitation and pollen (Fibras 1952) must more or less counter-balance the drainage losses and accumulation of this element of nitrogen in the bog peat. This conclusion is supported by studies of Danish bogs where the estimated accumulation of nitrogen over a 2500-year period can be accounted for by the amount supplied in precipitation over the same period (Moore and Bellamy 1973, p. 108).

The richer types of peatland have a greater productivity, hence greater nitrogen requirements. In some poor mire *Narthecium ossifragum* communities in Sweden the annual uptake of nitrogen can excede 50 kg ha^{-1} (Malmer 1962); and in spruce sedge forests of the Vologda region of the USSR P'yavchenko (1960) measured the uptake by the entire vegetation at 43 kg ha^{-1} of which 7 kg ha^{-1} were retained; the corresponding figures for the moss and sedge layer of this system were 19 and 4 kg ha^{-1} respectively. In these Vologda spruce–sedge peatlands the amount of soluble nitrogen in the rooting zone decreased from 27 to 3 kg ha^{-1} during the course of the season (P'yavchenko, op. cit.), and in many European peatlands the amount of soluble inorganic nitrogen in fen peats during the peak growing season is about 2 kg ha^{-1} (Waughman and Bellamy, unpublished data). In some rich fen communities with annual net production in excess of 30 t ha^{-1}, the nitrogen demand is even greater (Moore and Bellamy 1973, p. 90). In all types of fen the inflowing water will contribute nitrogen to the system; however, loss of nitrogen in the outflow is also greater. Furthermore, the influx of nitrogen from flowing water from many catchments, may not be very large: Bormann and Likens (1969) calculated that the nitrogen loss in stream water from a forest watershed in the North East United States (which could become a peatland inflow) was only 2 kg ha^{-1} a^{-1}, although the influx from peat covered catchments may be greater. Losses from blanket peat have been estimated at 18 kg ha^{-1} a^{-1} (Crisp 1966). Therefore, considering that over much of the land surface where fen peat formations occur the nitrogen contributed by precipitation is less than 12 kg ha^{-1} a^{-1} (Eriksson 1952), and that some nitrogen is entrapped with the growing peat mass, the amounts of biological nitrogen fixation in fens (upwards of 20 kg ha^{-1} a^{-1}, Table 5.12) are consistent with the greater nitrogen demands of this part of the peatland ecosystem.

5.6 Nitrogen fixation in sand dune systems

These environments occupy much smaller areas than the others discussed in this chapter. However, coastal dunes have long interested biologists due, in part at least, to the fact that in many of these areas a chronological sequence of soil and vegetation types is easier to identify and study than in many other ecosystems.

Dune soils in the early stages of formation provide an unstable substrate for vegetation, and may contain high concentrations of salts. The concentration of soluble nitrogen in the young dune soils is usually less than 2 p.p.m. and total nitrogen less than 0.01 per cent (Gorham 1958; Hassouna and Wareing 1964; Waughman 1972*b*; Olsson 1974), and there is experimental evidence that in these situations nitrogen shortage is a severe limiting factor to vegetation growth (Willis and Yemm 1961; Pemadasa and Lovell 1974). During the early phases of succession on sand dunes there is a progressive increase of both total and soluble nitrogen in the soil, and it is possible that this accumulation has a marked influence on many of the vegetation changes which occur (Olson 1958; Olsson 1974). Several workers have provided data suggesting that the nitrogen accumulation in dune systems is greater than can be accounted for by precipitation (e.g. Ovington 1951; Olson 1958; Stevenson 1959; Dommergues 1963).

Legumes may be abundant on young non-acid dunes and dune pastures prior to the leaching of bases from the soil, and whether planted or natural, their effect on the nitrogen balance may be considerable (Gadgil 1971; Waughman 1972*b*). However, it is the occurrence of nodulated non-legumes which often has the greatest influence on both the accumulation of nitrogen, and succession on the sand dunes. *Hippophaë*, which grows on many fixed dunes throughout Eurasia (Skogen 1972), possesses nodules that fix nitrogen in soils within the pH range 5.4–7.0 (Bond, MacConnel, and McCallum 1956). The activity of these nodules may give rise to concentrations of soil nitrogen which are much greater than in the younger dune soils (Stewart and Pearson 1967; Waughman 1972*b*). In England it has been estimated that four- and thirteen-year-old stands fix 4.4 and 5.7 kg N ha^{-1} a^{-1} respectively, and that the rate of nitrogen fixation declines under stands of 16 years and older (Stewart and Pearson 1967). The soil nitrogen beneath stands of *Hippophaë* is enhanced both by leaf fall, and by direct exudation of nitrogenous material from the nodulated roots (Skogen 1972). Where this nitrogen-fixing species occurs, it has a marked effect on succession, growing as closed thickets and causing an increase in the number of associated nitrophilous lichens, and vascular species such as *Solanum dulcamara*, *Humulus lupulus*, *Rubus idaeus*, and *Filipendula ulmaria* (Géhu and Ghestem 1965; Skögen 1972; Waughman 1972*b*).

Acetylene-reduction tests have indicated a nitrogen-fixing potential in root nodules of *Myrica pensylvanica*, one of several *Myrica* species growing on coastal dunes of the Eastern United States (Morris,

Eveleigh, Riggs, and Tiffney 1974). *M. pensylvanica* has been attributed with a key successional role in both dune and non-dune systems, speeding the conversion of various pioneer communities into the later shrub phases (Morris *et al.*, op. cit.; Blizzard 1931). *Caasurina* ranges widely throughout tropical Asia and Australasia with a number of species growing on coastal dunes. *Casuarina equisetifolia* planted on dunes near Dakar in West Africa was estimated to fix an average of 58 kg N ha^{-1} a^{-1} over a thirteen-year period (Dommergues 1963).

Results of both acetylene reduction and culturing experiments suggest that activity by free-living nitrogen-fixing bacteria is generally very low in dune soils (e.g. Tchan and Beadle 1955; Line and Loutit 1969; Hassouna and Wareing 1964; Akkermans 1971; Waughman 1972*b*); however, the bacterial activity under vegetation is greater than in free sand (Webley, Eastwood, and Gimmingham 1952; Akkermans 1971), and culturing experiments have shown that *Azotobacter* species are abundant in this rhizosphere flora (Hassouna and Wareing 1964; Line and Loutit 1969). The results of experiments by Hassouna and Wareing (op. cit.) suggest that a nitrogen-fixing rhizosphere flora is an important aspect of the nutrition of *Ammophila arenaria*; they also found that an exogenous carbon source was required in order to stimulate nitrogen-fixing activity. Nitrogen-fixing bacteria have also been reported in root sheaths of the grasses *Oryzopsis hymenoides* growing in semi-deserts (Wullstein *et al.* 1972), and *Agropyron juneiforme* growing on fore-dune in Ireland from which Murphy (1975) isolated *Bacillus polymyxa*. The external oxygen concentration appears to have little effect on the rate of nitrogenase activity by the rhizosphere flora provided that the soil–root system is intact (Döbereiner, Day, and Dart 1972; Harris and Dart 1973). Therefore a depletion of oxygen within the rhizosphere of sand dune species may create a suitably low oxygen environment, as required by free-living nitrogen-fixing bacteria, in what is otherwise a well aerated soil. In addition, root exudates in the rhizosphere may provide the energy needed by heterotrophic bacteria.

Nostoc, Anabaena, and *Tolypothrix* occur in dune systems (John 1942; Stewart 1965), and the transfer of nitrogen fixed by these algae to higher plants colonizing sand-dune slack regions has been demonstrated using ^{15}N (Stewart 1967*a*, *b*); pigmented bacteria, which may fix nitrogen, are also found in these low areas (Stewart 1967*a*; Henriksson 1971). However, the general surveys of Lund (1947) and Granhall and Henriksson (1969) suggest that blue–green algae are of very restricted distribution on sand dunes. On many dune systems the lichen *Peltigera canina* is abundant. This species, which fixes nitrogen (Millbank 1972), often covers large areas of stabilized grey dune, and where it occurs it may make an important contribution to the build up of nitrogen in the dune soil.

The later stages of succession on dunes may take the form of an acid heathland on the highly leached sandy soil. In these acid conditions the numbers of bacteria are much smaller than in the earlier dune phases (Webley *et al.* 1952), and John (1942) found blue–green algae to be absent from all the heath soils with a reaction lower than pH 6 that he examined. One source of nitrogen on acid soils may be provided by *Clostridium butyricum* which is very common on sandy heaths (Boswell 1955). Legumes, such as *Ulex* spp and *Genista* spp are frequent in vegetation of the heath phase of temperate dune systems, but the role of these species in the nitrogen balance remains to be established. In Australia, heath vegetation may contain an abundance of both nodulated legumes and *Casuarina* spp (Hannon 1956; Sprecht and Rayson 1957).

5.7 Nitrogen fixation in relation to ecological succession

Much of the foregoing account has illustrated the role of nitrogen fixation during pioneer stages of succession. This aspect of the subject can be conveniently studied on cooled lava flows and cinder cones which present a sterile substratum to the colonizing species. Treub (1888) observed that blue–green algae were amongst the most abundant plants colonizing the lava and ash on Krakatau after the 1883 eruption. Blue–green algae were likewise amongst the earliest colonizers of the new island of Surtsey following submarine eruptions (Schwabe 1972) and nitrogen-fixing bacteria have been isolated from cinder cones on Reception Island (Cameron and Beniot 1970). Mosses such as *Funaria, Grimmia,* and *Rhacomitrium* were the first macrophytes to be found on Surtsee, and on the new lava flows of the nearby island of Heimaey (Henriksson *et al.* 1972; Englund, 1976); although these genera do not possess a nitrogen-fixing symbiont, they have all been reported as possessing an epiphytic-fixing microflora of either bacteria or blue–green algae (Henriksson, Henriksson, and Pejler 1972; Schwabe 1972; Snyder and Wullstein 1973*a*, *b*; Englund 1976).

Reviewing the problem of the nitrogen supply to vascular plants in pioneer situations, Stevenson (1959) concluded that many such plants obtain part of their nitrogen from the atmosphere, possibly through associated microflora. Since the development of the acetylene-reduction method, there has been a rapid accumulation of evidence for nitrogen fixation in the rhizosphere. (See reviews by Döbereiner (1978). In this context it is particularly interesting to note that many workers attempting to detect free-living bacteria in soil, pioneer or otherwise, report their presence, but most of these authors also conclude that the estimated number, or the measured activity, is too low to make a significant contribution to the pool of biologically fixed nitrogen. Therefore it is possible that free-living nitrogen-fixing bacteria may actively fix nitrogen mostly when in the proximity of the rhizosphere, where the supply of energy rich exudates and the reduced oxygen levels will provide suitable conditions for bacterial fixation.

The rhizosphere, however, may not always provide a desirable location for nitrogen-fixing organisms. Leuck and Rice (1976) found that bacteria from the rhizosphere of *Aristida oligantha* were generally inhibitory to both *Azotobacter* and *Rhizobium*, and nitrogenase activity was found by Rychert and Skujins (1974) to be reduced beneath the desert shrubs *Atriplex confertifolia*, *Eurotia lanata*, and *Artemisia tridentada*. Phenolic compounds produced by numerous vascular plants in the early stages of succession on range soils are inhibitory to nitrogen-fixing bacteria, these allelopathic reactions, according to Rice and his co-workers, prevent the build up of nitrogen in the soil, and so delay development of the tall grass prairie. This is a later successional stage with several species having a greater nitrogen requirement than those of the earlier stages (Rice, Penfound, and Rohrbaugh 1960). Allelopathic inhibition has been demonstrated for nitrogen-fixing bacteria (Rice 1964), blue-green algae (Parks and Rice 1969), and legumes (Rice 1968). The ideas concerning the role of nitrogen fixation in range succession have received additional support from the more recent acetylene reduction tests which gave yearly averages of 23 and 15 pmol C_2H_4 g^{-1} h^{-1} for soil from the first and second seral stages compared with 83 pmol C_2H_4 g^{-1} h^{-1} for soil from a tall grass prairie site; these differences could not be accounted for by differences in available substrate (Kapustka and Rice 1976).

Another allelopathic reaction during the later stages of succession is the suppression of nitrifying bacteria by chemicals from higher plants (see reviews by Harmsen and Van Schrevan (1955) and Rice (1977)), this would explain the relative paucity of nitrifying bacteria in the later stages of succession which has been noted by several workers (e.g. Berlier, Dabin, and Leneuf 1956; Greenland 1958; Nye and Greenland 1960; Smith, Bormann, and Likens 1968). Low levels of nitrifying activity will cause an accumulation in the soil of the ammonium ion rather than the nitrate ion. This has been observed in old field succession (Rice and Pancholy 1972; Hains 1977) and in Scandinavian conifer woodlands (Viro 1963); earlier examples of this feature are quoted in the review by Harmsen and Van Schrevan (op. cit.).

In addition to allelopathic reactions, the accumulation of nitrogen in the later stages of succession might be expected to lead to a reduction of nitrogen fixation both through the direct effect of the ammonium ion on nitrogenase production, and indirectly through competition favouring plant species with a high nitrogen requirement, but which do not themselves fix nitrogen. Such a sequence has been well documented for vegetation on the morains deposited on the receding ice at Glacier Bay, Alaska. Plants colonizing these morraines show symptoms of nitrogen deficiency which can be eliminated by the addition of nitrogen fertilizer; during the next stage of succession a thicket develops in which the nitrogen-fixing species *Alnus crispa*, *Dryas drumondii*, and *Shepherdia canadensis* are of common occurence. During this stage nitrogen accumulates within the ecosystem at a rate of 62 kg ha^{-1} a^{-1}. Subsequently the thicket is replaced by a woodland of *Picea sitchensis* and *Tsuga* spp. During this stage there is a slight loss of nitrogen from the system (Crocker and Major 1955; Lawrence 1958; Lawrence, Schoenike, Quispel, and Bond 1967). Newton *et al.* (1968) found that increments of soil nitrogen due to fixation by alder ceased when the trees were about 20 years old, and Tarrent, Lu, Bullen, and Franklin (1969) estimated that the total nitrogen beneath a stand of *Alnus rubra* in the north-west United States could not be accounted for by the then current rates of fixation, and concluded that the rates of nitrogen fixation must have been greater during the early stages of woodland development. Similarly, Stewart and Pearson (1967) estimated that the amount of fixed nitrogen by a stand of *Hippophaë* in eastern England was 0.44, 4.19, and 5.75 g m^{-2} a^{-1} for stands aged 3, 11, and 13 years respectively, but fell to 0.15 g m^{-2} a^{-1} for a 16-year-old stand. In parts of California the accumulation of nitrogen in mud flows reportedly occurs mostly during the early stages of succession when nitrogen-fixing species of *Ceanothus* and *Purschia* are abundant in the vegetation (Dickson and Crocker 1953). In southern Germany, peat containing 37 μg $NH_4^+{-}N$ g^{-1} beneath an unwooded, early seral stage of fen vegetation reduced acetylene at the rate of 173 nmol C_2H_2 g^{-1} day^{-1}, compared with 3.5 nmol C_2H_2 g^{-1} day^{-1} for a nearby peat containing 72 μg $NH_4^+{-}N$ beneath a later, fen woodland stage of vegetation (Waughman and Bellamy 1980).

It is self-evident that nitrogen accumulates within terrestrial ecosystems during ecological succession from pioneer to climax, therefore in the context of this present review, some problems which need to be resolved are: to what extend does nitrogen fixation affect the rate of succession in various ecosystems; is the decline of nitrogen fixation a characteristic of the later stages of succession to woodland; are the nitrogen budgets of semi-arid grassland climax ecosystems out of balance in the short term due to a continued relatively high rate of nitrogen fixation, and do these ecosystems achieve a steady state with regard to nitrogen in the long term (see p. 153), and is allelophathic inhibition of nitrogenase activity a characteristic of seral stages in these ecosystems? Some examples of nitrogen fixation during several different types of succession are presented in Table 5.13.

TABLE 5.13

Some examples of nitrogen fixation in relation to succession

Succession stage when estimate was made	Location	Main organism involved	Estimate of nitrogen accumulated or fixed (kg ha^{-1} a^{-1})	Reference
Succession on glacial moraine: *Dryas*, *Populus*, and *Salix* scrub phase	Glacier Bay, Alaska	*Dryas drummondii*	12[c]	Crocker and Major (1955)
Succession on glacial moraine: *Dryas*, *Populus*, and *Salix* scrub phase	Glacier Bay, Alaska	*Dryas drummondii*	10[f]	Lawrence, Schoenike, Quispel, and Bond (1967)
Succession on glacial moraine: *Alnus* thicket stage	Glacier Bay, Alaska	*Alnus crispa*	62[b]	Crocker and Major (1955)
Pioneer on rocky shore rocks	Scotland	*Calothrix scopulorum*	25 (^{15}N)	Stewart (1967*a*)
Dune sand	Holland	Rhizosphere bacteria	1 (C_2H_4)	Akkermans (1971)
Scrub vegetation	Gainsville, Florida	*Myrica cerifera*	3.4 (C_2H_4)	Silver and Mague (1970)
Ammophiletum (artificial)	New Zealand	*Lupin* spp (planted)	160[a]	Gadgil (1971)
Dune scrub	Holland	*Hippophaë rhamnoides* (1–2 years old)	2 (C_2H_4)	Akkermans (1971)
Dune scrub	Lincolnshire, England	*Hippophaë rhamnoides* (3 years old)	4.4 (^{15}N)	Stewart and Pearson (1967)
Dune thicket	Holland	*Hippophaë rhamnoides* (10–15 years old)	15 (C_2H_4)	Akkermans (1971)
Dune thicket	Lincolnshire, England	*Hippophaë rhamnoides* (13 years old)	58 (^{15}N)	Stewart and Pearson (1967)
Dune thicket	Lincolnshire, England	*Hippophaë rhamnoides* (16 years old)	1.5 (^{15}N)	Stewart and Pearson (1967)
Dune thicket (artificial)	Cap-Vert, W. Africa	*Casuarina equisetifolia* (planted)	58[a]	Dommergues (1963)

TABLE 5.13

Some examples of nitrogen fixation in relation to succession

Succession stage when estimate was made	Location	Main organism involved	Estimate of nitrogen accumulated or fixed (kg $ha^{-1} a^{-1}$)	Reference
Wet pioneer situations	New Zealand	*Gunnera/Nostoc*	72 (C_2H_4)	Silvester and Smith (1969)
Unwooded bog peatland	S. Germany	Bacteria	0.7 (C_2H_4)	Waughman and Bellamy (1980)
Unwooded bog peatland	Scotland	Bacteria	0.3 (C_2H_4)	Waughman and Bellamy (1972)
Mixed mire	N. Sweden	Bacteria	0.3–1.5 (C_2H_4)	Granhall and Sellander (1973)
Unwooded poor fen	Scotland	Bacteria	12 (C_2H_4)	Waughman and Bellamy (1972*b*)
Unwooded fen	S. Germany	Bacteria	21 (C_2H_4)	Waughman and Bellamy (1980)
Fen thickets	Montreal Island	*Alnus rugosa*	165[b]	Daly (1966)
Dried pond bed	Connecticut	*Alnus rugosa*	85[c]	Voigt and Steucek (1969)
Abandoned farmland	Oregon	*Alnus rubra*	137[c]	Tarrant *et al.* (1969)
Early phase of succession on mud flows	California	*Ceanothus* spp and *Purschia* spp	62[b]	Dickson and Crocker (1953)
Range land, pioneer weed stage	Oklahoma	Blue–green algae	1.2 (C_2H_4)	Kapustka and Rice (1976)
Range land, annual grass stage	Oklahoma	Bacteria	0.7 (C_2H_4)	Kapustka and Rice (1976)
Range land, tall grass prairie	Oklahoma	Bacteria	3.5 (C_2H_4)	Kapustka and Rice (1976)

Estimate of nitrogen fixed by accreation unless indicated by ^{15}N or C_2H_4 in column 4.

[a] Net amount of biological nitrogen fixation based on accreation method.

[b] Net accreation in the soil.

[c] Net accreation in soil and vegetation.

5.8 Nitrogen fixation in animals

Introduction

Many workers have reviewed biological nitrogen fixation in various ecosystems (e.g. Delwiche 1970; Bergersen 1971; Postgate 1974; Burris 1974, 1976; Broughton and Parker 1978; Bolin and Arrhenius 1977; Neyra and Döbereiner 1977) and commented on prospects for increasing nitrogen inputs (Hardy 1976; Evans and Barber 1977; Bolin and Arrhenius 1977) but there is scant information on the role of nitrogen fixation by micro-organisms associated with insects and other animals in the various ecosystems. In this section an attempt at synthesizing the main findings regarding nitrogen fixation in animals and insects is presented, with particular emphasis on the ecological significance of nitrogen fixation to these groups.

Nitrogen fixation in animals other than insects

In this section the term animals includes all phyla in the Animalia, except Insecta. The following is an overview of the limited experimental evidence published on nitrogen fixation in animals.

Using nitrogen-free culture medium inoculated with bacteria from the rumen of herbivorous farm animals, Tóth (1948, 1949) concluded that nitrogen fixation was occurring. However, it was not until the development of the acetylene-reduction assay that Hardy, Holsten, Jackson, and Burns (1968) demonstrated the nitrogen-fixing ability of ruminants. They found that rumen contents from a fistulated steer reduced acetylene. Ethylene formation was ten-fold greater under anaerobic than aerobic conditions and methane formation was markedly decreased in the presence of acetylene. The nitrogen-fixing activity of a rumen calculated on the basis of the anaerobic results was 10 mg N fixed per rumen per day. Hungate (1966) considered that rumen provided suitable conditions for nitrogen fixation because of its carbohydrate content, anaerobic milieu, near neutral pH, and presence of nitrogen. Potential nitrogen-fixing bacteria present in the rumen were *Clostridium* and sometimes *Azotobacter*, although Elleway, Sabine, and Nicholas (1971) considered the presence of *Azotobacter* unlikely because these highly aerobic bacteria would not survive under these conditions. Experiments incorporating ^{15}N alone into the rumen of a cow failed to demonstrate nitrogen fixation (Moisio, Kreula, and Virtanen 1969). Bergersen and Hipsley (1970) using this latter technique and acetylene reduction, reported the isolation of nitrogen-fixing bacteria from the faeces of humans, whose diets consisted mainly of sweet-potato (a protein-deficient food) and also from the intestinal contents of pigs and guinea pigs. The best nitrogen fixers appeared to be *Klebsiella aerogenes* (0.75 nmol N g^{-1} h^{-1} from diluted human faeces collected in Papua New Guinea), but at least three other genera occurred. These were provisionally identified as *Enterobacter cloacae*, from human faeces, and *Escherichia coli*, from pigs and guinea pigs.

A group of Mexican research workers (Ulloa, Herrera, and Lanza 1971) when investigating an Indian maize mash ('Pozol') found that its nitrogen content increased by 30–60 per cent during the 5 to 10 days fermentation period customary in that area, and they attributed this enrichment to the activity of nitrogen-fixing organisms. However, Ulloa *et al.* (1971) used nitrogen-free or nitrogen-deficient media in their study, and as Kleiner (1975) commented, this conclusion should be tested using other techniques such as acetylene-reduction assay or $^{15}N_2$-incorporation. Granhall and Ciszuk (1971) showed that the rumen and intestinal contents obtained from sheep, a goat, a rabbit, and a reindeer fed on different diets for a minimum of two weeks, indicated nitrogen fixation when measured *in vitro* by the acetylene-reduction method. They found values for rumen and rabbit caecum activity of the same order (0.0 to 1.4 nmol C_2H_4 g^{-1} h^{-1}), corresponding to about 1.1 mg N fixed day^{-1}. The sheep had a nitrogen requirement of the order of 10 mg day^{-1}. Even the amount of nitrogen fixed in the rabbit's caecum was considered much too low to contribute substantially to the nutrition needed by the animal.

Similar results were found with acetylene-reduction experiments on rumen microflora by Elleway *et al.* (1971). They estimated that about 1.0 mg N per bovine rumen per day was fixed *in vitro*, whereas Hardy *et al.* (1968) obtained a value of 10 mg N fixed per bovine rumen per day. Further Elleway *et al.* (1971) considered the rumen nitrogen fixation of negligible value to a sheep when fed a lucerne diet (28 g N kg dry wt^{-1}). They found also that acetylene reduction in rumen contents was only detected when the diet was supplemented with urea. This suggested that an adequate nitrogen intake is necessary for acetylene reduction. Thus, animals on a low-nitrogen diet are unlikely to benefit from nitrogen fixation by the rumen microflora. Cuthbertson (1970) also concluded that under usual conditions of ruminant feeding, nitrogen fixation by rumen micro-organisms does not occur, because the quantity of available fixed nitrogen is adequate. This is particularly true during the period soon after feeding, when nitrogen is normally in excess. During later periods, some nitrogen enters from the blood and salivary secretions, and some is released from the bodies of bacteria, both by endogenous metabolism and by utilization of microbial bodies as food by other bacteria and by protozoa.

On examining nitrogen fixation in the rumen of living sheep, Hobson, Summers, Postgate, and Ware (1973) found that the slight and variable ability of samples of rumen contents to reduce acetylene reflected the actual in-vivo conditions and reported an average rate of nitrogen fixation of 0.4 mg sheep^{-1} day^{-1}. This is the same order as for the in-vivo tests. Hobson *et al.* (1973) considered that the in-vivo fixation was probably due to free-living, nitrogen-fixing bacteria entering the rumen with food and from the surroundings, small numbers of such contaminants being always present in the rumen. This conclusion was confirmed by Jones and Thomas (1974). Using $^{15}N_2$ incorporation and acetylene-reduction techniques they estimated a maximum fixation rate for sheep at pasture of 0.8 mg N g rumen^{-1} day^{-1} or 32 mg sheep rumen^{-1} day^{-1}. The nitrogen-fixing bacteria were ingested when the sheep were at pasture or provided with fresh, non-sterile food. Jones and Thomas (1974) suggested that a permanent nitrogen-fixing microflora does not develop in the rumen. The addition of nitrogen-fixing bacteria to sheep rumen coupled with additional carbohydrates (10 per cent molasses) in the feed increased the nitrogen-fixing capacity from 30 mg N to 800 mg N fixed rumen^{-1} day^{-1}. Thus nitrogen fixation in the rumen could be of value when the diet is low in nitrogen, and under such conditions possibly accounts for 6 per cent of the total nitrogen intake. (This compares favourably with the rate of nitrogen intake in termites; see Rohrmann and Rossman (1978).) Salter (1973) also pointed out that with dietary regimes supplying abundant good quality protein, the microflora has only a marginal influence on the protein nutrition of the host (rumen). With severely restricted protein intakes and poor quality diets, or in starvation, it is possible that synthesis of non-essential amino acids from nitrogen released as ammonia from urea or partly digested proteins, or recycling of endogenous nitrogen may be of value. Recycling of urea nitrogen has been reported for ruminants (Harris and Mitchell 1941; Cuthbertson 1970; Nolan, Norton, and Leng 1973; Kinnear and Main 1975) and other mammals, such as rats and rabbits (Houpt 1963), Koala (Harrop and Degabriele 1976), snails (Friedl 1975), terebellid polychaetes (O'Malley and Terwilliger 1975), and reef coral (Muscatine and Porter 1977). However, the amount of nitrogen fixation in goats fed on low protein diets for a sustained period (16 weeks) was considered negligible (2.5 mg N goat^{-1} day^{-1}) by Pun and Satter (1975).

Citernesi, Neglia, Seritti, Lepidi, Filippi, Bagnoli, Nuti, and Galluzzi (1977) demonstrated that nitrogen fixation occurred in the gastroenteric cavity of soil animals such as a mollusc, *Helix aspersa*, and an earthworm, *Eisnia foetida*. Their data showed that the isolated bacteria (*C. beijerinckii*, *C. butyricum*, and *C. paraputrificum*) were genuine nitrogen-fixing organisms though they provided lower nitrogenase activity than cultures isolated from sources other than gastroenteric cavities.

McCoy and Seidler (1973) reported the presence of *Klebsiella*, *Enterobacter*, and *Citrobacter* spp, associated with small green pet turtles, and

Jackson and Fulton (1970) isolated *Enterobacter* and *Citrobacter* from other turtle genera. Thus these animals carry many representatives of enteric bacteria which are closely associated with warm-blooded animals. It is interesting that a *Citrobacter* (*C. freundii*) was the nitrogen fixer isolated from termite hindguts (French, Turner, and Bradbury 1976) and paper pulp mill ponds (Neilson and Sparell 1976).

Nitrogen fixation in insects

Several reports of nitrogen enrichment in culture by symbiotic organisms of insects have been published (Peklo 1946; Peklo and Satava 1949, 1950; Tóth 1952). Peklo (1946) considered that insects with well-developed mycetomes (e.g. Aphids, Homoptera) fixed free nitrogen much more energetically than insects which are devoid of them (e.g. Heteroptera). Aphids, Homoptera, and Heteroptera, Bark beetles (*Ips* spp) wood-destroying insects (*Anobium paniceum*), the grain beetle (*Sitophilus*), and larvae of the moth *Sitotroga cerealella* were found to assimilate free nitrogen. Peklo (1946) considered that this remarkable insect symbiosis corresponded in magnitude with that in Leguminosae and characterized the fixing bacterium as an *Azotobacter*. A yeast, *Torulopsis* (Fungi Imperfecti) associated with the ambrosia beetle, *Xylebous dispar*, and its primary symbiotic fungus *Ambrosiella hartigii*, when grown *in vitro* was considered to fix nitrogen (Peklo and Satava 1950). However, subsequent studies have found that yeasts do not fix nitrogen (Postgate 1971). Increases in the weights of *Anobium punctatum* larvae growing in wood was directly correlated with the nitrogen content of the material (Bletchly and Taylor 1964). Later, Baker, Laidlaw, and Smith (1970) studied the efficiency of utilization of wood as food by *A. punctatum* and found a discrepancy in the nitrogen budget which they could only attribute to additional inputs, presumably from microbial nitrogen fixation. Also, the intercellular symbionts of the grain beetle *Sitophilus oryzae* were found capable of fixing small quantities of nitrogen from air *in vitro* (Dang-Gabrani 1971).

Although all these reports suggested nitrogen fixation by micro-organisms in insects, it was not until Dilworth (1966) and Schollhorn and Burris (1967) showed that acetylene was reduced to ethylene by nitrogen-fixing systems, and the subsequent development of a sensitive acetylene-reduction assay for nitrogen-fixing activity (Koch and Evans 1966; Stewart, Fitzgerald, and Burris 1967; Hardy *et al.* 1968) that unequivocal evidence was presented indicating that animals fixed atmospheric nitrogen. Using this assay, Breznak, Brill, Mertins, and Coppel (1973) demonstrated that the termites *Coptotermes formosanus*, *Reticulitermes flavipes*, *Zootermopsis* spp, and *Cryptotermes brevis* can fix atmospheric nitrogen; Benemann (1973) confirmed this work independently while investigating the termite *Kalotermes minor*. He concluded that nitrogen fixation must be considered a major source of nitrogen for at least some termites under some circumstances. Breznak *et al.* (1973) showed that nitrogen fixing ability increased significantly when termites were fed on diets low in nitrogen compounds (e.g. cellulose) and was decreased by adding salts containing nitrogen. This showed that dietary nitrogen intake affects nitrogen fixation. Both these studies used whole insects in their assays. However, an anaerobically grown bacterium, *Citrobacter freundii*, isolated from the hindgut contents of three Australian termite workers, *Coptotermes lacteus, Mastotermes darwiniensis,* and *Nasutitermes exitiosus* was nitrogenase-positive as assayed by the acetylene-reduction assay (French *et al.* 1976). Nitrogen fixation confirmed with ^{15}N, was highest in the isolate from *M. darwiniensis* (5.5 nmol C_2H_4 termite^{-1} h^{-1}). This represents about 9 mg N fixed g^{-1} a^{-1}, with the workers weighing 50–5 mg fresh. Estimates of nitrogenase activity (0.03 nmol C_2H_4 termite^{-1} h^{-1}) by live termites of comparable size (10–15 mg fresh wt worker^{-1}) from northern (Breznak *et al.* 1973; Benemann 1973) and southern hemispheres (French *et al.* 1976) were similar (see Table 5.14). Potrikus and Breznak (1977) isolated two strains of the facultative anaerobic nitrogen-fixing bacterium *Enterobacter agglomerans* from the gut of *C. formosanus*. Recently, anaerobic bacterial isolates of bacteriaceal present in hindguts of *Reticulitermes flavipes* included *Citrobacter* and *Enterobacter cloacae* (Schultz and Breznak 1978).

There are some anomalous reports in the literature with respect to nitrogen fixation within the same species of insect. For instance, Jurzita (1970) reported that the evidence from cultural studies indicated that the grain beetle *Lasioderma serricorne* fixed atmospheric nitrogen, whereas Breznak *et al.* (1973) found no fixation in this insect using the sensitive acetylene-reduction method. Dissimilar results have been reported when both researchers have used the acetylene reduction assay. Breznak *et al.* (1973) examined two mealworm beetles *Tenebrio molitor*, and found no fixation, yet Citernesi *et al.* (1977) reported fixation from whole body specimens, both of larvae and adults, of *T. molitor* though the number of individuals used in the assay was not mentioned. Benemann (1973) and Breznak *et al.* (1973) said that under certain circumstances nitrogen fixation may occur, and implied at other times it may not. The acetylene-reducing ability of *C. formosanus* can vary over 200-fold, with high rates being exhibited by young, growing larvae. In fact, it was estimated that the amount of nitrogen fixed by young larvae could, over the period of a year, allow the termites to double their nitrogen content if the fixation rates remained constant (Potrikus and Breznak 1977). Breznak (1975) reported that immature termites at certain stages have high rates of nitrogen fixation which may

TABLE 5.14

Nitrogen fixation in insects (whole body measurements) using acetylene reduction assays

Insect	Insect form	Organism involved	Amount nitrogen fixed (mg N g^{-1} $individual^{-1}$ day^{-1})	Reference
Kalotermes minor	worker	?	0.01	Benemann (1973)
Cryptotermes brevis	worker	?	0.0003	Benemann (1973)
Zootermopsis angusticollis	worker	?	0.0003	Benemann (1973)
Coptotermes formosanus	worker	?	0.0002	Breznak *et al.* (1973)
C. formosanus	soldier	?	0.000 03	Breznak *et al.* (1973)
Z. angusticollis	worker	?	0.000 06	Breznak *et al.* (1973)
C. brevis	worker	?	0.0004	Breznak *et al.* (1973)
C. lacteus	worker	*Citrobacter freundii*	0.0008	French *et al.* (1976)
Mastotermes darwiniensis	worker	*C. freundii*	0.90	French *et al.* (1976)
Nasutitermes exitiosus	worker	*C. freundii*	0.000 05	French *et al.* (1976)
C. formosanus	worker	*Enterobacter agglomerans*	0.003	Potrikus and Breznak (1977)
Tenebrio molitor	adult	*Bacillus macerans, Clostridium aurantibutyricum*	0.000 0005	Citernesi *et al.* (1977)
Cotonia aurata	adult	*B. macerans, C. aurantibutyricum*	0.000 0005	Citernesi *et al.* (1977)
Trinervitermes trinervoides	worker	?	0.01	Rohrmann and Rossman (1978)
T. trinervoides	soldier	?	0.004	Rohrmann and Rossman (1978)
Cubitermes sp	worker	?	0.0002	Rohrmann and Rossman (1978)

permit them to meet the higher nitrogen demands during periods of rapid growth. However, Rohrmann and Rossman (1978) found that none of the immature termites examined in their study fixed nitrogen.

The flexible nature of the nitrogen-fixing mechanism in insects was further illustrated when diapausing adults of the scolytid beetle, *Xyleborus dispar*, failed to reduce acetylene (French and Roeper 1973). Yet when non-diapausing *X. dispar* beetles were cultured with their symbiotic fungus and yeast complex, nitrogen fixation occurred (Peklo and Satava 1950).

A variety of insects when tested with the acetylene-reduction assay have shown no nitrogen fixation (Breznak *et al.* 1973). The wood-destroying beetles *Lyctus planicollis* and *Xyletinus peltatus* gave negative results; also the Australian dung beetle, *Onthophagus gazella* and the Argentine ant *Iridomyrmex humilis*, when examined, failed to reduce acetylene (French, unpublished data). The wood-eating cockroach, *Cryptocercus punctulatus*, reduced acetylene at rates comparable to those of termites on a body weight basis (Breznak 1975). As with termites, Breznak (1975) showed that antibacterial drugs quickly abolished the nitrogen fixing activity of bacteria (but not protozoa) from the gut of the cockroach.

Recently, Bismanis (1976) characterized several yeast-like endosymbionts of the wood-eating beetle, *Sitodrepa panicea*, as *Torulopsis buchnerii*. He mentioned that this endosymbiont readily obtained nitrogen and carbon from the haemolymph of the insect.

Nodules of the desert sagebush, *Artemisia ludoviciana*, were shown to reduce acetylene (Farnsworth 1975). On opening up these nodules larvae of the gall midge *Rhopalomyia subhumilis* (Cecidomyiidae) were found. Farnsworth (1975) hypothesized that this midge may act as a 'vector of infection', and while ovipositing eggs in the leaves, stems or roots of the sagebush also injects a nitrogen-fixing bacterial organism. Though evidence needs to be gathered to evaluate this hypothesis, other cases have been reported of insect-host bacterial associations. The bacterium, *Klebsiella pneumoniae*, was isolated from the faecal detritus (frass) of the sugarcane borer, *Diatraea saccharalis*, and found to be fixing nitrogen (Nunez and Colmer 1968). *K. pneumoniae* has also been isolated from the surfaces of vegetables and from within living trees, although the predominant organism in these habitats is *Enterobacter* (Brown and Seidler 1973). Thus, it is conjectured that insects feeding in these habitats are highly likely to ingest nitrogen-fixing bacteria. Indeed, Potrikus and Breznak (1977) isolated two strains of the facultative anaerobic nitrogen-fixing bacteria, *E. agglomerans* from the gut of the termite *C. formosanus*. An *Enterobacter* was found to be fixing nitrogen when isolated from decay in living white fir trees (Seidler *et al.* 1972).

Citernesi *et al.* (1977) indicated that anaerobic nitrogen-fixing organisms are usually present among bacteria isolated from the digestive tracts of some soil animals feeding on organic matter and that nitrogen fixation occurred in their intestines. For instance, nitrogen-fixing bacteria such as *Bacillus macerans*, *Clostridium aurantibutyricum*, and *C. butyricum* were isolated from both larvae and adults of the scarab *Cetonia aurata*; and *C. beijerinckii*, *C. butyricum*, and *C. paraputificum* from the millipede *Glomeris cingulata*. The rates for both animals were low, averaging 5.0×10^{-7} mg N fixed $g^{-1} h^{-1}$ (Table 5.14). They isolated the nitrogen-fixing bacteria *B. macerans*, *C. aurantibutyricum*, *K. pneumoniae*, and *Enterobacter cloacae* from the beetle *T. molitor*. Bergersen and Hipsley (1970) also confirmed the nitrogen fixing ability of *E. cloacae* isolated from the faecal material of natives in the Morobe Territory of Papua New Guinea, whose main diet was sweet potato. Recently, *E. cloacae* was isolated from within living elm wood, *Ulmus procera*, and when transferred to nutrient agar slopes attracted inflight European elm-bark beetles, *Scolytus multistriatus* (French, unpublished data).

There is a suggestion that aphids, whilst not actually shown to fix nitrogen themselves, excrete sugars which are quickly utilized by nitrogen-fixing bacteria in the soil beneath the trees on which the aphids are feeding (Owen and Wiegert 1976). The faeces of the larvae of many species of psyllids which feed on the phloem sap of the host plant contain honey-dew or lerps (which is predominantly starch composed of amylose and amylopectin); other psyllids produce both lerps material and honey-dew, and many only lerp material. The enzymes necessary for the synthesis of lerps are probably produced by micro-organisms in the psyllids mycetome (White 1972). This mechanism serves as a protection cover for the insect, yet at the same time, faecal droppings and dead lerp material probably 'energize' the nitrogen-fixing organisms both under and within host plants. It has been shown that artificially adding sugars, such as mannitol, sucrose, and glucose to soils increases the rate of nitrogen fixation (Delwiche 1970; Bahadur and Tripathi 1975).

On examining nitrogen fixation in deteriorating wood Sharp (1975) found that a strain of *B. circulans* gave appreciable acetylene reduction, whilst two strains of a non-fluorescent *Pseudomonas* species had less activity. Machado (1971) suggested a close symbiosis among bees of the subfamily *Meliponinae* (Apidae) and some species of bacteria of the genus *Bacillus*. A species of *Bacillus*, relatively close to *B. subtilis*, but with morphological and physiological characteristics different from both *B. subtilis* and *B. pumilus*, was found present in the tract of old larvae and adults, in stored pollen and in brood food. Although the role of nitrogen fixation was not examined by Machado (1971), he showed that this *Bacillus* was highly sensitive to streptomycin. A colony of bees died out in 30 days after this treatment, suggesting some form of obligatory symbiosis with micro-organisms. Breznak (1975) demonstrated that antibacterial drugs caused a loss of nitrogen-fixing activity in cockroaches. *B. subtilis* and *B. pumilus* have recently been isolated from fresh elm wood, together with *E. cloacae*, *E. aerogenes*, *Flavobacterium* sp, and a *Corynebacterium* sp (French, unpublished data).

5.9 Ecological significance of nitrogen fixation in animals

Introduction

Accepting that one can not, with only one or two plausible explanations, fully understand complex ecosystems, a fruitful approach is to develop multiple hypotheses as to what may be happening. Thus, a number of hypotheses relating to animal abundance and food quality have been selected in order to form a conceptual framework in which to fit the many simple and disparate pieces of information gathered in the previous sections relating to nitrogen fixation in insects and animals, and an attempt made to comment on the ecological significance of such a mechanism in impoverished environments.

White (1978) proposed that for many if not most animals, both herbivore and carnivore, vertebrate and invertebrate, the single most important factor limiting their abundance is a relative shortage of nitrogenous food for the very young. Any component of the environment of a plant, by affecting growth, may affect the amount of adequately nutritious plant tissue available to herbivores, hence the abundance of food through all subsequent trophic levels (Gosz, Holmes, Likens, and Bormann 1978); in this regard weather may be important more often than is immediately obvious. The hypothesis proposes that animals live in a variably inadequate environment wherein many are born but few survive, and leads to a concept of populations being 'limited from below' rather than 'controlled from above'. Herbivores are seen to be limited to green plant food which, although photosynthetic, may have a nitrogen content which is often too low to provide a young herbivore with its minimum protein requirement; the young will eat it, but will not survive on it. White (1978) considered that this in turn removes the problem of how the system is kept in balance. Balance is an unnecessary idea when we are thinking about natural populations—for the usual and variable shortage of food provides an alternative explanation. Mattson and Addy (1975), Owen and Wiegert (1976), Mattson (1978), and Springett (1978) hypothesized that consumers, like pollinators, have a mutualistic relationship with plants: they suggested that plants regulate, and even encourage, a spectrum of consumers whose overall effect is to maximize plant fitness. For instance,

Owen and Wiegert (1976) pointed out that the supply of nitrogen is often the main limiting factor for plant productivity and reproduction. The large quantity of sugary honey-dew excreted by aphids may stimulate free nitrogen fixation beneath the plant from which the sugar is extracted. Hence, provided a plant can produce 'surplus' sugar, and provided that this can be fed to the soil through the action of the aphids, both plants and aphids could benefit.

Tundra and taiga

Owing to the severity of the physical environment, the species of insects (Downes 1965) and animals (Ehrlich, Ehrlich, and Holdren 1977) are few in number; this ecosystem is both simple, lacking in integration, and with a poorly developed soil fauna. The ecology of the Arctic, and plant aspects of nitrogen fixation within this ecosystem are dealt with on pages 142–8. The highest rates of nitrogenase activity by blue–green algae and bacteria occur in the mesic meadows which constitute about 20 per cent of the lowlands, and are the principal life-support areas for Arctic wildlife (Jordan *et al.* 1978). In the brief Arctic summer, an astonishing diversity of birds (especially waterfowl) migrate to the tundra to breed, feeding on an ephemeral bloom of insects and freshwater invertebrates. The aquatic Chironomid larvae are the most numerous insect fauna, feeding on algae and micro-organisms that fix nitrogen. Adult Chironomids are terrestrial, and in temperate latitudes the adults of some Chironomids feed on nectar, but in the Arctic they do not feed at all (Downes 1962); the possibility exists of this adult insect partially surviving on nitrogen being fixed endogenously by bacteria ingested at the larval stage.

Caribou, musk-oxen, and rabbits are among the mammals found in the tundra, with wolves, foxes, and polar bears being their main predators. Animal populations are subject to dramatic oscillations in size. Best known are the lemming cycles, which induce cycles in the owls, jaegars, (predatory gulls) and other predators that feed on them (Ehrlich *et al.* 1977). As photoautotrophic nitrogen fixation is considered more important in these regions than heterotrophic fixation (Jordan *et al.* 1978) the suggestion is offered that climatic conditions may severely limit the 'growing season' of blue–green algae, which in turn limits the amount of available nitrogenous food for herbivores, and subsequently the carnivores. Thus the lemming cycles may be directly attributed to the amount of nitrogen available to them in these mesic areas.

In-vitro experiments with rumen and intestinal contents of rabbit and reindeer gave measured rates of nitrogenase activity which were considered to be too low to 'contribute substantially' to the nutrition of these animals (Granhall and Ciszuk 1971); the rate of, and necessity for nitrogen fixation by these animals in the wild may, however, show quite a different pattern. Cuthbertson (1970) recognized this phenomenon, and stated that a study of the selection of herbage and other plant material by wild ruminants at different seasons of the year should precede any physiological investigation of their digestive processes.

Savannahs

Many large herbivores, such as bison, buffalo, giraffes, zebras, antelopes, and kangaroos are typically found in grasslands. In the African savannahs heavy grazing by the migratory wildebeest population as it leaves the Serengeti Plains prepares the plant community (many members of which probably possess nitrogen-fixing ability) for subsequent dry season exploitation by gazelle (McNaughton 1976). Thus, rather than competition, there has evolved a facilitation of energy flow into the gazelle population by the wildebeest population through the impact of the latter on the plant community. The grazing wildebeest substantially improves forage quality. Nutrient content and digestibility are considerably greater in the rapidly growing grasses than in mature tissues. Also, insect populations probably benefit from this enhanced growth, which presumably increases the nitrogen content of the grassland via nitrogen fixation, for no phylum of animals is more widespread and more adaptable than insects. Every type of plant is used by some kind of insect. Plant food quality (nitrogenous content) varies with season and affects the response of insect grazers (Fox and Macauley 1977). Flowers and seeds, the highest food value (Westman and Rogers 1977), support a relatively large number of insects (prey and predators), but are available only for short periods of the year. Stems and roots provide a continuous food supply of lower nitrogen quality that is most abundant in the spring. However, termites are the most studied of all the insects for nitrogen fixation.

Termites are principally inhabitants of tropical, subtropical, and semi-arid regions. They feed on all forms of plant material from living (green) tissues to humidified material in soil. In savannahs, species feeding on living and fresh dead tissues are the dominant feeding group (Wood 1976; Wood and Sands 1977). The harvester termite, *Hodotermes mossambicus*, feeds on most of the indigenous grasses and shrubs in the pastoral regions of southern Africa. The termites compete with cattle and sheep for food and are often regarded as a serious pest of natural grazing (Nel, Hewitt, and Joubert 1970; Lee and Wood 1971). *H. mossambicus* feeds on dry plant litter in nature, preferring grasses rather than shrubs. Another grass-harvester termite, *Trinervitermes geminatus*, in the southern Guinea savannah, Nigeria, feeds almost exclusively on dry grass (Ohiagu and Wood 1976). The abundance of *T. geminatus* in Nigeria indicates that its biomass may exceed 10 mg m^{-2}, which is similar to the highest recorded biomass of grazing mammals (Lee and Wood 1971). Thus termites are efficient producers and must have a significant influence on the rate and direction of energy flow in grassland ecosystems, Lee and Wood (1971) considered that termites are influencing the flow of energy at both the herbivore and decomposer level. They felt that in such areas the relative respiration of herbivores and decomposers, and thus the fate of plant tissues (primary production), may be strongly influenced by the relative abundance of herbivorous and saprophagous termites. Although nitrogen fixation has not been examined in these termites, acetylene-reduction assays have been made of another harvester termite, *T. trinervoides* and a humivorous species, *Cubitermes bilobatus*, and both found positive (Rohrmann and Rossman 1978). A rate of 2.5 mg N fixed g $worker^{-1}$ a^{-1} by *T. trinervoides*, equivalent to 8–12 per cent of their total nitrogen content, was recorded. Rohrmann and Rossman (1978) considered that this amount of nitrogen, if available to termites, would be a significant contribution to overall nitrogen balance providing for colony growth and acting to counter nitrogen loss due to predation and alate production. The Macrotermitinae are a group of termites of Africa and Asia which construct and maintain a comb composed of organic matter, within which fungi, *Termitomyces*, are cultivated. Many areas of the Macrotermitinae range are arid but produce abundant vegetation subsequent to brief seasonal rains. In such climates, plant detritus decomposes slowly. Comb construction creates a damp-wood environment which allows fungal growth in such areas. The success of the symbiosis is evident by the fact that the Macrotermitinae are the most important detrivores in some areas and capable of creating the largest of all termite colonies (Rohrmann 1978). Of particular interest is the fact that these fungus-cultivating termites, such as *Macrotermes ukuzii*, did not fix nitrogen (Rohrmann and Rossman 1978). The fungus consumed by *M. ukuzii* comprises 38 per cent protein, and may account for the lack of nitrogen fixation, and because their diet is so rich in nitrogen the nitrogen-fixing organisms may not be induced.

Semi-arid regions

In areas where rainfall is less than 250 mm a year the world's deserts are found (Ehrlich *et al.* 1977). Plants and animals have solved the problem of water shortage and nitrogen deficiency in many diverse ways. Excretory systems of desert mammals (Main 1968; Cuthbertson 1970; Nolan *et al.* 1973; Kinnear and Main 1975), and presumably the insects (Moore 1969), have evolved to conserve water and recycle the excretory nitrogen products uric acid and urea. Though available nitrogen represses urease formation in many bacteria (Kaltwasser, Kramer, and Confer 1972) which are found in rumen (Wozny, Bryant, Holdeman, and Moore 1977). However, Kinnear and Main (1975) considered that the potential to recycle nitrogen in ruminants depends on the plant substrate, and plant nitrogen must be rate-limiting for the growth of the ruminant microbes. They demonstrated that the recycling potential of the wild tammar wallaby was fully exploited during the extended protein deficient periods of its life cycle. Karakul sheep feed on the sage bush, *Artemisia* sp, and other shrubs during autumn and winter until the spring ephemerous pastures are ready (Cuthbertson 1970). An *Artemisia* sp, was found to possess nodules, caused by a gall midge, which was fixing nitrogen (Farnsworth 1975), and the same mechanism may be possible elsewhere. Ingestion of nitrogen-fixing organisms by feeding sheep seems highly probable.

The mobility and browsing habits of many wild African ruminants make it doubtful if these animals are likely to suffer from a nitrogen deficiency at any particular season of the year unless they eat only grasses (Cuthbertson 1970). The same was recorded for Australian ruminants, such as the red kangaroo (Hume 1974). Berwick (1976) found that in the Gir Forest region of India 80 per cent of the woody plants began to grow and form leaves during the hot, dry season. This new, succulent growth is very high in nutrients such as proteins (10–15 per cent), whereas the grasses, dry at this time of year, were very low in nutrients (2 per cent). Thus cattle and buffalo, which are obligate grazers, were living below minimum maintenance standards, which range between 6 and 11 per cent protein for adult sheep and cattle, and 7 and 8 per cent for deer. The livestock's food must therefore be supplemented with cotton-seed and peanut-cake. The wildlife however, as browsers, did very well during the dry, hot season. Also, Berwick (1976) considered that predation by large carnivores kept herbivores below the point at which they would damage their forage plants.

Several researchers have examined the natural food preferences of desert termites (Haverty and Nutting 1975), their foraging behaviour (Haverty, Nutting, and La Fage 1975; Bodine and Ueckert 1975; La Fage, Haverty, and Nutting 1976), survival and food consumption (Spears and Ueckert 1976). Bodine and Ueckert (1975) indicated that desert termites accounted for 55 per cent of the dry mass of litter removed from the soil surface. They estimated the population densities of *Gnathamitermes tubiformans* to average 2139 m^{-2} in the upper 300 mm of soil in a shortgrass community in west Texas over a three-year period, and reached a peak of 9127 m^{-2}. Assuming that this termite was capable of fixing an equivalent amount of nitrogen to *T. trinervoides* (Rohrmann and Rossman 1978), the average would be 0.08 kg N fixed

$m^{-2} a^{-1}$; whereas *M. darwiniensis*, living north of the Tropic of Capricorn in Australia, was estimated to yield 9 kg N $colony^{-1} a^{-1}$. Probably one could expect a single colony per hectare, thus: the nitrogen fixed compares favourably with the estimate of global biological nitrogen fixation by Frissel (1977). However, Frissel (1977) made no specific mention of insect and animal nitrogen fixation.

Cloudsley-Thompson (1975) explained that seeds and dry vegetable matter provided the primary source of energy in the simple food webs of some smaller desert animals, principally insects; and a secondary supply appeared in the form of corpses, both of desert-dwelling animals and of migratory birds. However, it seems that many animals, including birds and reptiles, benefit from the high protein food source provided by alates (winged termites) during periods when they swarm (Watson, personal communication).

Decaying wood as an impoverished environment

Wood is decomposed by the interaction of a rich community of organisms (Swift 1977). These organisms include Basidiomycete fungi, and insects such as wood-destroying beetles and termites. The combined action of the microbial enzymes and those of the endosymbionts of the wood-destroying insects ensure the decomposition of available woody components. Also the termites of the Macrotermitinae cultivate fungi within their colonies (Rohrmann and Rossman 1978) from which they derive appreciable amounts of protein (38 per cent). Both the fungi (Seidler *et al.* 1972) and insects, particularly termites (see Table 5.14), fix atmospheric nitrogen. Thus, although wood is lower in nutrient content (especially nitrogen and phosphorus) than other plant materials, the ability to fix nitrogen overcomes this apparent disadvantage for such decomposers. In this manner, as suggested by White's (1978) hypothesis, there is an important regulatory role in the recycling of nutrients in the environment (in this case wood) 'from below'. The pattern of overall wood decay fluctuates in natural conditions and may be significantly inhibited by lack of nutrients (Swift 1977). In these circumstances the fresh input of nutrients by nitrogen fixation is most important ecologically. It may be concluded that wood which is capable of fungal or insect attack may not be an impoverished environment so much as a variably inadequate environment, with nitrogen-fixing bacteria involved in wood decomposition (Sharp 1975).

5.10 Future for improving impoverished environments with animals

Our knowledge of the role of insects and animals in biological nitrogen fixation is scant. Most researchers have concluded that the amount of nitrogen fixed was too low to contribute substantially to the nutrition of the animal (Bergersen and Hipsley 1971; Granhall and Ciszuk 1971; Elleway *et al.* 1971; Hobson *et al.* 1973). Though Jones and Thomas (1974) suggested that nitrogen fixation in the rumen could be of value when the diet is low in nitrogen. However, the ability of wildlife to breed in impoverished environments, such as the Arctic and desert regions, suggests adaptations in obtaining nitrogen that entail intestinal microbial nitrogen fixation, or ingestion of nitrogen-fixing organisms from host plant material, or probably a combination of both these pathways.

Several authors have written on the prospects of future nitrogen fixation (Delwiche 1970; Evans and Barber 1977; Hardy and Silver 1977; Frissel 1977; Söderlund and Svensson 1976; Phillips, Rains, Valentine, and Huffaker 1978); others have indicated areas offering promise for the future, such as the management of denitrification (Delwiche 1970; Huber, Warren, Nelson, and Tsai 1977; Bolin and Arrhenius 1977; Miller and Smith 1976), and new nitrogen-fixing processes (Chatt 1977). But as Rosswall (1976) indicated, most nitrogen flows occur within the soil and between soil and vegetation, which seems to support White's (1978) concept of populations being 'limited from below'. It was mentioned above that agriculture including forestry, is one of the most disruptive activities in the global nitrogen cycle. The removal of natural legumes is probably the most predominant feature of current large-scale agriculture (Wilken 1977) and forestry practices (Waring 1976; Bormann, Likens, and Melillo 1977). In addition food crops require increasing amounts of nitrogen fertilizers and insecticides both of which reduce biological fixation (Bolin and Arrhenius 1976; Neyra and Döbereiner 1977; Chi-Yang 1977; Aspiras, Halos, Tejada, and Jusgado 1977). In addition to the many excellent suggestions concerning the strategies of nitrogen fixation for the future by Bolin and Arrhenius (1977), Evans and Barber (1977) and others, may be added the suggestion of managing impoverished environments using the nitrogen-fixing abilities of insects, animals, and plants. Already a feasibility study is underway in the United States of America to use the ability of termites, microarthropods and plant microorganisms, which all possess nitrogen-fixing organisms, in order to reclaim mine spoils (or other abused, denuded areas) to a fertile condition (Whitford and Ettershank, personal communication). To this self-sustaining system, independent of soil moisture and thus not requiring irrigation, is added waste cellulose material as an additional carbon source. If this project proves viable, this will open up new options for the management of impoverished environments.

Citernesi *et al.* (1977) demonstrated that nitrogen fixation probably occurs within the gastro-enteric cavities of many species of soil fauna, not just termites. Also, goats (Dehority and Grubb 1977) and termites

(Eutick, O'Brien, and Slaytor 1978; Schultz and Breznak 1978) have similar intestinal bacteria, suggesting a probable common link via host-plant material. Jones and Thomas (1974) considered that a permanent nitrogen-fixing microflora did not develop in a rumen, suggesting that such microflora entered the rumen with food and from the surroundings. Thus, if one examines the present potential for animal production in arid areas, such as in Australia, prospects for greatly increased production of cattle and sheep are remote (Squires 1978). However, Winter (1978) considered that by improving cattle breeds and their fertility, better water supply and fencing, and by introducing low-phosphate demanding legumes into the native (tropical) grasslands, productivity of the zone could be increased. However, another approach could be to favour the utilization of Australian wildlife (ruminants), along the lines suggested by Cuthbertson (1970) for African wildlife, and for Indian wildlife by Berwick (1976). Thus, euros or hill red kangaroos and feral donkeys could produce the best crop in semi-arid Australia because they are best adapted to the marginal conditions. In contrast, sheep and cattle are the least efficient and most destructive herbivores. It would seem that the nitrogen fixation ability of insects, animal wildlife and plants could be theoretically combined to increase protein production for man's needs. Whether we utilize such nitrogen-fixing biological systems remains to be seen. But paraphrasing Schumacher (1974): 'If man selects for production, methods which are biologically sound, build up soil fertility and produce health, beauty, and permanence; then continuous productivity will look after itself!'

References

Abd-el-Malek, Y. (1971). *Pl. Soil* Special Volume, 423–42.

Aho, P. E., Seidler, R. J., Evans, H. J., and Aju, P. N. (1974). *Phytopathology* **64** 1413–20.

Alexander, V. (1974). In *Soil organisms and decomposition in tundra*, pp. 109–22. Tundra Biome Steering Committee, Stockholm.

—— and Kallio, S. (1976). *Rep. Kevo Subarctic Res. Stat.* **13**, 12–15.

—— and Schell, D. M. (1973). *Arctic Alpine Res.* **5**, 77–8.

—— Billington, M., and Schell, D. (1974). *Rep. Kevo Subarctic Res. Stat.* **11**, 3–11.

Akkermans, A. D. L. (1971). Ph.D. thesis, University of Leiden.

Aspiras, R. B., Halos, S. C., Tejada, F. S., and Jusgado, L. O. (1977). *N.S.D.B. Technol. J.* **2**, 74–81.

Bahadur, K. and Tripathi, P. (1975). *Zentbl. Bakt. ParasitKde. Abt. II. Bd.* **130**, 17–20.

Baines, E. F. and Millbank, J. W. (1978). In *Environmental role of nitrogen-fixing blue–green algae and asymbiotic bacteria* (ed. U. Granhall) *Ecol. Bull., Stockh.* **26**, 193–8.

Baker, J. H. (1974). *Oikos* **25**, 209–15.

Baker, J. M., Laidlaw, R. A., and Smith, G. A. (1970). *Holzforschung* **24**, 45–54.

Barsdate, R. T. and Alexander, V. (1975). *J. Environ. Qual.* **4**, 111–17.

Beadle, N. C. W. and Tchan, Y. T. (1955). *Proc. Lin. Soc. N.S.W.* **80**, 62–70.

Becker, V. E., Reeder, J., and Stetler, R. (1977). *Bryologist* **80**, 93–9.

BELLAMY, D. J. (1969). *Proc. 3rd Int. Peat Cong.*, Quebec, pp. 74–9.
BENEMANN, J. R. (1973). *Science, N.Y.* **181**, 164–5.
BERGERSEN, F. J. (1971). *A. Rev. Pl. Physiol.* **22**, 121–40.
—— and HIPSLEY, E. H. (1970). *J. gen. Microbiol.* **60**, 61–5.
BERLIER, Y., DABIN, B., and LENEUF, N. (1956). *Trans. 6th Int. Cong. Soil Sci. E*, pp. 499–502.
BERWICK, S. (1976). *Am. Scient.* **64**, 28–40.
BILLINGS, W. P. (1974). In *Arctic and alpine environments* (ed. J. D. Ives and R. G. Barry) pp. 403–43. Methuen, London.
BILINGTON, M. and ALEXANDER, V. (1978). In *Environmental role of nitrogen-fixing blue–green algae and asymbiotic bacteria* (ed. U. Granhall). *Ecol. Bull., Stockh.* **26**, 209–15.
BISMANIS, J. E. (1976). *Can. J. Microbiol.* **22**, 1415–24.
BLASCO, J. A. and JORDAN, D. C. (1976). *Can. J. Microbiol.* **22**, 897–907.
BLETCHLY, J. D. and TAYLOR, J. M. (1964). *J. Inst. Wood Sci.* **12**, 29–43.
BLISS, L. C. (1962). In *Die Stoffproduktion der Pflanzendecke* (ed. H. Licth) pp. 35–46. Gustav Fischer Verlag, Stuttgart.
—— (1966). *Ecol Monogr.* **36**, 125–55.
—— and CANTLON, J. E. (1957). *Am. Midl. Nat.* **58**, 452–69.
—— and WEIN, R. W. (1972). *Can. J. Bot.* **50**, 1097–109.
BLIZZARD, A. W. (1931). *Ecology* **12**, 201–31
BODINE, M. C. and UECKERT, D. N. (1975). *J. Range Mgmt* **28**, 353–8.
BOLIN, B. and ARRHENIUS, E. (1977). *Ambio* **6**, 96–105.
BOLYSHOV, N. N. and EVDOKINOV, T. I. (1944). *Pochvovedenie* 345–55.
BOND, G. (1951). *Ann. Bot., N.S.* **15**, 447–59.
—— (1952). *Ann. Bot., N.S.* **16**, 467–75.
—— (1961). *Z. allg. Mikrobiol.* **1**, 93–9.
—— (1976). In *Symbiotic nitrogen fixation in plants* (ed. P. S. Nutman) pp. 443–74. Cambridge University Press.
—— MACCONNEL, J. T., and MCCALLUM, A. H. (1956). *Ann. Bot., N.S.* **20**, 501–11.
BOOTH, W. E. (1941). *Ecology* **22**, 38–46.
BORMANN, F. H. and LIKENS, G. E. (1969). In *The ecosystem concept in natural resource management* (ed. G. M. Van Dyne) pp. 49–76. Academic Press, New York.
—— and MELILLO, J. M. (1977). *Science, N.Y.* **196**, 981–3.
BOSWELL, J. G. (1955). *New Phytol.* **54**, 311–19.
BOYD, W. L. (1958). *Ecology* **39**, 332–6.
—— and BOYD, J. W. (1962). *J. Bacteriol*, **83**, 429–30.
BREZNAK, J. A. (1975). In *Symbiosis* (ed. D. H. Jennings and D. L. Lee) Soc. Exp. Biol., Symposium Ser., No. 29, Cambridge University Press.
—— BRILL, W. J., MERTINS, J. W., and COPPEL, H. C. (1973). *Nature, Lond.* **244**, 577–80.
BROUGHTON, W. J. and PARKER, C. A. (1976). In *Global impacts of applied microbiology—State of the art, 1976* (ed. W. R. Stanton and E. DaSilva) pp. 113–22. UNEP/UNESCO/ICRO, Penerbit Universiti Malaya.
BROUZES, R., LASIK, J., and KNOWLES, R. (1969). *Can. J. Microbiol.* **15**, 899–905.
BROWN, C. and SEIDLER, R. J. (1973). *Appl. Microbiol.* **25**, 900–4.
BURGEFF, H. (1961). *Mikrobiologie des Hochmoores*. Gustav Fischer Verlag, Stuttgart.
BURRIS, R. H. (1974). *Pl. Physiol., Lancaster* **54**, 443–9.
CAMERON, R. E. (1972). In *Taxonomy and biology of blue–green algae* (ed. T. V. Desikachary) pp. 353–84. University of Madras.
—— and BENOIT, R. E. (1970). *Ecology* **51**, 802–9.
—— KING, J., and DAVID, C. N. (1970). In *Antarctic ecology* (ed. M. W. Holdgate) Vol. 2, pp. 701–16. Academic Press, London.

Cengia Sambo, M. (1923). *Atti Soc. ital. Sci. nat.* **62**, 226.
Chatt, J. (1977). *Phil. Trans. R. Soc.* **B281**, 243–8.
Chi-Yang, H. (1977). *Taiwania* **22**, 80–90.
Christenen, P. J. and Cook, F. D. (1970). *Can. J. Soil Sci.* **50**, 171–8.
Citernesi, U., Neglia, R., Seritti, A., Lepidi, A. A., Filippi, C., Bagnoli, G., Nuti, M. E., and Galluzzi, R. (1977). *Soil Biol. Biochem.* **9**, 71–2.
Clark, F. E. (1977). *Ecology* **58**, 1322–33
Cloudsley-Thompson, J. (1975). *Deserts and grasslands. Part I. Desert life* p. 144. Aldus Jupier, London.
Coats, R. N., Leonard, R. L. and Goldman, C. R. (1976). *Ecology* **57**, 995–1004.
Collins, V. G., D'Sylva, B. T., and Latter, P. M. (1978). In *Production ecology of British moors* (ed. O. N. Heal and D. F. Perkins) pp. 94–112. Springer Verlag, Berlin.
Cornaby, B. W. and Waide, J. B. (1973). *Pl. Soil* **39**, 448–55.
Corns, I. G. W. (1974). *Can J. Bot.* **52**, 1731–45.
Crisp, D. T. (1966). *J. appl. Ecol.* **3**, 327–48.
Crittenden, P. D. (1975). *New Phytol.* **74**, 41–9.
Crocker, R. L. and Major, J. (1955). *J. Ecol.* **43**, 427–48.
Cuthbertson, D. P. (1970). *Dietetics* **12**, 413–51.
Cyplenkin, E. J. and Schilin, D. G. (1936). *Chemisation Socialistic Agr. (U.S.S.R.)* **5**, 59.
Daly, G. J. (1966). *Can. J. Bot.* **44**, 1607–21.
Dang-Gabrani, K. (1971). *Experientia* **27**, 107.
Davies, J., Briarty, L. G., and Rieley, J. O. (1973). *New Phytol.* **72**, 167–74.
Dehority, B. A. and Grubb, J. A. (1977). *Appl. Environ. Microbiol.* **33**, 1030–40.
Delwiche, C. C. (1970). *Scient. Am.* **223**, 136–46.
—— (1977). *Ambio* **6**, 106–11.
Denison, R., Caldwell, B., Bormann, B., Eldred, L. Swanberg, C., and Anderson, S. (1977). *Water, Air Soil Pollut.* **8**, 21–34
Denison, W. C. (1973). *Scient Am.* **228**, 74–80.
Dickson, B. A. and Crocker, R. L. (1953). *J. Soil Sci.* **4**, 142–54.
Dilworth, M. J. (1966). *Biochim. biophys. Acta* **127**, 285–94.
Döbereiner, J. (1978). In *Environmental role of nitrogen-fixing blue–green algae and asymbiotic bacteria* (ed. U. Granhall). *Ecol. Bull., Stockh.* **26**, 243–52.
—— Day, J. M., and Dart, P. J. (1972). *J. gen. Microbiol.* **71**, 103–16.
Dommergues, Y. (1963). *Agrochimica* **7**, 435–40.
Dooley, F. and Houghton, J. A. (1973). *Br. Phycol. J.* **8**, 289–93.
Douglas, L. A. and Tedrow, J. C. F. (1959). *Soil Sci.* **88**, 305–12.
Downes, J. A. (1962). *Can. Ent.* **94**, 143–62.
—— (1965). *A. Rev. Ent.* **10**, 257–74.
Duffey, E. (1975). *Deserts and grasslands. Part 2. Grassland life.* p. 144. Aldus Jupiter, London.
Durrell, L. W. (1962). *Trans. Am. Microsc. Soc.* **81**, 267–73.
Edmisten, J. A. (1970). In *USABC Report IId-24270, Book J.* (ed. H. T. Odum and R. G. Pigeon) Chaps. 11–21, pp. H211–H215.
Ehrlich, P. R., Ehrlich, A. H., and Holdren, J. P. (1977). *Ecoscience: population, resources and environment,* p. 1051. W. H. Freeman, San Francisco.
Elleway, R. F., Sabine, J. R., and Nicholas, D. J. D. (1971). *Arch. Mikrobiol.* **76**, 277–91.
Englund, B. (1975). *Pl. Soil* **43**, 419–31.
—— (1976). *Oikos* **27**, 428–32.
—— and Meyerson, H. (1974). *Oikos* **25**, 283–7.
Eriksson, E. (1952). *Tellus* **4**, 215–32.
Eutick, M. L., O'Brien, R. W., and Slaytor, M. (1976). *J. Insect Physiol.* **22**, 1377–80.
—— (1978). *Appl. Environ. Microbiol.* **35**, 823–8.

Evans, H. J. and Barber, L. E. (1977). *Science, N.Y.* **197**, 332–9.
Farnsworth, R. B. (1975). *Nitrogen fixation in shrubs.* Paper given at USDA workshop and symposium on wildland shrubs. Brigham Young University, Utah.
——and Clawson, M. A. (1972). *Am. Soc. Agron. Abs.* 96.
—— and Hammond, M. W. (1968). *Proc. Utah Acad. Sci.* **45**, 182–8.
—— Romney, E. M., and Wallace, A. (1976). *Gt Basin Nat.* **36**, 65–80.
Fibras, F. (1952). *Veröff. geobot. Inst., Rübel* **25**, 177–200.
Flensburg, T. (1967). *Acta phytogeogr. suec.* **51**, 1–132.
—— and Malmer, N. (1970). *Bot. Notiser* **123**, 269–99.
—— and Sparling, J. H. (1973). *Can. J. Bot.* **51**, 743–9.
Fleschner, M. D., Delwiche, C. C., and Goldman, C. R. (1976). *Am. J. Bot.* **63**, 945–50.
Fletcher, J. E. and Martin, W. P. (1948). *Ecology* **29**, 95–9.
Fletcher, W. W. (1955). *Ann. Bot., N. S.* **19**, 501–13.
Fogg, G. E. and Stewart, W. D. P. (1968). *Br. Antarct. Surv. Bull.* **15**, 39–46.
Forman, R. T. (1975). *Ecology* **56**, 1176–84.
Fox, L. R. and Macauley, B. J. (1977). *Oecologia* **29**, 145–62.
French, J. R. J. (1975). *Mater. Org.* **10**, 1–13.
—— and Roeper, R. A. (1973). *J. Insect Physiol.* **19**, 593–605.
—— Turner, G. L., and Bradbury, J. F. (1976). *J. gen. Microbiol.* **95**, 202–6.
Friedl, F. E. (1975). *Comp. Biochem. Physiol.* **52A**, 377–9.
Frissel, M. J. (ed.) (1977) *Agro-Ecosyst.* **4**, 1–354.
Gadgil, R. L. (1971). *Pl. Soil* **35**, 113–26.
Garcia-Moya, E. and McKell, C. M. (1970). *Ecology* **51**, 81–8.
Géhu, J. M. and Ghestem, A. (1965). *Annls Inst. Pasteur, Paris* Suppl. **109** No. 3, 136–52.
Gessel, S. P., Cole, D. W., and Steinberger, E. C. (1973). *Soil. Biol. Biochem.* **5**, 19–34.
Goldman, C. R. (1961). *Ecology* **42**, 282–94.
Gorham, E. (1958). *J. Ecol.* **46**, 373–9.
Gosz, J. R., Holmes, R. T., Likens, G. E., and Bormann, F. H. (1978) *Scient. Am.* **238**, 92–103.
Granhall, U. and Ciszuk, P. (1971). *J. gen. Microbiol.* **65**, 91–3.
—— and Henriksson, E. (1969). *Oikos* **20**, 175–8.
—— and Hofsten, A. V. (1976). *Physiol. Pl.* **36**, 88–94.
—— and Lindberg, T. (1978). In *Environmental role of nitrogen-fixing blue–green algae and asymbiotic bacteria* (ed. U. Granhall). *Ecol. Bull., Stockh.* **26**, 178–92.
—— and Selander, H. (1973). *Oikos* **24**, 8–15.
Greenland, D. J. (1958). *J. Agric. Sci., Camb.* **50**, 82–9.
Haag, R. W. (1973). *Can. J. Bot.* **52**, 103–16.
Haines, S. G., Haynes, W. L., and White, G. (1978). *Soil Sci. Soc. Am. J.* **42**, 130–2.
Hains, B. L. (1977). *Oecologia* **26**, 295–304.
Hallgren, J. H. and Nyman, B. (1977). *Studia Forestalia suecica* **137**, 1–40.
Hannon, N. J. (1956). *Proc. Linn. Soc. N.S.W.* **81**, 119–43.
Hanssen, J. F., Lid-Tovsuik, V., and Tovsuik, T. (1973). *Microbiologi. In IBP i Norge.* Årsrapport 1972, 75–93. Norsk IBP, Oslo.
Hardy, R. W. F. (1976). *Ist Int. Symp. on Nitrogen fixation Proc. 1974*, Vol. 2, pp. 693–717. Washington State University.
—— and Silver, W. S. (ed.) (1977) *A treatise on dinitrogen fixation.* Wiley-Interscience, New York.
—— Burns, R. C., and Holsten, R. D. (1973). *Soil Biol. Biochem.* **5**, 47–81.
—— Holsten, R. D., Jackson, E. K., and Burns, R. C. (1968). *Pl. Physiol., Lancaster* **43**, 1185–207.
Harmsen, G. W. and Van Schrevan, D. A. (1955). *Adv. Agron.* **7**, 299–398.

HARRIS, D. and DART, P. J. (1973). *Soil Biol. Biochim.* **5**, 277–9.
HARRIS, L. E. and MITCHELL, H. H. (1941). *J. Nutr.* **22**, 167–82.
HARROP, C. J. F. and DEGABRIELE, R. E. (1976). *Aust. J. Zool.* **24**, 201–15.
HASSOUNA, M. G. and WAREING, P. F. (1964). *Nature, Lond.* **202**, 467–9.
HAUKE-PACEWICZOWA, T., BALANDREAU, J., and DOMMERGUES, Y. (1970). *Soil Biol. Biochem.* **2**, 47–53.
HAVERTY, M. I. and NUTTING, W. L. (1975). *Ann. Ent. Soc. Am.* **68**, 533–6.
—— and LA FAGE, J. P. (1975). *Environ. Ent.* **4**, 105–9.
HEILMAN, F. and EKUAN, G. (1973). *Forest Sci.* **19**, 220–4.
HENRIKSSON, E. (1971). *Pl. Soil* Special Volume, 415–19.
—— and SIMU, B. (1971). *Oikos* **22**, 119–21.
—— HENRIKSSON, L. E., and PEJLER, B. (1972). *Surtsey Res. Prog. Rep.* **6**, 66–8.
HOBSON, P. N., SUMMERS, R., POSTGATE, J. R., and WARE, D. A. (1973). *J. gen. Microbiol.* **77**, 225–6.
HOLM-HANSEN, O. (1964). *Phycologia* **4**, 43–51.
HORNE, A. J. (1972). *Br. Antarct. Surv. Bull.* **27**, 1–18,
HOUPT, T. R. (1963). *Am. J. Phys.* **205**, 1144–50.
HUBER, D. M., WARREN, H. L., NELSON, D. W., and TSAI, C. Y. (1977). *Bioscience* **27**, 523–31.
HUME, I. D. (1974). *Aust. J. Zool.* **22**, 13–23.
HUNGATE, R. E. (1966). *The rumen and its microbes.* Academic Press, London.
HUSS-DANELL, K. (1977). *Can. J. Bot.* **55**, 585–92.
JACKSON, C. G., Jr and FULTON, M. (1970). *Proc. Ann. Conf. J. Wildl. Dis.* **6**, 466–8.
JAIN, M. K. and VLASSAK, K. (1975). *Annls Microbiol. Inst. Pasteur, Paris* **126A**, 119–22.
JEPSON, W. L. (1936). *A flora of California.* California School Book Depository, San Francisco.
JOHN, R.P. (1942). *Ann. Bot., N. S.* **6**, 323–49.
JOHNSON, F. D. (1968). In *Biology of alder* (ed. J. M. Trappe, J. F. Franklin, R. F. Tarrant, and G. M. Hansen), N.W. Sci. Ass. 40th Ann. Meeting Symp. Proc., 1967, pp. 9–22.
JONES, K. (1970). *Ann. Bot.* **34**, 239–44.
—— (1974). *J. Ecol.* **62**, 553–65.
—— (1976). In *Microbiology of aerial plant surfaces* (ed. C. H. Dickinson and T. F. Preece) pp. 451–63. Academic Press, London.
—— (1978). In *Environmental role of nitrogen-fixing blue–green algae and asymbiotic bacteria* (ed. U. Granhall) *Ecol. Bull., Stockh.* **26**, 199–205.
—— and THOMAS, J. G. (1974). *J. gen. Microbiol.* **85**, 97–101.
—— KING, E., and EASTLICK, M. (1974). *Ann Bot.* **38**, 765–72.
JORDAN, D. C., McNICOL, P. J., and MARSHALL, M. R. (1978). *Can. J. Microbiol.* **24**, 643–9.
JURGENSON, J. R. and WELLS, C. G. (1970). *Agron. Abs.* 159.
JURGENSEN, M. F. and DAVEY, C. B. (1971). *Pl. Soil* **34**, 341–56.
JURZITA, G. (1970). *Arch. Mikrobiol.* **72**, 203–22.
KAILA, A. (1954). *J. Sci. agr. Soc. Finland* **26** 40–9.
—— SOINI, S., and KIVINEN, E. (1954). *Maataloust. Aikakausk.* **26**, 79–95.
KALININSKAYA, T. A. (1967). *Mikrobiologiya* 345–9.
KALLIO, P. and KALLIO, S. (1978). In *Environmental role of nitrogen-fixing blue–green algae and asymbiotic bacteria* (ed. U. Granhall) *Ecol. Bull., Stockh.* **26**, 225–33.
—— SUHONEN, S., and KALLIO, H. (1972). *Rep. Kevo Subarct. Res. Stat.* **9**, 7–14.
KALLIO, S. (1973). *Rep. Kevo Subarct. Res. Stat.* **10**, 34–42.
—— (1978). In *Environmental role of nitrogen-fixing blue–green algae and asymbiotic bacteria* (ed. U. Granhall) *Ecol. Bull., Stockh.* **26**, 217–24.
KALTWASSER, H., KRAMER, J., and CONFER, W. R. (1972). *Arch. Mikrobiol*, **81**, 178–96.

Kapustka, L. A. and Rice, E. L. (1976). *Soil Biol. Biochem.* **8**, 497–503.
—— —— (1978). *Soil Biol. Biochem.* **10**, 553–4.
Kinnear, J. E. and Main, A. R. (1975). *Comp. Biochem. Physiol.* **51A**, 793–810.
Kleiner, D. (1975). *Angew. Chem. Int. Ed.* **4**, 80–6.
Knowles, R. and O'Toole, P. (1975). In *Nitrogen fixation by free-living micro-organisms* (ed. W. D. P. Stewart). Cambridge University Press.
Koch, B. and Evans, H. J. (1966). *Pl. Physiol., Lancaster* **41**, 1748–50.
Krasilnikov, N. A. (1949). *Mikrobiologiya* **18**, 3–6.
La Fage, J. P., Haverty, M. I., and Nutting, W. L. (1976). *Sociobiology* **2**, 155–69.
Latter, P. M., Cragg, J. B., and Heal, O. W. (1967). *J. Ecol.* **55**, 445–64.
Lawrence, D. B. (1958). *Am. Scient.* **46**, 89–122.
—— Schoenike, R. E., Quispel, A., and Bond, G. (1967). *J. Ecol.* **55**, 793–813.
Lee, K. E. and Wood, T. G. (1971). *Termites and soil.* Academic Press, London.
Leuck, E. E. and Rice, E. L. (1976). *Bot. Gaz.* **137**, 160–4.
Lid-Torsvik, V. (1971). I.B.P. Norwegian tundra biome, annual report. [Quoted by Alexander (1974).].
Line, M. A. and Loutit. M. W. (1969). *N.Z.Jl. agric. Res.* **12**, 630–8.
—— —— (1973). *N.S. Jl. agric. Res.* **16**, 87–94.
Loub, W. (1963). *Bodenkulture* **14**, 189–208.
Lund, J. W. G. (1947). *New Phytol.* **46**, 35–60.
MacConnel, J. T. and Bond, G. (1957). *Pl. Soil* **8**, 378–88.
McCoy, R. H. and Seidler, R. J. (1973). *Appl. Microbiol.* **25**, 534–8.
Macgregor, A. N. (1972). *Soil Sci. Soc. Am. Proc.* **36**, 594–6.
—— and Johnson, D. E. (1971). *Soil Sci. Soc. Am. Proc.* **35**, 843–4.
Machado, J. O. (1971). *Ciênc. Cult., S Paulo* **23**, 625–33.
McNaughton, S. J. (1976). *Science, N.Y.* **191**, 92–4.
McVean, D. N. (1956*a*). *J. Ecol.* **44**, 321–30.
—— (1956*b*). *J. Ecol.* **44**, 219–25.
Magdoff, F. R. and Bouldin, D. R. (1970). *Pl. Soil* **33**, 49–61.
Main, A. R. (1968). *Proc. Ecol. Soc. Aust.* **3**, 96–105.
Makrinov, M. A. and Stepanova, M. L. (1930). *Arkh. biol. Nauk* **30**, 293–302.
Malmer, N. (1962). *Op. bot. Soc. bot. Lund.* **7**(2), 1–67.
Maltseva, N. N. (1973). *Mȳkrobiol. Zh.* **35**, 143–8.
Marthaler, H. (1939). *Jb. wiss. Bot.* **88**, 723–58.
Mattson, W. J. (1978). *Oecologia* **33**, 327–49.
—— and Addy, N. D. (1975). *Science, N.Y.* **190**, 515–22.
Mayland, H. F. and McIntosh, T. H. (1966). *Soil Sci. Soc. Am. Proc.* **30**, 606–9.
—— —— and Fuller, W. H. (1966). *Soil Sci. Soc. Am. Proc.* **30**, 56–60.
Millbank, J. W. (1972). *New Phytol.* **71**, 1–10.
—— (1978). In *Environmental role of nitrogen-fixing blue–green algae and asymbiotic bacteria* (ed. U. Granhall). *Ecol. Bull., Stockh.* **26**, 260–5.
Miller, R. J. and Smith, R. B. (1976). *J. Environ. Qual.* **5**, 274–8.
Mitchell, J. E., Waide, J. B., and Todd, R. L. (1975). In *Mineral cycling in southeastern ecosystems* (ed. F. G. Howell, J. B. Gentry, and M. H. Smith) pp. 41–57. Technical Information Center, Office of Public Affairs, US Energy, Research and Development Administration.
Moisio, T., Kreula. M., and Virtanen, I. (1969). *Acta chem. fenn.* **92**, 432–3.
Moore, A. W. and Becking, J. H. (1963). *Nature, Lond.* **198**, 915–16.
Moore, B. P. (1969). In *Biology of termites* (ed. K. Krishna and F. M. Weesner) Vol. 1, pp. 407–32. Academic Press, New York.
Moore, P. D. and Bellamy, D. J. (1973). *Peatlands.* Paul Elek, London.
Morris, M., Eveleigh, S., Riggs, C., and Tiffney Jr, W. N. (1974). *Am. J. Bot.* **61**, 867–70.

Mostafa, M. A. and Mahmoud, M. Z. (1946). *Nature, Lond.* **167**, 446–7.
Murphy, P. M. (1975). *Proc. R. Irish Acad.* **75B**, 453–64.
Muscatine, L. and Porter, J. W. (1977). *Bioscience* **27**, 454–60.
Neilson, A. H. and Sparell, L. (1976). *Appl. Environ. Microbiol.* **32**, 197–205.
Nel, J. J. C., Hewett, P. H., and Joubert, L. (1970). *Phytophylactica* **2**, 27–32.
Nelson, A. D., Barber, L. E., Jepkema, J., Russell, S. A., Powelson, R., Evans, H. J., and Seidler, R. J. (1976). *Can J. Microbiol.* **22**, 523–30.
Newton, M., El Hassan, B. A., and Zavitkoyski, J. (1968). In *Biology of alder* (ed. J. M. Trappe, J. F. Franklin, R. F. Tarrant, and G. M. Hansen) pp. 73–84. N. W. Sci. Ass. 40th Ann. Meeting Symp. Proc. 1967.
Neyra, C. A. and Döbereiner, J. (1977). *Adv. Agron.* **29**, 1–38.
Nolan, J. V., Norton, B. W., and Leng, R. A. (1973). *Proc. Nutr. Soc.* **32**, 93–110.
Nunez, W. J. and Colmer, A. R. (1968). *Appl. Microbiol.* **16**, 1875–8.
Nye, R. H. and Greenland, D. J. (1960). Tech. Commun. 51. Commonwealth Bureau of Soils, Harpenden, England.
Ohiagu, C. E. and Wood, T. G. (1976). *J. appl. Ecol.* **13**, 705–13.
Olson, J. S. (1958). *Bot. Gaz.* **119**, 125–69.
Olsson, H. (1974). *Acta phytogeogr. suec.* **60**, 1–176.
O'Malley, K. L. and Terwilliger, R. C. (1975). *Comp. Biochem. Physiol.* **52A**, 367–9.
Ovington, J. D. (1951). *J. Ecol.* **39**, 363–75.
—— (1954). *J. Ecol.* **42**, 71–80.
—— (1957). *New Phytol.* **56**, 1–11.
Owen, D. F. and Wiegert, R. G. (1976). *Oikos* **27**, 488–92.
Parks, J. M. and Rice, E. L. (1969). *Bull. Torrey bot. Club* **96**, 345–60.
Paul, E. A., Myers, R. J. K., and Rice, W. A. (1971). *Pl. Soil* Special Volume, 495–507.
Peklo, J. (1946). *Nature, Lond.* **158**, 795–6.
—— and Satava, J. (1949). *Nature, Lond.* **163**, 336–7.
—— —— (1950). *Experientia* **15**, 190–1.
Pemadasa, M. A. and Lovell, P. H. (1974). *J. Ecol.* **62**, 647–57.
Persson, A. (1962). *Op. bot. Soc. bot. Lund.* **6**, 1–100.
Phillips, D. A., Rains, D. W., Valentine, R. C., and Huffaker, R. C. (1978). *Cereal Foods World* **23**, 26–9.
Popova, L. S. (1961). In *Soil micro-organisms and organic matter*, pp. 98–118. Izd-vo Acad. Nauk SSSR, Moscow.
Porter, L. K. and Grable, A. R. (1969). *Agron. J.* **61**, 521–32.
Postgate, J. R. (1971). *The chemistry and biochemistry of nitrogen fixation*, p. 163. Plenum, London.
—— (1974). *J. appl. Bact.* **37**, 185–202.
Potrikus, C. J. and Breznak, J. A. (1977). *Appl. Environ Microbiol.* **33**, 392–9.
Pringsheim, H. (1909). *Zentbl. Bakt. ParasitKde* **23**, 300–4.
Pun, H. H. L. and Satter, L. D. (1975). *J. Anim. Sci.* **41**, 1161–3.
P'Yavchenko, N. I. (1960). *Pochvovedenie* **61**, 21–32.
Quispel, A. (1958). *Acta bot. neerl.* **7**, 191–204.
Raju, P., Evans, H. J., and Seidler, R. J. (1972). *Proc. natn. Acad. Sci., U.S.A.* **69II**, 3474–8.
Rehder, H. (1976*a*). *Oecologia* **22**, 411–23.
—— (1976*b*). *Oecologia* **23**, 49–62.
Remacle, J. (1970). *Bull. Soc. r. Bot. Belg.* **103**, 83–96.
—— (1977). *Oecol. Plant.* **12**, 23–43.
Rice, E. L. (1964). *Ecology* **45**, 824–37.
—— (1968). *Bull. Torrey bot. Club* **95**, 346–58.
—— (1977). *Biochem. Syst.* **5**, 201–6.
—— and Pancholy, S. K. (1972). *Am J. Bot.* **59**, 1033–40.

—— Paul, E. A., and Wetter, L. R. (1967). *Can. J. Microbiol.* **13**, 829–36.
—— Penfound, W. T., and Rohrbaugh, L. M. (1960). *Ecology* **41**, 224–8.
Richards, B. N. (1964). *Aust. For.* **28**, 68–74.
—— (1973). *Soil Biol. Biochem.* **5**, 149–62.
Rodin, W. A. and Bazilevich, N. I. (1967). *Production and mineral cycling in terrestrial vegetation.* [English translation ed. G. E. Fogg. Oliver and Boyd, London.]
Rohrmann, G. F. (1978). *Pedobiologia* **18**, 89–98.
—— and Rossman, A. Y. (1978). In press.
Rogers, R. W., Lang, R. T., and Nicholas, D. J. D. (1966). *Nature, Lond.* **209**, 96–7.
Rosswall, T. (1976). In *Nitrogen, phosphorus and sulphur—global cycles* (ed. B. H. Svensson and R. Söderlund) SCOPE Report 7. Ecological Bulletins, No. 22, Sweden.
Ruinen, J. (1974). In *The Biology of nitrogen fixation* (ed. A. Quispel) pp. 121–67. North Holland, Amsterdam.
Russell, R. S. (1940). *J. Ecol.* **28**, 269–88.
Rychert, R. C. and Skujins, J. (1974). *Soil Sci. Soc. Am. Proc.* **38**, 768–74.
Sabet, Y. S. (1946). *Nature, Lond.* **157**, 656–7.
Salter, D. N. (1973). *Proc. Nutr. Soc.* **32**, 6571.
Schell, D. M. and Alexander, V. (1973). *Arctic* **26**, 130–7.
Schollhorn, R. and Burris, R. H. (1967). *Proc. natn. Acad. Sci. U.S.A.* **58**, 213–16.
Schultz, J. E. and Breznak, J. A. (1978). *Appl. Environ. Microbiol.* **35**, 930–6.
Schumacher, E. F. (1974). *Small is beautiful.* Abacus, London.
Schwabe, G. H. (1972). In *Taxonomy and biology of blue–green algae* (ed. T. V. Desikachary) pp. 353–84. University of Madras.
Scope Report 7. *Ecological Bulletins No. 22*, Sweden.
Scott, G. D. (1956). *New Phytol.* **55**, 111–16.
Seidler, R. J., Aho, P. E., Raju, P. N., and Evans, H. J. (1972). *J. gen. Microbiol.* **73**, 413–16.
Sharp, R. F. (1975*a*). *Soil Biol. Biochem.* **7**, 3–14.
—— (1975*b*). *Mycopathologia* **55**, 185–192.
Shields, L. M., Mitchell, C., and Drouet, F. (1957). *Ecology* **38**, 661–3.
Shklyar, M. Z. (1956). *Dokl. Vashnil.* **8**, 32–7.
Silver, W. S. and Mague, T. (1970). *Nature, Lond.* **227**, 378–9.
Silvester, W. B. and Bennett, K. J. (1973). *Soil. Biol. Biochem.* **5**, 171–9.
—— and Smith, D. R. (1969). *Nature, Lond.* **224**, 1231.
Sjörs, H. (1950). *Oikos* **2**, 241–58.
Skogen, A. (1972). *K. norske Vidensk. Selsk. Skr.* **4**, 1–115.
Small, E. (1972). *Can. J. Bot.* **50**, 2227–33.
Smith, W., Bormann, F. H., and Likens, G. E. (1968). *Soil Sci.* **106**, 471–3.
Snyder, J. M. and Wullstein, L. H. (1973*a*). *Am. Midl. Nat.* **90**, 257–65.
—— —— (1973*b*). *Bryologist* **76**, 196–9.
Snyder, R. M. and Wyant, Z. N. (1932). *Tech. Bull. Mich. (St. Coll.) agric. Exp. Stn* **129**, 1–63.
Söderlund, R. and Svensson, B. H. (1976). In *Nitrogen, phosphorus and sulphur—global cycles* (ed. B. H. Svensson and E. Söderlund) pp. 23–74. SCOPE Report 7. Ecological Bulletins, No. 22, Sweden.
Spears, B. M. and Ueckert, D. N. (1976). *Environ. Ent.* **5**, 1022–25.
Sprecht, R. L. and Rayson, P. (1957). *Aust. J. Bot.* **5**, 103–14.
Springett, B. P. (1978). *Aust. J. Ecol.* **3**, 129–40.
Squires, V. R. (1978). *Proc. Aust. Soc. Anim. Prod.* **12**, 75–84.
Stevenson, G. (1959). *Ann. Bot., N.S.* **23**, 622–34.
Stewart, W. D. P. (1963). *Z. allg. Mikrobiol.* **3**, 152–6.
—— (1965). *Br. phycol. Bull.* **2**, 514.
—— (1966). *Nitrogen fixation in plants,* p. 68. Athlone Press, London.

—— (1967*a*). *Ann. Bot., N.S.* **31**, 385–407.
—— (1967*b*). *Nature, Lond.* **214**, 603–4.
—— and BOND, G. (1961). *Pl. Soil* **14**, 347–59.
—— and PEARSON, M. C. (1967). *Pl. Soil* **26**, 348–60.
—— FITZGERALD, G. P., and BURRIS, R. H. (1967). *Science, N.Y.* **158**, 536.
STUTZ, R. C. and BLISS, L. C. (1975). *Can. J. Bot.* **53**, 1387–99.
SUSAKI, T., TANABE, I., MATSUGUCHI, T., and ARARAGI, M. (1975). In *Nitrogen fixation and nitrogen cycle* (ed. H. Takahashi) pp. 51–60. JIBP, University of Tokyo Press.
SWIFT, M. J. (1977). *Sci. Prog., Lond.* **64**, 175–99.
TARRANT, R. F. and MILLAR, R. E. (1963). *Proc. Soil Sci. Soc. Am.* **27**, 231–4.
—— LU, K. C., BULLEN, W. B., and FRANKLIN, J. F. (1969). USDA Forest Serv. Res. Pap. PNW 76.
TCHAN, Y. T. and BEADLE, N. C. W. (1955). *Proc. Linn. Soc. N.S.W.* **80**, 97–104.
TIBBETTS, T. E. (1969). *Proc. 3rd Int. Peat Cong. Quebec*, pp. 8–21.
TICHOMIROV, B. A. (1957). *Bot. Zh.* **42**, 1691–717.
TODD, R. L., MEYER, R. D., and WIDE, J. B. (1978). In *Environmental role of nitrogen-fixing blye–green algae and asymbiotic bacteria* (ed. U. Granhall). *Ecol. Bull., Stockh.* **26**, 172–7.
—— WAIDE, J. B., and CORNABY, B. W. (1975). Significance of biological nitrogen fixation and denitrification in a deciduous forest ecosystem. In *Mineral cycling in south eastern ecosystems* (ed. F. G. Howell, J. B. Gentry, and M. H. Smith) pp. 729–35. Technical Information Center, Office of Public Affairs, US Energy, Research and Development Administration.
TÓTH, L. (1948*a*). *Hung. Acta biol.* **1**, 22–9.
—— (1948*b*). *Experientia* **4**, 395–6.
—— (1949). *Experientia* **5**, 474–5.
—— (1951). *Zool. Anz.* **146**, 191–6.
—— (1952). *Tijdschr. Ent.* **95**, 43.
TREUB, M. (1888). *Ann. Jarb. Bot. Buitenzorg* **7**, 221–3.
ULLOA, M., HERRERA, T., and LANZA, G. (1971). *Revta lat.-am. Microbiol.* **13**, 113–24.
VAVULO, F. P. (1958). *Liet. TSR Mokslu Akad. biol. Inst. Darb.* **7**, 139–46.
VIRO, P. J. (1963). *Soil Sci.* 24–30.
VLASSAK, K., PAUL, E. A., and HARRIS, R. E. (1973). *Pl. Soil* **38**, 637–49.
VOIGT, G. K. and STEUCEK, G. L. (1969). *Soil Sci. Soc. Am. Proc.* **33**, 946–9.
WALLACE, A. and ROMNEY, E. M. (1972). USAEC Monograph TIC-25954.
WARING, H. D. (1976). Official Hansard Report (18/10/76) pp. 4609–10.
WARREN-WILSON, J. (1966). *Ann. Bot., N.S.* **30**, 383–402.
WAUGHMAN, G. J. (1972*a*). *Pl. Soil.* 521–8.
—— (1972*b*). *Oikos* **23**, 206–12.
—— (1976). *Can. J. Microbiol.* **22**, 1561–6.
—— (1977). *J. exp. Bot.* 949—60.
—— (1980). *J. Ecol.* In press.
—— and BELLAMY, D. J. (1972*a*). *Proc. 4th Int. Peat Cong.*, Vol. 1, pp. 309–18. Helsinki.
—— —— (1972*b*). *Oikos* **23**, 353–8.
—— —— (1980). *Ecology*. In press.
WEBLEY, D. M., EASTWOOD, D. J., and GIMMINGHAM, C. H. (1952). *J. Ecol.* **40**, 168–78.
WESTMAN, W. E. and ROGERS, R. W. (1977). *Aust. J. Ecol.* **2**, 447–60.
WHITE, T. C. R. (1972). *J. Insect Physiol.* **18**, 2359–67.
—— (1978). *Oecologia.* **33**, 71–86.
WIGGINS, I. L. and THOMAS, J. H. (1962). *A flora of the Alaskan arctic slope*. University of Toronto Press.
WILKEN, G. E. (1977). *Agro-Ecosyst.* **3**, 291–302.
WILLIAMS, B. H. (1943). *Science, N.Y.* **97**, 441–2.
WILLIAMS, B. L. (1972). *Forestry* **45**, 177–88.

WILLIS, A. J. and YEMM, E. W. (1961). *J. Ecol.* **49**, 377–90.

WILSON, J. S. and WILSON, B. D. (1933). *Mem. Cornell Univ. agric. Exp. Stn* **184**, 3–15.

WINTER, W. H. (1978). *Proc. Aust. Soc. Anim. Prod.* **12**, 86–93.

WOOD, T. G. (1976). In *The role of terrestrial and aquatic organisms in decomposition processes* (ed. J. M. Anderson and A. Macfadyen) pp. 145–68 (17th Symp. Brit. Ecol. Soc.) Blackwell, Oxford.

—— and SANDS, W. A. (1977). In *Production biology of ants and termites* (ed. M. V. Brian). IBP Synthesis Volume. Cambridge University Press.

WOODMANSEE, R. G., DODD, J. L., BOWMANN, R. A., CLARK, F. E., and DICKINSON, C. E. (1978). *Oecologia* **34**, 363–76.

WOZNY, M. A., BRYANT, M. P., HOLDEMAN, L. V., and MOORE, W. E. C. (1977). *Appl. Environ. Microbiol.* **33**, 1097–104.

WULLSTEIN, L. H., BRUENING, M. L., and SNYDER, L. S. (1972). *Abs. Amer. Soc. Ecol. Western Section: Amer. Ass. Adv. Sci.*

6 Nitrogen fixation in waters

H. W. PAERL, K. L. WEBB, J. BAKER, AND
W. J. WIEBE

6.1 Introduction

Research on freshwater and marine nitrogen fixation has expanded dramatically in the past ten years. This is largely due to the advent of the easily executed acetylene reduction assay as well as the recognition of nitrogen fixation as an important process from a budgetary point of view.

Although much has been done, many questions regarding the quantitative significance of nitrogen fixation, validity of current measuring techniques, unexplored and potentially important nitrogen-fixing microenvironments remain unanswered. In this chapter our discussions are intended to present the available knowledge of nitrogen fixation in aquatic ecosystems, including an assessment of its role in community nitrogen supplies, and some of the current problems of in-situ research facing an aquatic ecologist.

6.2 Freshwater habitats

In general, information on the quantitative and qualitative aspects of nitrogen fixation is more extensive in fresh water than in marine ecosystems. It appears that nitrogen-fixing organisms are capable of achieving much higher biomass in freshwater systems than in marine (Fogg, Stewart, Fay, and Walsby 1973). However, if estuaries are included with marine systems, the diversity of nitrogen-fixing organisms is similar. As in marine systems, nitrogen fixation in fresh water is confined to prokaryotic micro-organisms, both photosynthetic and heterotrophic. Such micro-organisms can also function as symbiotic partners with higher plants and possibly animals. A summary of confirmed freshwater nitrogen-fixing organisms is given in Table 6.1.

Nitrogen fixation in fresh water is carried out by certain groups of

TABLE 6.1
Major freshwater bacterial and blue–green algal (cyanophycean) genera for which nitrogenase activity has been demonstrated

Types	Genera	General references
Anaerobic bacteria		
heterotrophic	*Clostridium*	Postgate (1971*a*)
	Desulfovibrio	Postgate (1971*a*)
chemoautotrophic	*Thiobacillus*	Stewart (1977*a*)
photosynthetic	*Chlorobium*	Postgate (1971*b*)
	Chromatium	Postgate (1971*b*)
	Ectothiorhodospira	Stewart (1973)
	Pelodictyon	Stewart (1968)
	Rhodomicrobium	Kamen and Gest (1949)
	Rhodomicrobium	Lindstrom, Lewis, and Pinsky (1951)
	Rhodopseudomonas	Lindstrom, Burris, and Wilson (1949)
Facultatively aerobic bacteria		
heterotrophic	*Bacillus*	Postgate (1971*a*)
	Klebsiella	Stewart (1969)
Aerobic bacteria		
heterotrophic	*Azotobacter*	Beijerinck and Van Delden (1902) Kuznetzov (1959)
	Azomonas	Postgate (1971*a*)
	Azotococcus	Postgate (1971*a*)
	Arthrobacter	Postgate (1971*a*)
	Beijerinckia	Postgate (1971*a*)
	Corynebacterium	Gogotov and Schlegel (1974)
	Methylobacter	Dalton (1974)
	Methylococcus	Dalton (1974)
	Methylocystis	Dalton (1974)
	Pseudomonas	Stewart (1977*a*)
	Spirillum	Postgate (1971*a*)
Blue–green algae		
unicellular	*Gloeocapsa*	Wyatt and Silvey (1969) Rippka, Neilson, Kunisawa, and Cohen-Bazire (1971)
non-heterocystous	*Lyngbya*	Kenyon, Rippka, and Stanier (1972)
	Oscillatoria	Kenyon *et al.* (1972)
	Phormidium	Stewart (1977*a*)
	Raphidiopsis	Stewart (1977*a*)
heterocystous filamentous	*Anabaena*	Bortels (1940) Fogg (1942)
	Anabaenopsis	Watanabe (1951)
	Aphanizomenon	Horne and Goldman (1972) Peterson and Burris (1976)
	Calothrix	Stewart (1970*a*)
	Chloroglea	Stewart (in preparation)

Gloeotrichia	Broughton and Parker (1978)
Mastigocladus	Billaud (1967) Stewart (1970*a*)
Nostoc	Bortels (1940) Herriset (1952)
Scytonema	Cameron and Fuller (1960)
Stigonema	Vankataraman (1961)
Tolypothrix	Watanabe (1951)
Westiellopsis	Pattnaik (1966)

bacteria and blue–green algae. The freshwater nitrogen-fixing bacteria are represented by heterotrophic and photosynthetic (non-oxygen-evolving) genera. Although bacterial nitrogen fixation is probably present in all lakes and streams, it usually constitutes only a minor fraction of biological nitrogen fixation (Waksman, Hotchkiss, and Carey 1933; Kuznetsov 1968); one exception is the case of meromictic lakes, where layers of photosynthetic and chemo-autotrophic bacteria can dominate the microbial biomass (Stewart 1968; Keirn and Brezonik 1971).

Three major groups of blue–green algae (Cyanophyta) dominate nitrogen fixation in lakes. They are unicellular, non-heterocystous filamentous, and heterocystous filamentous forms. All employ oxygen-evolving photosynthesis, and since oxygen is a strong inhibitor of nitrogen fixation (Schmidt-Lorenz and Rippel-Baldes 1957; Stewart and Pearson 1970), they have evolved a variety of structural, biochemical, and behavioural modifications, including associations with other organisms, to deal with the effects of oxygen evolution (Carr and Whitton 1973). The ecological significance of such modifications with respect to bloom formation in highly oxygenated surface waters is discussed on page 203.

Because lakes usually have defined measurable sources of nitrogen input, limnologists have been able to estimate the relative importance of nitrogen input by nitrogen fixation (Horne and Fogg 1970; Horne and Viner 1971; Horne, Dillard, Fujita, and Goldman 1972). The quantitative significance of nitrogen fixation as established by both the acetylene reduction and ^{15}N methods is outlined in Table 6.2. These results leave little doubt that nitrogen fixation can be an important source of combined nitrogen in fresh water. The blooms responsible for nitrogen fixation in lakes are often dominated by only one or two planktonic species. Most of this nitrogen fixation occurs in the euphotic regions of the water column (Stewart, Fitzgerald, and Burris 1967; Stewart 1977*a*). Since biomass measurements and taxonomic surveys are often easier to perform on planktonic as opposed to benthic or sediment micro-organisms, freshwater ecologists have been able to assess the numerical dominance and diversity of nitrogen-fixing organisms with relative ease. However, it is known that (i) significant rates of nitrogen fixation

TABLE 6.2

The significance of nitrogen fixation in diverse lakes expressed as nitrogen fixation in (a) (kg ha^{-1} $^{-1}$) or (b) (μg l^{-1} day^{-1}) and per cent of annual fixed nitrogen input (where applicable)

Lake/location	Principal nitrogen-fixing organism	Rate	% of fixed nitrogen input	References
Sanctuary Lake, Pennsylvania, USA	*Anabaena flos-aquae* *A. circinalis* *A. spiroides*	(b) 125	—	Dugdale and Dugdale (1962)
Lake Mendota, Wisconsin, USA	*Gloeotrichia echinulata*	(b) 8.5	—	Goering and Neess (1964)
Lake Wingra, Wisconsin, USA	*Anabaena* sp	(b) 12	—	Goering and Neess (1964)
Lake Mary, Wisconsin, USA	Anoxic bacteria	(a) 1.2–4.6	—	Brezonik and Harper (1969)
Esthwaite Water and Windermere, UK	*Anabaena flos-aquae* *A. solitaria*	(a) 0.4–2.9	2–72	Horne and Fogg (1970)
Lake Erken, Sweden	*A. flos-aquae* *A. spiroides* *Aphanizomenon flos-aquae* *Gloeotrichia echninulata*	(a) 2.5–13.2	40	Granhall and Lundgren (1971)
Lake George, Uganda	*Anabaena* spp *Anabaenopsis* spp	(a) 40–150	62–86	Horne and Viner (1971)
Clear Lake, California, USA	*Anabaena circinalis* *Aphanizomenon flos-aquae*	(a) 18	43	Horne and Goldman (1972)
Lake Rotongaio, New Zealand	*Anabaena circinalis* *A. oscillarioides*	(b) 4–6	20–40	Paerl and Kellar (unpublished data)

occur at night (Dugdale and Dugdale 1962; Goering and Neess 1964) or in aphotic waters (Horne and Viner 1971; Peterson, Friberg, and Burris 1977), and (ii) sediments often support abundant nitrogen-fixing organisms (Brezonik and Harper 1969; Howard, Frea, Pfister, and Dugan 1970). Although these sources often do not contribute the majority of a lake's fixed nitrogen, they must be considered in budgets and predictive models of nitrogen inputs. Much work, therefore, remains for aquatic ecologists, since current techniques for measuring nitrogen fixation in benthic regions and underlying sediments are difficult to apply.

Other freshwater nitrogen fixation is largely confined to microbial associations with macrophytes (Stewart 1977*a*). One of the most common is the symbiotic association of the heterocystous cyanophyte *Anabaena azollae* and the aquatic fern *Azolla*. *Azolla* can dominate surface waters in small tropical impoundments (Becking 1976) and temperate ponds (Kellar 1978), and has been used as a nitrogen source in rice paddys (Watanabe, Shirota, Endo, and Yamamoto 1971).

Macrophyte leaf surfaces as well as rhizosphere regions can also provide microzones rich in excreted organic matter which can support heterotrophic nitrogen-fixing bacteria (Stewart 1977*a*; see Chapter 2). Areas bordering particles, organisms and strata of micro-organisms where strong gradients of dissolved oxygen, carbon dioxide, nutrients, pH, and light exist can give rise to increased microbial activity (Jannasch and Pritchard 1972), including carbon dioxide and nitrogen fixation (Lange 1971; Paerl and Kellar 1978). Large-scale physical and chemical analyses often fail to detect such regions. Minute gradients formed by oxygen consumption near particles may create microzones conducive to anaerobic nitrogen-fixing organisms or even denitrifying and ammonifying bacteria. Such gradients cannot be detected by measuring ambient dissolved oxygen levels with a probe.

In any ecological study, the following question invariably arises: how does the organism optimize its growth given a particular set of environmental conditions? Through laboratory experiments on axenic cultures, inhibition, regulation, and stimulation of aquatic nitrogen fixation by environmental parameters have been examined. In-situ investigations or laboratory experiments with naturally occurring microbial associations examining these environmental effects have been rare (Whitton 1973; Fogg *et al.* 1973). The interaction of physical, chemical, and biological processes in lakes results in a complex interplay of shifting environmental variables. Such factors as fluctuating climatic conditions, nutrient fluxes from catchment and precipitation dynamics, dissolved oxygen and carbon dioxide, pH, and light regimes within the lake pose certain ecological problems which must be dealt with by aquatic organisms. Thus, if strategies that nitrogen-fixing micro-organisms employ to optimize growth in specific aquatic niches are to be understood, more emphasis must be placed on relating laboratory findings to environmental constraints occurring in lakes. In this chapter, ecological strategies of freshwater nitrogen-fixing organisms will be examined by combining current knowledge of biochemistry and physiology with some environmental extremes which dominant genera face in lakes. The bloom-forming cyanophycean genera will receive particular attention.

Nitrogen fixation is a highly reductive process which is restricted to prokaryotic micro-organisms. The process was probably present in the early stages of evolution of micro-organisms, presumably in oxygen-free systems (Schopf 1970; Stewart 1977*b*). Nitrogen fixation is powered by light in all photosynthetic micro-organisms (Fogg *et al.* 1973), but can proceed in darkness through chemo-autotrophic reactions (Kuznetzov 1959) or utilization of energy-rich carbon compounds (heterotrophy) (Allison, Hoover, and Morris 1937; Fay 1965). The anaerobic photosynthetic bacteria, as exemplified by *Rhodospirillum* and *Chromatium*, can fix nitrogen using hydrogen sulphide as an electron donor in highly reduced conditions. Assuming that photosynthetic bacteria were forerunners of oxygen-evolving photosynthetic cyanophytes (Stewart 1977*b*; Stanier and Cohen-Bazire 1977), their ability to fix nitrogen indicates that nitrogen fixation predated oxygen-evolving photosynthesis. In addition to the photosynthetic bacteria, anaerobic decomposers such as *Clostridium* and *Desulfovibrio* can fix nitrogen (Mulder 1975; Stewart 1977*a*) in similar anaerobic environments, though at lower fixation rates than in photosynthetic bacteria.

The rise of cyanophycean nitrogen-fixing algae represented an important step in the evolution of nitrogen fixation, since they are the only nitrogen-fixing organisms with photosynthetic machinery similar to higher plants (Calvin and Lynch 1952; Fogg 1968). Oxygen-evolving photolysis of water (as opposed to hydrogen sulphide utilization in photosynthetic bacteria) generates electrons which reduce nitrogen to ammonia. Oxygen evolution posed a major problem during cyanophyte evolution, forcing this group to accommodate oxygen-inhibited nitrogen fixation contemporaneously.

Consequently, freshwater and marine cyanophyceans have evolved a myriad of biochemical and structural strategies to cope with the presence of oxygen, while the photosynthetic bacteria (and all other obligate anaerobes) have remained confined to anaerobic hypolimnia, anoxic sediments, and reduced microzones. The cyanophycean adaptation to oxygen has allowed such common bloom-forming genera as *Anabaena*, *Aphanizomenon*, and *Gloeotrichia* unrestricted use of the radiant energy-rich epilimnia of lakes, while the photosynthetic bacteria have been forced to remain in the radiant energy-poor anaerobic hypolimnion. As a result, nitrogen fixation in eutrophic lakes supporting both groups of photosynthetic organisms is dominated by cyanophycean species (Dugdale and Dugdale 1962; Kuznetzov 1968), so amplifying their evolutionary success.

Photosynthetic bacteria have optimized their hypolimnetic existence to some extent. They are capable of photosynthetic growth at lower light levels than cyanophyceans (Pfenning 1967; van Gemerden 1974). Furthermore, the bacteriochlorophyll and accessory carotenoids of photosynthetic bacteria are adapted to utilize light not absorbed by overlying cyanophytes and other algae. These pigments are particularly effective in absorbing low-energy red to far-red light sources (Pfennig 1967), which are usually the predominant wavelengths penetrating cyanophycean and higher algal surface blooms. Little is known, however, of the efficiency of nitrogen fixation by photosynthetic bacteria at the low light levels. Hence, assessments of nitrogen input by these organisms are very sparse at present.

The non-photosynthetic heterotrophic anaerobic nitrogen-fixing bacteria face less rigid niche constraints than photosynthetic bacteria. The former groups can occupy the entire deoxygenated hypolimnion being mainly limited by available carbon and hydrogen sulphide toxicity. These genera have low efficiencies of nitrogen fixation per amount of carbon consumed (Stewart 1977*a*), a possible explanation for the dominance of planktonic cyanophycean nitrogen fixation versus hypolimnetic and benthic bacterial nitrogen fixation in eutrophic lakes showing hypolimnetic oxygen depletion.

In the oxygenated epilimnia of lakes where nitrogen fixation occurs, cyanophyceans dominate nitrogenase activity when compared to smaller size fractions, indicating low abundance of planktonic nitrogen-fixing aerobic or micro-aerophilic bacteria. The presence of bacterial nitrogen fixation can be examined by prefiltering lake water through either 3 or 5 μm Nuclepore filters, followed by acetylene-reduction assays on the filtrate versus unfiltered lake water (Table 6.3).

Ecological strategies employed by nitrogen-fixing cyanophyta

There are distinct advantages to an organism of utilizing atmospheric nitrogen during nitrogen limited periods, a topic which has been widely discussed previously (Frank 1889; Beijerinck and Van Delden 1902; Fogg *et al.* 1973). This advantage is probably one of the main reasons for the persistence and, at times, dominance of nitrogen-fixing genera in planktonic, epiphytic, and benthic communities. Combined nitrogen deficiencies, particularly in the epilimnia during thermally stratified summer months in eutrophic lakes, are often followed by blooms in nitrogen-fixing cyanophytes (Fogg *et al.* 1973). But Nauwerck (1963) and others (Reynolds 1971; Horne *et al.* 1972) have observed that

TABLE 6.3

Acetylene reduction (nitrogen fixation), photosynthetic $^{14}CO_2$ *fixation and heterotrophic* [^{14}C]*-glucose assimilation activities of* > 5, > 3, *and* > 0.2 μm *sized micro-organisms in Lake Ngahewa, New Zealand waters. Lake water was filtered through respective porosity filters (5, 3, and 0.2* μm*) prior to acetylene-reduction assays. Organisms for* $^{14}CO_2$ *fixation and* [^{14}C]*-glucose assimilation experiments were filtered through respective filters after tracer addition and incubation. During these experiments* A. spiroides *and* A. circinalis *were the predominant cyanophycean nitrogen-fixing algae*

Analysis	Activities of size fractions (relative to 0.2 μm) (%)		
	>5 μm	>3 μm	>0.2 μm
Acetylene reduction	98	100	100
$^{14}CO_2$ fixation	65	90	100
[^{14}C]-glucose assimilation	22	36	100

non-nitrogen-fixing cyanophytes, such as *Microcystis* and *Oocystis*, often become established simultaneously. Fogg *et al.* (1973) have noted that the co-appearance of non-nitrogen-fixing cyanophytes may depend on the ability to store assimilated nitrogen in the form of cyanophycin granules (phycocayanin), which can be utilized during nitrogen-limited growth periods. Diatoms and green algae often co-exist during nitrogen-fixing *Anabaena* blooms in Lakes Rotorua, Rotongaio, and Tutira, New Zealand. Horne *et al.* (1972) observed the diatom *Melosira* in both *Anabaena* and *Aphanizomenon* blooms in Clear Lake, California. In the New Zealand lakes, autoradiography revealed that diatoms could be heterotrophic preferring amino acids, which are excretion products of *Anabaena* (Walsby 1974; Nalewajko, Dunstall, and Shear 1976). The evidence indicates that non-nitrogen-fixing genera have, to a limited extent, developed ways of coping with nitrogen deficiencies, including nitrogen storage and the utilization of nitrogen-containing excretory products from nitrogen-fixing genera.

Although nitrogen deficiency often promotes the appearance of nitrogen-fixing cyanophytes, nitrogen abundance often conversely fails to eliminate nitrogen fixation. Stewart (1969, 1973) has reviewed contradictory evidence claiming that combined nitrogen either fails to inhibit, partially inhibits, or completely inhibits nitrogenase activity in photosynthetic micro-organisms. He points out that much of the disagreement in findings could be the result of diverse sources, concentrations of combined nitrogen and differing exposure periods to combined nitrogen. He concludes that the critical compound is ammonia, which when added at 3 mM to cultures of *Anabaena cylindrica* can halve the nitrogenase activity while doubling the growth rate. However, lakes with such high ammonia concentrations are rare. In addition, it is not unusual to find both significant quantities of ammonia and nitrogen fixation simultaneously (Dugdale and Dugdale 1962; Goering and Neess 1964; Granhall and Lundgren 1971). Flett (1976), Kellar (1978), and others have found only marginal relations between ambient combined nitrogen (as ammonia or nitrate) and nitrogenase activity in cyanophycean dominated lakes. Flett (1976) points out that the ratio of combined nitrogen to phosphorus may be more important than absolute quantities of nitrogen in predicting the appearance of nitrogen fixers. It appears that the decline in nitrogenase activity is often related to factors other than ambient combined nitrogen concentration; a complex of limiting factors, including trace element availability (Goldman 1964; Holm-Hansen, Gerloff, and Skoog 1954), phosphorus (Stewart and Alexander 1971), iron (Horne *et al.* 1972), and carbon dioxide availability (Lange 1971) may be responsible for a decrease in nitrogenase activity. Unfavourable climatic and turbulent conditions, the excretion of toxins (Vance 1966), metabolites (Jakob 1954), and allelopathic control (Hutchinson 1957) may also result in a decline of nitrogenase activity or of a bloom in general. In fact, nitrogenase activity in the symbiotic fern *Azolla*, which contains *Anabaena*, appears to be completely independent of ambient ammonia concentrations and is probably more closely regulated by physical and climatic factors.

The earliest freshwater studies attempting to find relationships between physical-chemical parameters and the presence of nitrogen-fixing organisms focused on the abundance of organic substances. Pearsall (1932), for example, found a direct relationship between the concentration of dissolved organic matter (DOM) and the abundance of blue–green algae in the plankton of 11 lakes in the English Lake District. Singh (1955) and Lange (1971) accumulated similar evidence linking cyanophyte blooms with the appearance of both high DOM and dissolved organic nitrogen (DON). It is difficult to conclude, however, that high DOM or DON concentrations are mainly responsible for cyanophycean dominance, since enhanced phosphorus loading (Schindler 1974), increases bacterial populations (Lange 1971; Paerl and Kellar 1978), increases pH and lowers available carbon dioxide (Shapiro 1973), all of which are conducive to cyanophycean growth. Also these changes accompany high DOM concentrations. Furthermore, it is not clear whether elevated DON concentrations are actually caused by cyanophytes through the excretion of nitrogen-fixation products (Fogg *et al.* 1973). In short, although high DOM and DON concentrations in lake waters are *a priori* reasons for suspecting cyanophycean dominance, the actual causative factors remain unclear.

Nitrogenase activity appears to be closely related to phosphorus availability. This is not surprising since nitrogen fixation is an energy demanding process, met by cleavage of phosphate bonds from adenosine triphosphate (Bulen and Lecomte 1966). For a limited time, cyanophyceans can support nitrogenase activity by utilizing stored phosphorus in the form of polyphosphate bodies (volutin granules) (Kuhl 1968). These granules are formed during periods of adequate phosphorus supply and are used during phosphorus-deficient conditions (Batterton and van Baalen 1971; Volk and Phinney 1968). The maintenance of nitrogenase activity by use of polyphosphates was demonstrated during phosphorus starvation experiments with *Anabaena* (Stewart and Alexander 1971). When transferred to a phosphorus-free medium, nitrogenase activity continued for several days. We (Paerl *et al.*, in preparation) have repeated these experiments and find that after the exhaustion of stored phosphorus pools, both nitrogenase activity and growth cease.

In general, good direct relationships exist between phosphorus loading and the establishment of nitrogen-fixing cyanophytes. Vollenweider (1968) has quoted baseline phosphorus levels above which blooms are likely to occur in lakes. Lakes which have received accelerated phosphorus loading, including Lake Erie of the North American Great Lakes; Lake Erken, Sweden; Lake Mendota, Wisconsin; Clear Lake, California; and Lake Rotorua, New Zealand, have shown marked upsurges in cyanophycean dominance in response to high phosphorus loading; several of these lakes, most notably Lake Erie, have shown recent declines in both phosphorus loading and cyanophycean (especially nitrogen-fixing organisms) dominance. An excellent example of cyanophycean response to phosphorus loading is Lake 304 of the experimental lakes area in Canada, where Schindler (1974) monitored the increased dominance of *Anabaena flos aquae* after controlled phosphorus loading. The dominance by nitrogen-fixing organisms was minimized to some extent by nitrate supplementation.

The abundance of micronutrients, particularly iron, appears to be an important limiting factor at times with regard to overall photosynthetic production of lakes. In particular, nitrogen-fixing organisms have a high demand for iron (Fogg *et al.* 1973) and, to a lesser extent, molybdenum (Bortels 1940), since both are components of nitrogenase. Cobalt has also been found to be an important micronutrient (Holm-Hansen *et al.* 1954), while trace concentrations of boron, copper, manganese, and zinc are also essential for cyanophycean growth (Fogg *et al.* 1973). Early experiments by Goldman (1961) at Castle Lake, California, indicated that molybdenum could act as a limiting nutrient under natural conditions. Since then, many cases of micronutrient limitation of phytoplankton spread have been documented, but few have focused on cyanophycean-dominated lakes.

Both freshwater and marine phytoplankton appear capable of secreting micronutrient-binding organic compounds, the ecological significance of which is not clear. The binding properties of iron-specific chelators (siderochromes) have received most attention (Price 1968; Barber and Ryther 1969). Recently, Murphy, Lean, and Nalewjko (1976) have demonstrated the presence of a powerful iron-hydroxymate chelating agent in the filtrate of *A. flos aquae* cultures. Such chelators afforded a competitive advantage to *Anabaena* when grown in the presence of a eukaryotic alga (*Scenedesmus*). They concluded that iron deprivation induced the synthesis of hydroxymate chelators. These results indicate that certain cyanophyceans have developed chemical mechanisms for dealing with iron shortages during periods of iron demand by nitrogen fixation. The mechanisms appear to function at extremely low iron concentrations and are more efficient than chelating or direct uptake systems present in non-nitrogen-fixing algae.

Cyanophytes often regulate their buoyancy and vertical movement through gas vacuole formation (Walsby 1968). Diurnal movements have been postulated as a mechanism to satisfy specific nutrient demands, allowing planktonic algae to move from depleted to nutrient-rich depths. Gas vacuolation and buoyancy regulation are especially evident during summer stratification periods, when nutrient depletion is most profound in surface layers. During such periods cyanophycean surface blooms often migrate to nutrient-rich deeper waters during midday and return to the surface later in the afternoon. Nutrient depletion is probably only one of many reasons for diurnal vertical movement. Other factors, including high surface light intensities, carbon dioxide depletion, pH shifts, and oxygen supersaturation (leading to photorespiration), are also operative.

As mentioned previously, nitrogen-fixing cyanophytes produce oxygen by the photolysis reaction of photosystem II while fixing nitrogen through the oxygen inhibited, nitrogenase complex. To fix nitrogen the group has evolved spatial, temporal, and biochemical mechanisms to separate these processes. The study of these mechanisms has revealed some ecological strategies leading to cyanophycean domination in fresh water.

Filamentous nitrogen-fixing cyanophyceans including *Anabaena*, *Aphanizomenon*, *Gloeotrichia*, and *Nostoc* have developed highly specialized cells with a reductive intracellular environment, heterocysts, bordering highly oxygenated photosynthetic cells (Lang and Fay 1971). Heterocysts harbour most of the nitrogenase (Fay, Stewart, Walsby, and Fogg 1968; Wolk and Wojciuch 1971; Wolk 1973; Tel-Or and Stewart 1976), and appear to be the main sites of nitrogen fixation

(Wolk, Thomas, Shaffer, Austin, and Gabonski 1976). They possess thickened cell walls (Lang and Fay 1971), and multiple layers of extracellular mucilage, which shield against oxygen intrusion. Heterocysts are also biochemically specialized, being characterized by high respiration rates (Fay and Walsby 1966), and the absence of oxygen-evolving photosystem II (Thomas 1970). The enzyme ribulose 1, 5-diphosphate carboxylase is also absent (Winkenbach and Wolk 1973), making the heterocyst a non-carbon-fixing and non-oxygen-evolving cell. Photosystem I, however, is active in heterocysts (Cox and Fay 1969) providing photoreductant and cyclic photophosphorylation to power nitrogenase. Important cytochromes and reductant-transferring enzymes are also present (Stewart 1977*b*), assuring electron transfer to the photosystem I reaction centre followed by transfer to ferredoxin and NAD^+ ($NADP^+$), the consequent electron donors for nitrogenase.

Despite such assurance against oxygen inactivation of nitrogen fixation, short-term oxygen supersaturation, which readily occurs during surface blooms, can still have a marked inhibitory effect (Stewart and Pearson 1970). The biochemical and structural adaptations discussed above are not entirely adequate and supplemental oxygen-scavenging mechanisms are needed. Several workers (Sirenko, Stentsenko, Arendarchuk, and Kuzmenko 1968) have stated that mucilagenous material and secreted sheaths contain terminal sulphydryl groups, which are capable of binding oxygen. In certain *Anabaena* species thickened mucilagenous excretions often cover the heterocyst. Simpson and Neilands (1976) also reported that during hydroxymate synthesis by certain species of *Anabaena*, oxygen is utilized. Thus, the synthesis of these compounds acts as a direct oxygen sink.

Bacterial associations with nitrogen-fixing cyanophyceans also appear to be instrumental in removing oxygen from the vicinity of the heterocysts. We (Paerl and Kellar 1978) have observed specific associations of heterotrophic bacteria with the heterocysts of naturally occurring *Anabaena* and *Aphanizomenon*, being particularly common in junctions between the heterocysts and vegetative cells (Paerl 1976). The attached bacteria profit by assimilating *Anabaena* excretions in this region (Paerl 1978). The heterocyst–vegetative cell junction is probably the 'weak link' in the barrier excluding oxygen from the heterocysts (Figs. 6.1 and 6.2). Fixed carbon and nitrogen, reductant, and ATP are transported to and from the heterocyst through a minute 'pore channel' (Lang and Fay 1971) which forms the constricted polar regions of the heterocysts. It is this region which is most susceptible to oxygen invasion, due to the absence of thick cell walls surrounding the pore channel. The suspected site of nitrogen fixation in the heterocysts directly borders this channel. The inhibitory effect of oxygen is minimized by associated bacteria, probably through bacterial oxygen con-

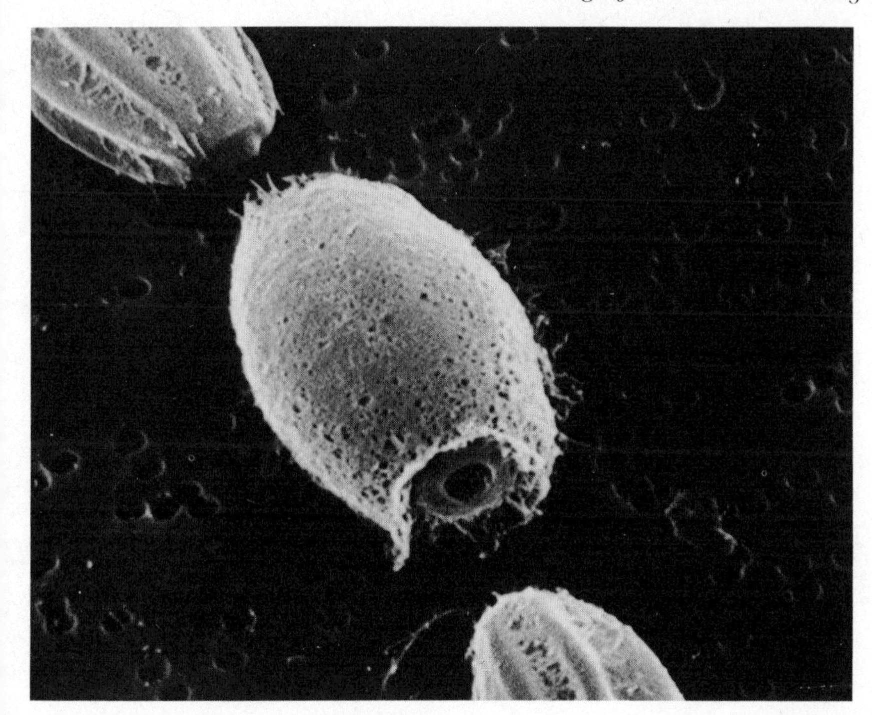

Fig. 6.1. Scanning electron micrograph of exposed 'pore channel' region of a detached heterocyst of *Anabaena oscillarioides*. Note the envelope of mucilage covering the heterocyst. Pore size of the Nucleopore filter supporting the specimen is 0.4 μm. Adjacent ribbed cells are vegetative cells.

sumption in these microzones. Tetrazolium salts, which produce coloured insoluble crystals under reduced conditions, have been used to demonstrate the abilities of blue–green algae and bacteria to produce reduced microzones in highly aerobic environments (Fay and Kulasooriya 1972; Paerl 1979). When applied to cultures of *Anabaena oscillarioides* containing bacteria, such salts are commonly deposited near sites of bacterial attachment.

In addition to forming associations with bacteria, cyanophytes have also formed symbiotic associations with higher plants in fresh water. The aquatic fern *Azolla* harbours *Anabaena* in the dorsal lobes of its leaflets. In this association *Anabaena* provides fixed nitrogen for the host (Peters and Mayne 1974) while *Azolla* supplies at least some of the photosynthetically fixed carbon compounds to *Anabaena* and probably provides protection from high oxygen levels. Phosphorus is also transferred via the *Azolla* roots to the nitrogen-fixing endosymbiont.

Cyanophytes have also developed endogenous means of protecting nitrogenase from oxygen inhibition. It has been shown that after oxygen

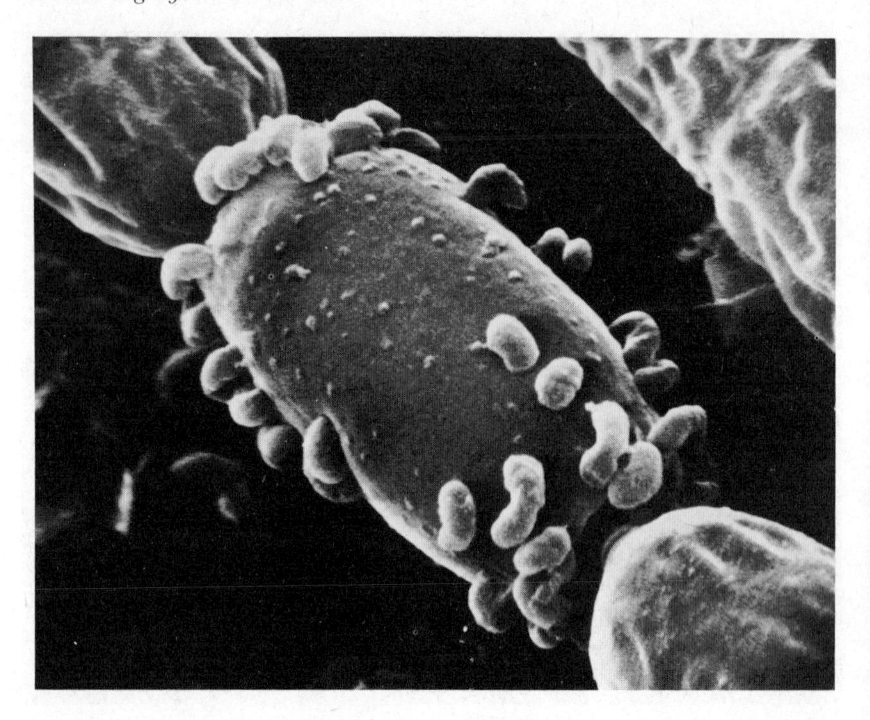

FIG. 6.2. A heterocyst of *Aphanizomenon flos aquae* in freshly filtered Clear Lake, California water during a spring bloom. Note the abundance of attached bacteria, which show specific attraction for the heterocysts as opposed to vegetative cells of this cyanophyte.

supersaturation, axenic *Anabaena* exhibit inhibition of nitrogenase activity (Stewart and Pearson 1970). However, after continued oxygen supersaturation (from two to six hours after initial oxygen supersaturation), nitrogenase activity shows full recovery to control (90–100 per cent oxygen saturation) levels (Paerl 1978). This recovery only occurs during daylight (in darkness oxygen supersaturation produces sustained inhibition) and is only marginally limited by 10^{-5} M dichloromethylurea (DCMU), a potent inhibitor of oxygen-evolving photosystem II. During nitrogenase recovery carbon dioxide fixation and oxygen evolution remain suppressed (Fig. 6.3). Furthermore, Paerl (1979) showed, by the use of tetrazolium reduction experiments, that the source of reductant powering the recovery probably resides in the vegetative cells of *Anabaena.* In summary, this indicates that during periods of oxygen supersaturation, which are common at midday hours during blooms, oxygen-evolving photosynthesis is more severely inhibited than nitrogen fixation. Photosystem I (cyclic or non-oxygen-

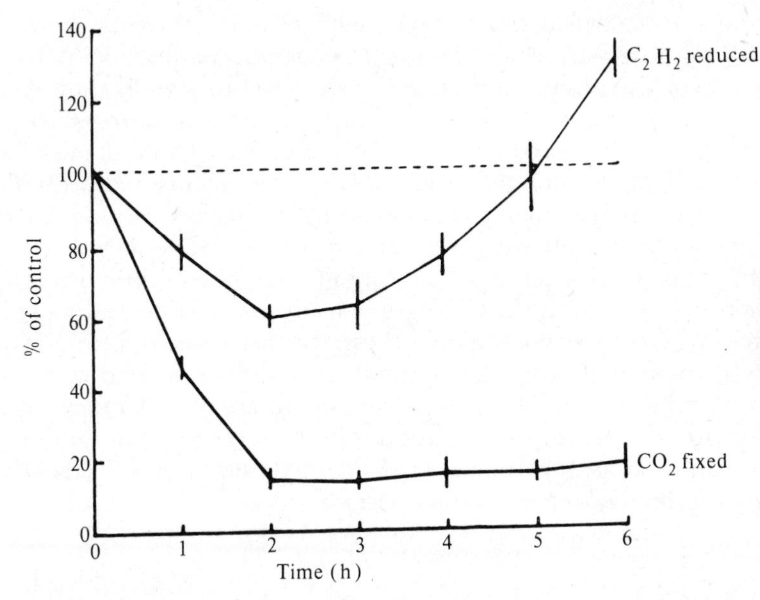

FIG. 6.3. Acetylene reduction and carbon dioxide fixation rates of *Anabaena oscillarioides* in response to oxygen supersaturation. Results are graphically compared to control (100 per cent oxygen saturation) cultures. Note that acetylene-reduction rates show a marked recovery after initial inhibition, while carbon dioxide fixation remains severely inhibited throughout the experimental period. (Results from Paerl (1978).)

evolving) appears to be the main source of reductant during these periods.

It is not known how *Anabaena* directly protects its nitrogenase during the recovery. It has been noted though that nitrogenase, in addition to fixing nitrogen, also evolves hydrogen, the functional significance of which is not clear. It would appear that hydrogen evolution is a wasteful process since photogenerated reductant is expelled without cellular utilization (Shubert and Evans 1976). Recently, however, Fujita, Ohama, and Hattori (1964) and Tel-Or, Luick, and Packer (1977) have reported an inducible hydrogenase (hydrogen-utilizing enzymes) in aerobically grown cultures of *Nostoc muscorum* and *A. cylindrica*. These findings seem to confirm original suggestions by Dixon (1972) and Shubert and Evans (1976) that: (i) hydrogenase may aid nitrogen fixation by recirculating hydrogen back through a series of electron transfer steps, and (ii) hydrogenases may function to inactivate oxygen near the nitrogenase complex. It is clear that many biochemical strategies designed to cope with optimizing nitrogen fixation in highly oxygenated environments are yet to be discovered.

From an ecological perspective, such recovery systems represent highly adaptive strategies designed to avoid competition between oxygen-evolving photosynthesis and reductive nitrogen fixation. With photosynthesis more inhibited than nitrogen fixation, nitrogenase can act as a sink for photoreductant during periods of high input of radiant energy. Evidence obtained from in-situ experiments indicates that nitrogenase activity can in fact act as a sink for available radiant energy during midday and afternoon oxygen-supersaturated periods (Fig. 6.4). Such periods tend to show strong inhibition of photosynthetic carbon fixation in surface waters. Owing to effective protection and recovery mechanisms, nitrogenase activity can be maintained during such periods, thereby allowing for optimal use of light operated reductant during daylight. Aside from representing an adaptive strategy in response to oxygen supersaturation, the sequential optimization of carbon and nitrogen fixation assures adequate supplies of fixed carbon prior to optimal nitrogen-fixation periods.

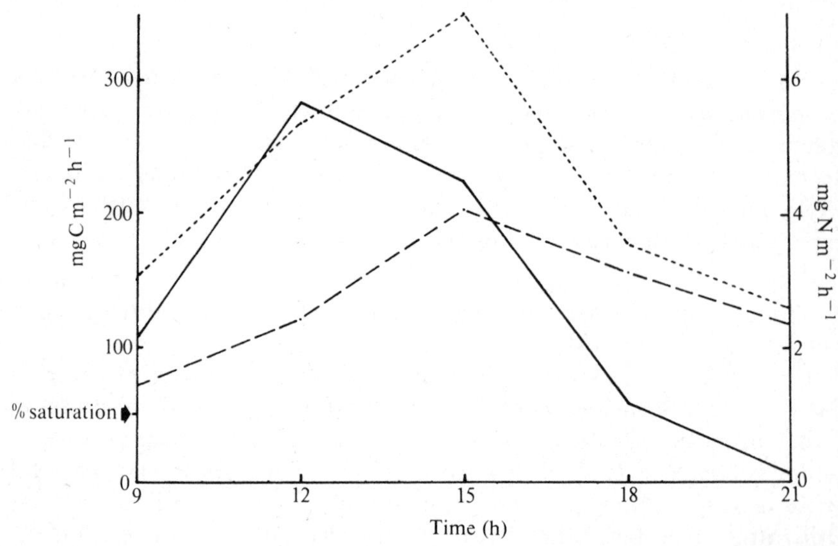

FIG. 6.4. Distribution per square metre of carbon fixation (as measured with the ^{14}C method) (solid line), nitrogen fixation (as measured by the acetylene-reduction method) (short dashes) and dissolved oxygen saturation (long dashes) in Lake Rotongaio, New Zealand. These results demonstrate that maximum carbon fixation tends to precede maximum nitrogen fixation on a diurnal basis. Nitrogen-fixation rates are related to oxygen buildups in the water column while carbon fixation rates appear suppressed during peak oxygen saturation in the water column. Data from this diurnal study were integrated throughout a 10-m water column (containing seven sampling depths).

Other genera have evolved somewhat different means of overcoming oxygen toxicity during nitrogen-fixing periods. *Oscillatoria* for example, possesses no apparent structural means of isolating nitrogen from carbon dioxide fixation. As a result, any nitrogen fixation (at present it is uncertain whether *Oscillatoria* fixes nitrogen in nature) occurs only in micro-aerophilic environments or at night time, when respiratory activity can deplete oxygen; it may also occur in anaerobic microzones formed by associated bacteria (see p. 203).

Finally, among the aerobic unicellular nitrogen-fixing cyanophytes, namely the genus *Gloeocapsa*, the coexistence of photosynthesis and nitrogen fixation is particularly difficult to reconcile since no obvious structural specialization has been found. It has been suggested that *Gloeocapsa* temporarily separates photosynthesis from nitrogen fixation, thereby avoiding direct competition for reductant between the two processes (Gallon, Kurz, and LaRue 1975).

6.3 Estuarine and coastal habitats

Biomes considered herein are land zones subjected directly to ocean water or indirectly to saltwater intrusion along estuaries, so that salinity and tidal inflow are constant variables. The most severe environments are devoid of higher plant life, characterized by low diversity of animal forms (i.e. sandy beach, rocky shore, dune slack) while zones with vegetation provide increased sediment stability, available energy sources, and ecological niches to organisms (i.e. sea grass, marsh, mangrove).

Where terrestrial drainage or tidal deposition yields land accretion, a primary producing community will develop eventually barring extreme physical factors. Since ocean waters are usually too nutrient poor to support plant assemblages alone, nitrogen-fixing species may be important as early colonizers of barren sand and sediments, building nitrogen resources. Populations of blue–green algae have been cited as one of the first inhabitants of south Texas bay flats (Wood 1965), as the primary recolonizers of rocky shores after the July algal felt peel (Stewart 1967*a*), and as the pioneer species of sand dune slack systems (Stewart 1967*b*).

The sources and amounts of nitrogen fixation and recycling in estuarine and coastal environments are poorly understood. The majority of recent studies involve salt marshes with fewer examinations of sandy beaches, rocky shores, sand dune slack, mangrove, and sea grass communities. The marsh habitat is used as the principal example of a potential nitrogen budget, including fixation, inflows, outflows, and withinflows, of a marginal salt water biome.

Species

Estuarine and coastal marine nitrogen-fixing species that have been identified are listed in Table 6.4; not all species shown have been verified to fix nitrogen in pure culture but are members of established nitrogen-fixing genera. Species range from free-living heterotrophic bacteria to photosynthetic organisms to symbiotic complexes. Table 6.5 gives available in-situ rates published for several species; the record is small as most acetylene-reduction experiments, especially in marsh studies, reflect community fixation values rather than individual species values.

TABLE 6.4
Nitrogen-fixing species identified in estuarine or coastal habitats

Habitat	Organism		Reference (see footnote)
Marsh	Free living		
	Heterotrophic		
		Bacillus	1,2
		Clostridium	1,2,3
		Desulphovibrio	1,2,3
	Photosynthetic		
	Pseudomondales		2,4,5
		Chloropseudomonas	3
		Rhodopseudomonas	3,4
	Cyanophyta		
		Anabaena sp	2,3,6
		A. torulosa†	7,8,9
		A. variablis†	8,9
		A. oscillarioides	9
		Calothrix sp	3
		C. aeruginea	10
		C. contrarenii	5
		C. crustacea	9
		C. membranacea	7
		C. scopulorum	7,11,12,13
		Cylindrospermum licheniforme	9
		Microchaete sp	10
		Nodularia sp	2
		N. harveyana	7,8,9
		N. spumigina	8,9,10,13
		Nostoc sp	2
		N. commune	8,9,13
		N. ellipsosporum‡	9
		N. entophytum	12,13
		N. linkia	13
		N. muscorum‡	9

		Rivularia atra	7,10
		R. biasolettiana	10
		R. nitida	7
		Scytonema hofmanii	9
		Spiruli subsalsu	8
		Tolypthrix tenius	7
Dune slack		*Microcoleus chthonoplastes*§	13
		Anabaena sp	14
		Calothrix sp	14
		Nodularia harveyana	14
		Nostoc sp	14,15
		N. entophytum	14
		Scytonema sp	14
	Symbiotic		
		Peltigera refuscens-Nostoc	16
		P. polydactyla–Nostoc	16
Mangrove		*Pseudomondales*	17
		Rhodospirillaceae	17
		Scytonema sp	18
Rocky shore		*Calothrix crustacea*	14
		C. fasciculata	14
		C. scopulorum	14
		Nodularia harveyana	14
		Scytonema sp	18
		Lichina confinis-Calothrix	16
Sandy beach		*Pseudomondales*	18
		Calothrix crustacea	18
		Rivularia sp	18
		Scytonema sp	18
		Hyella balani§	18
		Microcoleus chthonplastes§	18
		Pleurocapsa–Chroococcus§	18
Sea grass		*Calothrix* sp	19

† May be different life stages of *A. oscillarioides* (Ralph 1975).
‡ May be different life stages of *N. commune* (Ralph 1975).
§ Non-heterocystous.

References: (1) Kator and Herwig (1976); (2) Jones (1974); (3) Hanson (1977*b*); (4) Wynne-Williams and Rhodes (1974*a*); (5) Van Raalte *et al.* (1974); (6) Stacey *et al.* (1977); (7) Webber (1967); (8) Thomson (1977); (9) Ralph (1975); (10) Stewart (1973); (11) Stewart (1967*a*); (12) Stewart (1962); (13) Stewart and Pugh (1963); (14) Stewart (1965); (15) Stewart (1967*b*); (16) Hitch and Stewart (1977); (17) Gotto and Taylor (1976); (18) Potts and Whitton (1977); (19) Goering and Parker (1972).

Factors controlling nitrogen fixation

Analysis of nitrogen fixation by the acetylene-reduction technique has usually yielded high variability between spatially different sediment cores of the same time period in the marsh (Jones 1974; Marsho,

TABLE 6.5
In-situ nitrogen-fixation rates of pure cultures of blue–green algae and algal associations

Habitat	Organism	Rate	Reference
Sandy beach	*Calothrix crustacea*	26.41 g N m^{-2} a^{-1}	1
	Hyella balani	2.16–20.42 g N m^{-2} a^{-1}	1
	Microcoleus chthonoplastes	1.13–22.14 g N m^{-2} a^{-1}	1
	Pleurocapsa–Chroococcus	1.03 g N m^{-2} a^{-1}	1
	Rivularia sp	60.87 g N m^{-2} a^{-1}	1
	Scytonema sp	22.34–40.3 g N m^{-2} a^{-1}	1
Rocky shores	*Calothrix scopulorum*	2.5 g N m^{-2} a^{-1}	2
		10.2 μg N mg N^{-1} day^{-1}	3
	Lichina confinis–Calothrix	0.4 μg N mg N^{-1} algae day^{-1}	4
	Nostoc entophytum	11.7 μg N mg N^{-1} day^{-1}	3
	Scytonema sp	19.88 g N m^{-2} a^{-1}	1
Mangrove	*Scytonema* sp	22.93–102.6 g N m^{-2} a^{-1}	1
Dune slack	*Peltigera rufescens–Nostoc*	4.1 μg N mg N^{-1} algae day^{-1}	4

References: (1) Potts and Whitton (1977); (2) Stewart (1967*a*); (3) Stewart (1965); (4) Hitch and Stewart (1977).

Burchard, and Fleming 1975; Whitney, Woodwell, and Howarth 1975; Thomson 1977) and sea grass bedding (McRoy, Goering, and Chaney 1973). Marsho *et al.* (1975) and Thomson (1977) have suggested very patchy loci of fixation exist in the sediment; heterogenous sediment composition and environmental conditions could cause clumped distributions of microbe populations and/or nitrogen-fixing activity. General factors regulating nitrogen fixation can be described for pure culture, but species growth and nitrogen fixation activity in mixed populations in rapidly changing environments is less predictable especially when factors are non-additive and not correlated seasonally.

Heterotrophic bacteria have been characterized as inefficient energy users (Chapter 2), requiring greater carbohydrate supplies than found in most soils. The marsh (Stewart 1969; Jones 1974; Hanson 1977*a*,*b*), sea-grass (Patriquin and Knowles 1975), and mangrove (Zuberer and Silver 1978) sediments also seem to be carbon-limited habitats to these species. Additions of raffinose, arabinose, and mannose increased fixation rates for mixed heterotroph populations in marsh soil over other simple sugars (Hanson 1977*a*). Within plant systems, root exudates are thought to provide carbohydrates to microbes in exchange for fixed nitrogen (Jones 1974; Hanson 1977*b*). In addition to energy limitations, nitrogen fixation is repressed by combined nitrogen compounds. The degree and type of inhibition are reported to be concentration-dependent (Stewart 1969). Patriquin and Knowles (1975) suggest that the concentrations of ammonia required for repression are positively correlated with population densities of organisms. In pure culture, nitrogen compounds do not directly interfere with fixation but inhibit nitrogenase production over time (Stewart 1969). However, in sediment slurries amended with ammonium chloride and sodium nitrate at 25 to 100 times the endogenous concentration, Hanson (1977*a*) found immediate reduction of nitrogen fixation capacity, while amino nitrogen did not reduce nitrogen fixation.

The nitrogen-fixing cyanophyta require adequate light and are often limited by phosphate and trace elements (Stewart 1964, 1969). Sediment algal production is greater in areas of lesser higher plant cover (Van Raalte, Valiela, and Teal 1976), as a result of the canopy effect. Maximum nitrogen fixation has been observed *in situ* at temperatures of 18–30 °C (Stewart 1967*a*, 1970*b*; Jones 1974), pH of 7.5–9.5 (Fogg 1956; Stewart 1970*b*), and a salinity of 0.086 M NaCl (Jones and Stewart 1969), although a species of *Anabaena* has been isolated from a shallow Texas marine environment that has an optimum temperature of 42 °C for growth (Stacey, Van Baalen, and Tabita 1977). Greater salinities decrease nitrogen fixation of submerged algae; however, floating algae show little inhibition to 0.086 M NaCl (Jones 1974). Generally, nitrogen fixation is greater in free water subaerial conditions than submerged (Jones 1974). Combined nitrogen effects on nitrogen fixation by cyanophyta are concentration dependent (Stewart 1969). The factors controlling fixation of symbiotic lichens and blue–green algae are influenced by the physiology of both organisms. Hitch and Stewart (1977) have found moisture to be the most important requirement, and temperature more important than light as nitrogen fixation continued into darkness.

Generally, in temperate latitudes, peaks and troughs of nitrogen fixation for bacterial and algal soil populations and epiphytes have been correlated with the seasons; increases are synchronous with warming temperatures (early spring) and plant carbohydrate exudation (late summer, early autumn) (Jones 1974; Marsho *et al.* 1975; Haines, Chalmers, Hanson, and Sherr 1977; Hanson 1977*b*; Thomson 1977). Inhibition often occurs in early–mid-summer particularly among algal felts and is attributed to high temperatures, desiccation, and increased salinity (Stewart 1967*a*).

As indicated previously, microbial fixation is a potential source of nutrient nitrogen to non-nitrogen-fixing algae and higher plants in brackish water systems. Jones and Stewart (1969), Stewart (1967*a*), and Jones (1974) have documented uptake by higher plants of $^{15}N_2$ fixed by algae in salt marshes, dune slack, and rocky shore habitats. One method of evaluating the possible sufficiency of in-situ nitrogen fixation is by constructing a ratio of nitrogen fixed annually to primary production nitrogen demand for a system.

Nitrogen system budgeting

Environments relying solely on ocean water nutrient levels are poor in inorganic nitrogen nearly all year round unless nitrogen fixation or some other source of nitrogen is available. Estuarine habitats benefit from terrestrial nitrogen runoff. Table 6.6 provides a balance of fixation versus primary production demand for salt marshes and to a lesser extent rocky shore, dune slack, sea grass, and mangrove environments.

The marsh nitrogen pools which are large and relatively stable seasonally are present in soil (organic nitrogen and clay-bound ammonia) and subsurface macro-organic matter (Haines *et al.* 1977).

TABLE 6.6

Production and fixation nitrogen fluxes in various ecosystems

Habitat	System	Flux ($g N m^{-2} a^{-1}$)	References and notes
Marsh	Organism N demands		
	Benthic algae	10.55–28.73	1,2,3; C/N=10 (2)
	Angiosperm	1.5–53.34	2,4,5,6,7,8,9,10,11, 12,13; 0.5–1.85 %N of g dry wt (2,3,14)
	Roots	1.13–9.33	15; %N of g dry wt × production (12)
	minus		
	N withdrawn into rhizomes	0.53–20.86	2; live–dead %N of g dry wt × angiosperm production
		13.65–70.54	
	Denitrification	7–12	2,16
		20.65–82.54	
	Fixation		
	Soil depth	4.42–51.4	2,17,18
	Surface	0.22–1.2	2,17
	Epiphyte	0.18–0.87	2,17,19
		4.82–53.5	
	plus		
	Blue–green mat	4.13–176	19,20,21
		1.14–46.17	20,21
		× % of area covered, very variable	

Balance: Excluding mat and pool contributions, fixation accounts for a minimum of 23–65% of flux through production and denitrification and 35–76% through production alone

Rocky shore	Organism N demand		
	Algae	6.1	22; 2.5 as 41% of total algae N
	Fixation		
	Mat (algae, some bacteria)	0.996–19.88	22,23
	Lichen–algae	?	24
	Balance: Excluding lichen–algae contribution, Stewart (22) figured fixation accounted for 41% of algae annual production		
Dune slack	Fixation		
	Mat (algae, some bacteria)	4.58–5.89	23; based on 7–9 mos. activity (22)
	Lichen–algae	?	24
	Balance: Excluding lichen–algae contribution, fixation accounts for 21% of algae production (22)		
Sea grass	Organism N demand		
	Zostera (temperate zone)	73.00–109.5	25
	Thalassia (tropical zone)	3.14–25.9	26,27,28
	Fixation		
	Zostera sediment	0	25
	Thalassia sediment	0.0106–170	25,29; sparse to very dense beds
	epiphytes	110.4	27
		0.0106–280.4	
	Balance: No fixation was found for temperate species; a wide range of fixation values for *Thalassia* indicated negligble to over 100% satisfaction of N production requirements		
Mangrove	Organism N demand		
	Florida mangrove	26	30
	Fixation		
	Mangrove sediments	7.8	30
	Balance: Root associated fixation exceeds that of sediments, > 60% of N requirements met by fixation (30)		

References: (1) Van Raalte *et al.* (1976); (2) Haines *et al.* (1977); (3) Gallagher (1971); (4) Hanson (1977*a*); (5) Gallagher, Linthurst, and Pfeiffer (1977); (6) Mendelssohn and Marcellus (1976); (7) Squiers and Good (1974); (8) Valiela and Teal (1974); (9) Kowalski (1978); (10) Nixon and Oviatt (1973); (11) de la Cruz and Gabriel (1974); (12) Valiela, Teal, and Persson (1976*a*); (13) Gallagher, Reimold, and Thompson (1972); (14) Keefe and Boynton (1973); (15) Gallagher (1975); (16) Valiela, Vince, and Teal (1976*b*); (17) Hanson (1977b); (18) Tjepkema and Evans (1976); (19) Thomson (1977); (20) Jones (1974); (21) Whitney *et al.* (1975); (22) Stewart (1967*a*); (23) Stewart (1965); (24) Hitch and Stewart (1977); (25) McRoy *et al.* (1973); (26) Patriquin (1972); (27) Goering and Parker (1972); (28) Capone and Taylor (1977); (29) Patriquin and Knowles (1972); (30) Zuberer and Silver (1978).

Seasonal fluctuations are expressed in the smaller pools, exchangeable ammonia, nitrate, and nitrite of pore water. Their concentrations and turnover rates appear to determine the nitrogen available to primary producers (Haines *et al.* 1977; Hanson 1977*a*). Experimental subjection of marsh to sewage, urea nitrogen, and fertilizer loading has affirmed suspected nitrogen limitation for angiosperm growth (Valiela, Teal, and Sass 1973; Sullivan and Daiber 1974; Valiela and Teal 1974; Gallagher 1975; Valiela, Teal, and Sass 1975). Especially in the mid-high marsh, aerial biomass of short *Spartina alterniflora* was increased by nitrogen addition, though utilization of available nitrogen was thought inhibited by increasing salinities (Christian, Bankroft, and Wiebe 1975; Chalmers 1977). Of the *Spartina* plants, the rhizomes are the largest and perennial portion of underground biomass; measurements of dry weight have remained seasonally constant and insensitive to increased nitrogen (Valiela and Teal 1974; Gallagher 1975; Valiela, Teal, and Persson 1976). The rhizomes can store nitrogen from mature dying leaf and stem translocations (Haines *et al.* 1977); this accumulation can be increased by sediment fertilization and may delay a direct response of aerial biomass to nitrogen loadings during a growing season (Valiela and Teal 1974). Root production is seasonal and after late summer has been indicated to be lower than normal in enriched areas (Valiela, Teal, and Persson 1976). Marsh benthic primary production by algae and photosynthetic bacteria is not depressed by nitrogen supplements, though soil nitrogen fixation is lowered until added nitrogen has been absorbed by plants or removed by tides (Van Raalte, Valiela, Carpenter, and Teal 1974; Hanson 1977*b*).

Several species have been isolated from mangrove sediments and their *in situ* nitrogen fixation rates determined. Rates for a *Scytonema* sp mat have been expressed in terms of area^{-1} time^{-1} (Table 6.6) by Potts and Whitton (1977) and estimated for the whole mangrove community by Zuberer and Silver (1978). Light-dependent nitrogen fixation has also been observed on floating, decaying mangrove leaves (Odum and de la Cruz 1967; Gotto and Taylor 1976).

Sandy beaches are generally sterile, harsh environments with reduced organic sediment zones beneath the surface. Where beaches are protected from tidal energy (lagoon), significant blue–green and bacterial nitrogen fixation have been measured by Potts and Whitton (1977). Rates range from 1–61 g of nitrogen m^{-2} a^{-1} (Table 6.6); no biomass production values were available. However, even if one assumes a primary production of twice the marsh benthic algal upper limit (i.e. 57 g of nitrogen m^{-2} a^{-1}), algal fixation in this lagoon system would meet the nitrogen demand.

Algal colonies are the principal nitrogen-fixing entities in both rocky shore and sand dune slack regions with peaks in spring and fall (Stewart

1965; Stewart 1967*a*). Jones and Stewart (1969) found that algae released a minimum of 40 per cent of fixed nitrogen in the form of peptides and amino acids. Lichen–algae symbioses also contribute nitrogen to these systems, fixing most nitrogen from October to January (Hitch and Stewart 1977).

Documentations of sea-grass communities are controversial, the data being highly variable and not indicative of the role of fixation in supplying nitrogen to developing grasses. The temperate grass, *Zostera*, has not revealed associated sediment, rhizome, or epiphyte nitrogen fixation (McRoy *et al.* 1973), and while the tropical *Thalassia* has sediment, rhizome, and epiphyte nitrogen fixation (McRoy *et al.* 1973; Goering and Parker 1972; Patriquin and Knowles 1972), the total varied from zero to rates exceeding 100 per cent sufficiency for the production. The work of Capone and Taylor (1977) provides intermediate values for *Thalassia* while their continued efforts hopefully will go a long way towards resolving this controversy.

6.4 Nitrogen fixation in the ocean

Stewart (1971) has thoroughly reviewed marine nitrogen fixation up to 1970; thus, in this section, the work that has occurred since his review is emphasized. We will first examine the regions of the world's oceans in which nitrogen fixation has been reported. Then we will discuss the types of organisms involved, how much nitrogen is fixed, whether this fixation is of quantitative importance to the total nitrogen budget, and what factors control the rates of fixation. Finally, we will examine nitrogen fixation in two systems where it appears to be of quantitative importance: coral reefs and blooms of *Oscillatoria* (*Trichodesmium*).

Location and rates of nitrogen fixation

A summary of results from investigators since Stewart's review (1971) in the various regions of the oceans is given in Table 6.7. We can draw several conclusions. First, despite the recent explosion of interest in measuring nitrogen fixation in the biosphere, there have been few investigations in the sea, and the majority of these have dealt with blooms of *Oscillatoria*. (We will use the revised classification scheme of Drouet (1968) who classified all *Trichodesmium* species as *Oscillatoria eurythraea*.) Secondly, water and sediments in aphotic environments, only a little removed from land, show either no, or diminishingly small, rates of nitrogen fixation regardless of the region of the ocean investigated. Thirdly, quantitative assessments, by various authors, of nitrogen fixation in *Oscillatoria* blooms appear somewhat conflicting; however, we believe that Carpenter and Price (1976, 1977) have resolved many of the problems, as will be discussed. Fourthly, bacteria do

TABLE 6.7

Nitrogen fixation in waters and sediments of the world's oceans

	Habitat	Location	Probable nitrogen-fixing organisms	Estimated rate and importance to ecosystem	Reference
Bacteria	Sediment	Arctic—Beaufort Sea	Unknown	25 mg N m^{-2} a^{-1}; unimportant	Knowles and Wishart (1977)
	Sediment	Suruza and Sagami Bays, Japan	Unknown	Unknown	Maruyama *et al.* (1970)
		Tokyo Bay, Japan	Unknown bacteria	Unknown	Maruyama *et al.* (1974)
	Open ocean	Lat. 40°N/5°N, Long. 155°W	Unknown	Unimportant	Maruyama *et al.* (1970)
	Bay water	Maizuru Bay, Japan	Unknown	Unknown	Sugahara, Sawado, and Kawai (1971)
	Water and sediment	Japan and China Sea	Unknown	Place counts of nitrogen-fixing organisms only	Kawai and Sugahara (1971*a*,*b*, 1972)
	Sediments	Scottish lochs	Unknown	0.08–0.5 nM C_2H_4 g^{-1} h^{-1}: high rate	Hartwig and Stanley (1978)
	Sediments	Offshore, deep sea (4800 m), N. Atlantic	Unknown	Very low or zero; unimportant	Hartwig and Stanley (1978)
	Shallow carbonate sediments	Off Barbados	Unknown	Very low without organic substrate addition	Patriquin and Knowles (1975)
	Sediment	Iceland fjord mud, Aberystwyth harbour	Athiorhodaceae (cultured from mud)	Low rates	(Wynne-Williams and Rhodes (1974*a*)

	Seawater	3 km off Cardigan Bay	Aerobic bacteria (no *Azotobacter*)	Probably unimportant	Wynne-Williams and Rhodes (1974*b*)
	Coral reef flat	Kaneohe Bay, HA	Identified types: *Rhodopseudomonas*, *Thiocystis*, *Pelodictyon*, *Clostridium*, *Desulfovibrio* *Azotobacter*	2–10 ng N_2 g^{-1} h^{-1}; possibly important	Hanson and Gundersen (1976*a*)
	Sediment	Kingoodie Bay, Scotland	Identified types: *Azotobacter*, *Klebsiella*, *Enterobacter*, *Desulfovibrio*, *Clostridium*	Maximum rate 1.84 ng g^{-1} dry wt sed. h^{-1}; possibly important	Herbert (1975)
Blue–green algae	Open ocean	S.W. Sargasso Sea	*Oscillatoria thiebautii*	Average 0.044 μg N $colony^{-1}$ day^{-1}; significant addition	Carpenter (1973)
	Ocean water	Baltic Sea	*Nodularia spumigena*	Potentially very important	Ostrom (1976)
	Open ocean water	W. Sargasso Sea	*Oscillatoria thiebautii*	24 μg N m^{-2} day^{-1}; appears negligible	Carpenter and McCarthy (1975)
	Open ocean water	Caribbean Sea	*Oscillatoria* spp	1.3 ng N m^{-2} day^{-1}; may be major input	Carpenter and Price (1977)
	Open ocean water	Central North Pacific Ocean	*Oscillatoria* spp	33 μg at N m^{-2} day^{-1}; may meet 100% of requirements	Mague *et al.* (1977)
	Coral reef flat	Kaneohe Bay, HA	*Calothrix* spp, *Nostoc* spp	Unknown but potentially important	Hanson and Gundersen (1976*b*)
	Coral reef flat	Lizard Island, Australia	*Scytonema* spp	6.8–30.6 kg N ha^{-1} a^{-1}; potentially significant	Burris (1976)

Habitat	Location	Probable nitrogen-fixing organisms	Estimated rate and importance to ecosystem	Reference
Coral reef flat	Enewetak, MI	*Calothrix* spp, *Nostoc* spp, *Hormothamnion enteromorphoides*	Up to 400 kg ha^{-1} a^{-1}; great importance	Mague and Holm-Hansen (1975)
Coral reef flat	Enewetak, MI	*Calothrix crustaceae; Hormothamnion enteromorphoides* *Rivularia* spp	Average 1.8 kg ha^{-1} day^{-1}; great importance	Wiebe *et al.* (1975)
Intertidal lagoon	Aldabra Atoll	*Hyella balani*, *Scytonema* spp, *Calothrix crustaceae*, *Rivularia* sp, *Microcoleus chthonoplastes*	Probably great importance	Potts and Whitton (1977)

not appear important as nitrogen-fixing organisms in the ocean with the exception of perhaps localized zones. Finally, most authors who have attempted to construct nitrogen budgets recognize that until recycling rates are known, the importance of the nitrogen fixation to an organism or community cannot be accurately estimated. We will discuss some of the factors that control nitrogen fixation rates subsequently.

When examining the reports of nitrogen fixation by prokaryotes associated with eukaryotic organisms, a different picture emerges, as shown in Table 6.8. Included here are several coastal assemblages recently described. Only three of the 15 reports deal with bacteria, and of these only the paper by Carpenter and Culliney (1975) deals with bacteria in the open ocean. While they speculated that nitrogen fixation by bacteria associated with deep oceanic animals may be of nutritional importance, to date no one has made such investigations. It is certainly an area that should be investigated. Guerinot, Fong, and Patriquin (1977) demonstrated that the rate of nitrogen fixation found in the sea urchin gut depended upon the substrate being grazed, indicating that nitrogen fixation in animals may be more complicated than anticipated. Also, while they calculated that this nitrogen fixation could account for 8–15 per cent of the animal's growth requirements, they specified that no estimate of assimilation was made. They point out that in order for the fixed nitrogen to be shown to affect the animal's nutrition, assimilation studies are necessary. The finding that nitrogen fixation takes place in an animal does not necessarily prove that the animal benefits from the activity.

Most of the eukaryotic–prokaryotic associations in Table 6.8 represent epiphytic growth by blue–green algae on macro-algae. In six of the 12 reports the organisms responsible are not specified. Only three of the papers deal with open-ocean assemblages; the rest are found on coral reefs or in coastal sediments. The quantitative significance of nitrogen fixation ascribed by the authors, is in all cases speculative.

In summarizing the data in both Tables 6.7 and 6.8 we see that most of the papers involve *Oscillatoria* blooms and that reports of other nitrogen-fixing organisms are confined to a handful of observations. Even from these few data, however, we can conclude that in most areas of the ocean, nitrogen fixation appears of limited nutritional importance, that blue–green algae are responsible for most of the activity and that the local importance of the process, with the exception of coral reefs, remains to be established.

Factors affecting the measured rates of nitrogen fixation

Stewart (1971) discussed the advantages and disadvantages of measuring nitrogen fixation in aquatic environments by the ^{15}N and acetylene-reduction techniques. In this section we will not review the general

TABLE 6.8

Nitrogen fixation in eukaryote–prokaryote associations

Eukaryote community	Prokaryote responsible	Location	Importance of nitrogen Fixation	Reference
Codium fragile	*Azotobacter*	New England beaches, USA	Potentially important locally	Head and Carpenter (1975)
Marine shipworms	*Spirillaceae*	Coastal and Sargasso Sea	Potentially important for worm nutrition	Carpenter and Culliney (1975)
Strongylocentrotar aroebochiensis	Unspecified bacteria in gut	St. Margaret's Bay, Nova Scotia	Unknown assimilation (8–15% of requirement for growth)	Guerinot and Patriquin (1977)
Sargassum sp	*Dichothrix fasicoli*	Western Sargasso Sea	Possibly important	Carpenter (1972)
Colonial hydroid	*Oscillatoria*	Subtropical Atlantic Ocean	Unknown	Geiselman (1977)
Rhizosolenia styliformes var. *longispina*	*Richelia intracellaris*	Central North Pacific Ocean	4% of phytoplankton requirement	Mague *et al.* (1977)

R. cylindricus	*Richelia intracellaris*	Central North Pacific Ocean	4% of phytoplankton requirement	Mague *et al.* (1977)
Thalassia phyllosphere	*Calothrix*	South Florida, USA	Unknown	Capone and Taylor (1977) Capone (1978)
Microdictyon sp *Dictyota* sp *Padina* sp	Unspecified blue–green epiphyte	Coral reef near Grand Bahama Island Coral reef near Grand Bahama Island	Possible importance to reef Possible importance to reef	Capone *et al.* (1977) Capone *et al.* (1977)
Halimeda sp	epiphyte	Coral reef near Grand Bahama Island	Possible importance to reef	Capone *et al.* (1977)
Jania decussatodichotoma	*Hormothamnion enteromorphoides*	Coral reef, Enewetak, MI	High rates; potential importance	Mague and Holm-Hansen (1975)
Rhizoclonium	Unspecified	Coral reef, Enewetak, MI	Low rates	Mague and Holm-Hansen (1975)
Dictyosphaeria cavernosa	*Calothrix crustaceae*	Coral reef, Enewetak, MI	High rates Unknown	Webb *et al.* (1975)

arguments, but confine ourselves to discussing the factors that investigators recently have identified in field and laboratory studies. A list of these factors is found in Table 6.9 with at least one author who has discussed the effect. The list is not meant to be all-inclusive but rather illustrative of the problems.

The first three factors deal with problems encountered using the acetylene-reduction technique. The partial pressure of acetylene in the vessel affects ethylene production rates up to some plateau level. Some investigators (e.g. Crossland and Barnes 1976) may have chosen too low a partial pressure of acetylene, in their case 5000 Pa. The work of MacRae (1977) is included to demonstrate the wide range of acetylene concentrations that affect different organisms.

No less a problem is the length of incubation used. Here there are in fact two problems. First, as Mague, Weare, and Holm-Hansen (1974), among others discovered, acetylene can sometimes rapidly affect the organisms in question. Hanson and Gundersen (1977) noted that *Calothrix crustaceae* ceased nitrogen fixation after about six hours incubation. The second effect caused by various incubation periods concerns the kinetics of the reaction. Often only initial and final values are taken. If the rate is linear this is adequate, but there are reports that rates may not be linear, e.g. Wiebe, Johannes, and Webb (1975) for *C. crustaceae*, Patriquin and Knowles (1975) for mannitol-amended sediments. Changes in oxygen concentration and inorganic or organic nutrients can take place during incubation and these too could affect the rates. It is necessary to run at least some kinetic studies on any new system.

The most perplexing of the acetylene-related problems concerns the ratio used to convert ethylene produced to nitrogen fixed. As has been explained in many papers, the theoretical ratio is 3:1. However, when this ratio has been examined by the simultaneous use of $^{15}N_2$ and acetylene, the ratio has been shown to vary greatly, not only in the field, e.g. Burris (1976) and Mague *et al.* (1974), but also in pure cultures, e.g. MacRae (1977). Ratios higher than 3:1 mean that proportionately more ethylene is formed than N_2 is actually fixed. There are a variety of possible causes for this anomaly. For example, acetylene diffuses more rapidly in water than does nitrogen, and in compact sediments $^{15}N_2$ might diffuse less rapidly into the sites of fixation. Ratios lower than 3:1 could result from nitrogenase response to environmental factors other than nitrogen; e.g. binding capacity to specific fractions of the enzyme complex (Shah, Davis, and Brill 1975). Mague, Mague, and Holm-Hansen (1977) suggested that *in situ* in the presence of active nitrogen metabolism, reduction of nitrogen may be preferred to that of acetylene. Whatever the cause, the result is that using acetylene reduction alone to measure nitrogen fixation potentially can introduce large errors in calculating rates. For example, if Burris's (1976) average value of 1.9:1

TABLE 6.9

Some factors that affect measured rates of nitrogen fixation in the sea

	Factor	Effect	Reference
Method			
	Amount of acetylene used	*Calothrix* increased rate to 0.12 atm	Burris (1976)
		Beijerinckia indica increased to 0.8 atm	MacRae (1977)
	Incubation time	*Oscillatoria* stopped C_2H_4 production after 2 h	Mague *et al.* (1974)
		Calothrix sp stopped C_2H_4 production after 6 h	Hanson and Gundersen (1977)
	Ratio of C_2H_4 produced to N_2 fixed	Ratios of 1.9–20.7:1 (pure cultures)	MacRae (1977)
		Ratios of 1.5–8.4:1; mean 1.9:1 (coral reef material)	Burris (1976)
		Mean ratio 10:1 (open ocean)	Mague *et al.* (1974)
		Mean ratio 6.3:1 (open ocean)	Carpenter and Price (1977)
Physical	Turbulence	Reduced *Oscillatoria* N_2 fixation if O_2 present	Carpenter and Price (1977)
		Stimulated *Hormothamnion* N_2 fixation	Wiebe *et al.* (1975)
		Reduced N_2 fixation in mangrove leaves	Gotto and Taylor (1977)
		No effect on *Calothrix*	Wiebe *et al.* (1975)

Factor	Effect	Reference
Oxygen effect	Decreased N_2 fixation in carbonate sand	Patriquin and Knowles (1975)
	Decreased *Oscillatoria* N_2 fixation	Carpenter and Price (1977)
	Decreased N_2 fixation in sediments	Bohlool and Wiebe (in press)
	In micro amounts stimulated N_2 fixation in pure cultures of aerobes	Wynne-Williams and Rhodes (1974b)
Light regime: light stimulation	*Oscillatoria* N_2 fixation greatest near light saturation and rapid decline at lower light	Mague *et al.* (1977)
	Oscillatoria N_2 fixation greatest near surface and 9% at 50 m	Carpenter and McCarthy (1975)
	Calothrix N_2 fixation stimulated in high light	Wiebe *et al.* (1975)
Light regime: dark stimulation	*Hyella balani*—purple bacteria N_2 fixation in sediments	Potts and Whitton (1977)
Temperature	For *Calothrix crustaceae* no N_2 fixation at 24 °C, maximum at 37 °C	Wiebe *et al.* (1975)
	N_2 fixation rates stimulated in temperate sediments	Maruyama *et al.* (1974)
	For *Calothrix scopulorum* good rates at 1–10°C	Stewart (1971)

Salinity	2–40 p.p.t. S had no effect on *Calothrix crustaceae* N_2 fixation rate	Wiebe *et al.* (1975)
	45 p.p.t. S reduced N_2 fixation of *Calothrix scopulorum* to 1/2 that at 5 p.p.t.	Jones and Stewart (1969)
Season	*Hormothamnion enteromorphoides* prevalent in Guam only Nov–March	Tsuda (in press)
	Sediment rate greatest in summer	Wynne-Williams and Rhodes (1974*b*)
Nutrients		
Fixed nitrogen	Reduced N_2 fixation rates in carbonate sand when added	Patriquin and Knowles (1975)
	No effect in *Oscillatoria* at natural concentrations	Goering *et al.* (1965) Stewart (1971)
PO_4^{2-}	Addition stimulated N_2 fixation rate	Mague *et al.* (1974)
Fe^{3+}	Addition stimulated N_2 fixation rate	Stewart (1971)
Reduced carbon	Addition stimulated N_2 fixation rate in carbonate sand	Patriquin and Knowles (1975)
	Addition stimulated N_2 fixation rate in coastal sediments	Maruyama *et al.* (1974)
	Oscillatoria bloom—no effect upon addition	Carpenter and McCarthy (1975)

were the correct value for the work of Wiebe *et al.* (1975) who used 3:1, they would have underestimated the real rate by 37 per cent. Variability in ratios have been documented even under pure culture conditions. MacRae (1977) showed that two *Beijerinckii* sp ratios ranged from 1.9–20:1. Wynne-Williams and Rhodes (1974*b*) expressed concern over the unexplainable variability in ratios of pure cultures. This factor appears to be the source, potentially, of very serious errors in establishing rates. There is a need to establish the reasons for these variations if the acetylene-reduction technique is to be used for quantitative field studies.

The next set of factors deal with physical features of the environment. It has long been known that light levels affect the rate of nitrogen fixation in blue–green algae. Carpenter and Price (1977) found that *Oscillatoria* nitrogen-fixation rates per cell decreased progressively from the surface in the Sargasso Sea and from 15 m down in the Caribbean Sea. Mague *et al.* (1977) examined the light responses in isolated *Oscillatoria* colonies and found that acetylene reduction increased steadily up to 84–97 per cent of light saturation but that carbon fixation was maximum at from 32–58 per cent of maximum light intensity.

Stewart (1971) discussed the effect of temperature on nitrogen fixation in several blue–green algae. Wiebe *et al.* (1975) showed that fixation by *C. crustaceae* in an encrusted coral reef flat pavement doubled when the temperature was raised from 27 °C to 36 °C and that at 39 °C, after a brief erratic period of nitrogen fixation, the rate dropped dramatically. At 24 °C, nitrogen fixation ceased. Jones and Stewart (1969) reported that after 12 days of incubation at 40 °C, cultures of *C. scopulorum* did not recover when transferred to 25 °C. Maruyama, Taga, and Matsuda (1970) and Maruyama, Suzuki, and Otobe (1974) noted that increasing the incubation temperatures of Sagami and Tokyo Bay sediments dramatically stimulated nitrogen-fixation rates by bacteria; maximum rates were at 20 °C (in-situ temperature 2.5 °C) and 30 °C (in-situ temperature 16 °C) for the two bays respectively. They concluded that at in-situ temperatures little nitrogen could be fixed.

There appears to be some latitudinal adjustment in blue–green algae to different temperature regimes. Rates of fixation on a *Calothrix*-dominated rocky shore in Scotland (Stewart 1967*a*) were appreciable in February when temperatures averaged 0–10 °C, while Wiebe *et al.* (1975) found that nitrogen fixation ceased when temperatures became as low as 24 °C on a *Calothrix*-dominated coral reef flat.

Contradictory effects of salinity on nitrogen-fixation rates have been reported for *Calothrix* sp. Jones and Stewart (1969) found that cultures of *C. scopulorum* growing at different salinities attained maximum fixation rates at 0.086 M (the lowest examined) and that this was reduced to one-half after 12 days in 0.77 M sodium chloride. The effect was time-related; after three days incubation there was virtually no difference in the rates. Wiebe *et al.* (1975), using *Calothrix*-dominated coral reef flat rocks, found no effect on nitrogen-fixation rates between 0.05 and 0.77 M. In both of these studies the *Calothrix* was taken from intertidal or only slightly subtidal (reef flat) regions, where at low tide there can be essentially fresh water during heavy rains, or high salinities when evaporation increases salinity.

Carpenter and Price (1976) showed that, above a low threshold, increasing turbulence decreased the rate of nitrogen fixation in *Oscillatoria*. The effect appeared related directly to an increase of oxygen in the central portion of colonies where most of the nitrogen fixation was suspected of taking place and was caused by mechanically breaking up the bundles of cells. Wiebe *et al.* (1975) found that stirring the water during incubation of reef flat substrates dominated by *Calothrix* had no effect on the rate of nitrogen fixation, but that vigorous stirring increased the rate of fixation in the filamentous and tufted *Hormothamnion enteromorphoides* by about 50 per cent. They suggested that the effect might be related to increasing diffusion of nutrients into the tightly clumped filaments; the reef substrate contained a thin film of algae which presumably were not diffusion limited. Gotto and Taylor (1976) noted that turbulence under oxygenated conditions decreased the fixation rate of epiphytes on mangrove leaves. They showed that this was an oxygen effect; turbulence under anaerobic conditions did not affect the nitrogen-fixation rate.

There has been some work on the effect of the gaseous atmosphere composition in incubation chambers on nitrogen fixation. Patriquin and Knowles (1975), Carpenter and Price (1976), and Bohlool and Wiebe (1979) all showed that oxygen in the incubation atmosphere decreased the rate of fixation. Wynne-Williams and Rhodes (1974*b*) reported that small quantities of oxygen-stimulated nitrogen fixation of pure cultures of marine aerobic heterotrophs. Where oxygen depressed activity, the effect has been ascribed to the irreversible inhibition by oxygen on the nitrogenase complex as discussed in Chapters 1 and 2. There exists disagreement concerning the effect of nitrogen on the rates of acetylene reduction. Some workers (e.g. Wynne-Williams and Rhodes 1974*a*,*b*) sparge with argon or helium before running the acetylene assay. Others, e.g. Stewart, Fitzgerald, and Burris (1970), have shown that with adequate acetylene, the nitrogen in air does not compete with acetylene. As noted previously, Mague *et al.* (1977) cite some evidence for differential effects of nitrogen on acetylene reduction. Thus there are two different techniques for incubation: nitrogen-free and air. Both types of experiments appear necessary for any new system.

The third set of factors involve nutrient availability. *Oscillatoria* nitrogen-fixation rates have been shown to be stimulated by phosphate addition (Mague *et al.* 1974). Stewart (1971) noted stimulation with blue–green algae by addition of available iron. A variety of investigators have observed that organic substrates stimulate nitrogen fixation by bacteria, including Patriquin and Knowles (1975) with carbonate sands; Maruyama *et al.* (1974) for shallow and deep sediments; and Wynne-Williams and Rhodes (1974*a*) for coastal sediments.

One final factor that can affect the rates of nitrogen fixation is not easy to categorize: seasonality. We know of no seasonal studies on fixation in any non-estuarine environment except that of Wynne-Williams and Rhodes (1974*b*) and Hanson and Gundersen (1976*a*, 1977). In addition to contending with methodological, physical, and nutrient effects, our perception of the importance of nitrogen fixation in an environment may be affected by the seasonal occurrence of organisms as well. It is important that seasonal studies of nitrogen fixation be conducted in diverse marine environments.

Importance of nitrogen fixation in marine nitrogen budgets

Stewart (1971) in his summary of work on marine nitrogen fixation stated: 'Quantitative *in situ* data on marine nitrogen fixation are available only for *Trichodesmium* (*Oscillatoria*), *Calothrix* and *Nostoc*' [p. 559]. To a large extent this statement holds true today, and given the factors that affect the measurement of nitrogen-fixation rates, it is uncertain how truly quantitative many of the data are. Most workers who have calculated nitrogen budgets conclude that nitrogen added by fixation contributes from a few to perhaps 30 per cent of the calculated ecosystem needs. Most authors conclude their papers by stating that this amount of nitrogen fixation *may* provide an important source of nitrogen for marine communities. Only Mague *et al.* (1977) and Webb, DuPaul, Wiebe, Sottile, and Johannes (1975) found that the measured nitrogen-fixation rates could meet virtually 100 per cent of the requirements of their ecosystems. Ostrom (1976) attempted to use remote sensing techniques to calculate the effect of blooms of the nitrogen-fixing blue–green algae *Nodularia spumigena* on fertilization of the Baltic Sea. The blooms were massive and he related their occurrence to, possibly, the industrial and waste water release of phosphorus. He estimated approximately 2000 T N a^{-1} could be added to the Baltic Sea by this one organism. (He very carefully explained the speculative nature of this number.) This organism could contribute up to 20 per cent of the total nitrogen entering the Baltic Sea.

Two systems for which some nitrogen budget information is available are *Oscillatoria* blooms and coral reefs. Results for these systems are briefly summarized.

Oscillatoria

Dugdale, Menzel, and Ryther (1961) first measured nitrogen fixation in blooms of *Oscillatoria* off Bermuda using $^{15}N_2$. Dugdale, Goering, and Ryther (1964) studied blooms in the Arabian Sea, Indian Ocean, and Atlantic Ocean. They found occasional rates that were up to ten times greater than the rate of ammonia removal and concluded nitrogen fixation was responsible for considerable addition of nitrogen to the ecosystem. They, and subsequently others, e.g. Maruyama *et al.* (1970); Qasim (1970) cautioned that until bacteria-free cultures of *Oscillatoria* were examined, it was possible that bacteria, rather than *Oscillatoria*, were responsible for the nitrogen fixation. Indeed, Maruyama *et al.* (1970) found that almost all of the bacteria isolated from colonies of *Oscillatoria* were capable of growth on non-nitrogenous media. However, Carpenter and McCarthy (1975), using the same medium as Maruyama *et al.* (1970), isolated 50 cultures from *Oscillatoria* blooms and found that none showed positive nitrogen fixation. Taylor, Lee, and Bunt (1973) isolated bacteria from *Oscillatoria* blooms and concluded that there were far too few cells to account for the fixation rates. To date only one report of the laboratory cultivation of *Oscillatoria* has been published (Ramamurthy 1974) and other workers have not been successful in reproducing the results.

Recently, Carpenter and Price (1976) have explained how *Oscillatoria* can fix nitrogen without heterocysts; they showed that little carbon fixation took place in the middle of *Oscillatoria* colonies and thus little oxygen was generated locally. Also colony disruption in the presence (but *not* in the absence) of oxygen depressed the nitrogen-fixation rate. Previously, Fogg (1974) had speculated that nitrogen fixation might occur only in the centre of bundles. The Carpenter and Price (1976) study presents a strong case that *Oscillatoria* fixes nitrogen rather than associated bacteria. In addition, Carpenter and Price (1977) noted that only 2.2 per cent of the *Oscillatoria* trichomes in the Caribbean contained noticeable bacteria. Probably, however, the controversy will not end until axenic cultures of *Oscillatoria* are shown to fix nitrogen.

A variety of conclusions have been reached concerning the importance of *Oscillatoria* nitrogen fixation to the sea. Carpenter and McCarthy (1975) summarized the literature until the end of 1974, and stated that: 'as yet there exist no comprehensive data on the input of nitrogen to a large area of the sea via N_2 fixation' [p. 389]. They examined the nitrogen fixation and the incorporation of nitrate and ammonium into phytoplankton in the western Sargasso Sea. They concluded that 'NO_3^- and NH_4^+ are about 2000 times more significant than N_2 fixation by *O. thiebautii*' [p. 399] and that while *Oscillatoria* blooms may be important on a small scale, they exert a negligible effect on the Sargasso Sea as a whole. Subsequently, Carpenter and Price

(1977) re-examined nitrogen fixation in the western Sargasso Sea and also studied it in the Caribbean Sea. Their conclusions about its importance in the Sargasso Sea were identical to that just noted. However, in the Caribbean Sea, they found that *Oscillatoria* nitrogen fixation accounted for 8–20 per cent of the daily total phytoplankton requirement, comprised more than 50 per cent of the total chlorophyll *a* in the upper 200 m and was responsible for 20 per cent of the primary production. They commented that *Oscillatoria* spp were most abundant in surface slicks, most likely caused by Langmuir cell formation. (In this regard, it is interesting to examine the picture of long, wavy lines of *Oscillatoria* shown in Baas-Becking (1951). These discontinuities in local abundance must be accounted for in any quantitative sampling scheme.) Dunstan and Hosford (1978), working in the Georgia Bight of the south-eastern Atlantic Ocean, calculated that *Oscillatoria* could provide no more than 5 per cent of the new nitrogen required for overall primary production. Mague *et al.* (1977) examined *Oscillatoria* nitrogen fixation in the central North Pacific and found amounts up to about 33 μg of nitrogen m^{-2} day^{-1}, while Carpenter and Price (1977) estimated an average of about 100 μg of nitrogen m^{-2} day^{-1} in the Caribbean. These studies point out the potential danger of extrapolating data from one region to another.

Carpenter and Price (1976) found a very significant relationship between low wind speed and high population densities. Baas-Becking (1951) and Qasim (1970) similarly observed blooms in calm weather: Qasim (1970) noted that when the weather became cloudy, the bloom disappeared.

In summary, nitrogen fixation by *Oscillatoria* spp appears to be of variable importance for oceanic nitrogen budgets; major differences in rates between water masses have been documented. While axenic cultures have not been established, it is most likely that the *Oscillatoria* spp, not bacteria, are responsible for nitrogen fixation.

Coral reefs

The presence of heterocystous blue–green algae on coral reefs was noted by Baas-Becking (1951) and Wood (1963). Wood (1963) speculated that there might be a great deal of nitrogen fixation on reefs. Bunt, Cooksey, Heeb, Lee, and Taylor (1969, 1970) could find little fixation in Florida, Bahamas, or Virgin Island reefs. Their observations were all taken at 20 m or greater depth. Webb *et al.* (1975) examined the nitrogen flux across two transects of a windward coral reef at Enewetak Atoll, Marshall Islands. Ammonium, nitrate, dissolved organic nitrogen, and particulate organic nitrogen were measured. On one transect (II), the seaward half was largely an algal pavement while the downstream half was visually dominated by corals. In the other (Transect

III), benthic algae covered the entire length with a few scattered corals downstream. There was a significant export of nitrogen from this system, although the products released differed. On one Transect (II), net rates of combined export were 0.8 mM of nitrogen m^{-2} h^{-1} while Transect III yielded 0.29 mM of nitrogen m^{-2} h^{-1}. The rate of nitrogen release on Transect II implies a fixation rate of about 1000 kg ha^{-1} a^{-1}. Wiebe *et al.* (1975) subsequently investigated the rate of nitrogen fixation on this same transect and estimated an average nitrogen fixation rate of about 650 kg of nitrogen fixed ha^{-1} a^{-1}. Mague and Holm-Hansen (1975) in a nearby area of Enewetak measured rates that yielded up to 400 kg of nitrogen $ha^{-1}a^{-1}$. In both cases, *Calothrix crustaceae* was identified as the dominant organism. Potts and Whitton (1977) measured *C. crustaceae* fixation rates at Aldabra Atoll in the Indian Ocean and found rates comparable to those of Wiebe *et al.* (1975) and Mague and Holm-Hansen (1975). Both authors used a C_2H_4:N_2 conversion ratio of 3:1. Hanson and Gundersen (1976*b*) examined C_2H_4:N_2 ratios using $^{15}N_2$ and found a ratio of 3.3:1 of material on a dead coral head. Recently Burris (1976) reported an average ratio of 1.9:1 for benthic samples at Lizard Island, Great Barrier Reef. If this ratio is used to convert the Webb *et al.* (1975) data, the average value becomes 1050 kg nitrogen fixed ha^{-1} a^{-1}, exactly that predicted by the input–output study. Regardless of the exact value, it is clear that high fixation rates, approaching the highest ever reported for any community, occur on at least some coral reefs.

Lower rates for at least one coral reef have been reported. Burris (1976) calculated 6–30 kg nitrogen fixed ha^{-1} a^{-1}, an order of magnitude lower than Mague and Holm-Hansen (1975) and Wiebe *et al.* (1975). The dominant nitrogen-fixing organism was identified as *Scytonema.* The latter two groups of investigators had incubated whole rock communities, while Burris (1976) scoured the organisms off the rock, and resuspended them in water. Whether the difference in organisms, sample preparation, or seasonal factors contributed to the differences in values is not known.

In addition to these reports there have been other investigations of coral reef nitrogen fixation. Crossland and Barnes (1976) found light-stimulated acetylene reduction activity associated with skeletal portions of corals and on the surface of macro-algae. Activity in fine sediments was not light-stimulated. Hanson and Gundersen (1977) measured active nitrogen fixation in Kaneohe Bay, Hawaii. They found that blue–green algae on dead coral heads, predominantly *Calothrix* and *Nostoc*, could fix 760 kg nitrogen ha^{-1} a^{-1} and that in sediments, heterotrophic bacteria fixed 6 kg nitrogen ha^{-1} a^{-1}. They concluded that in the south bay, where sewage pollution increased the combined nitrogen in the water, nitrogen fixation was not affected. Capone, Taylor, and

Taylor (1977) measured nitrogen fixation associated with coral reef macro-algae in the Bahamas and found that a variety of blue–green algal–plant associations yield appreciable rates of nitrogen fixation, with the maximum rates being associated with *Microdictyon* sp. Capone and Taylor (1977) reported fixation rates in the phyllosphere of *Thalassia testudinum* could supply from 4–38 per cent of the nitrogen required for leaf production. Goldner (1978) has documented appreciable rates of nitrogen fixation on reef flats and macro-algae assemblages in the Bahamas and Jamaica.

6.5 Conclusions

In the numerous freshwater systems examined thus far it appears that nitrogen fixation can occur under diverse environmental conditions and constraints. These range from radiant-energy-rich oxygenated euphotic epilimnia to completely de-oxygenated hypolimnia and benthic regions.

With certainty, freshwater nitrogen fixation is limited to autotrophic (photosynthetic and chemo-autotrophic) and heterotrophic prokaryotes. Overall, the blue–green algae are quantitatively dominant, although anaerobic photosynthetic and heterotrophic bacteria and certain aerobic and facultatively aerobic bacteria may dominate nitrogen fixation when favourable conditions prevail.

The importance of nitrogen fixation in locally reduced microzones in and on particles, live organisms, and sediments residing in oxygen-rich waters remains to be established. Such microzones may yield ample energy supplies and oxygen-scavenging properties to allow nitrogen fixation to proceed. They are reminiscent of known symbiotic nitrogen-fixing associations where adequate energy sources and protection from oxygen inhibition are provided by a host organism.

The occurrence and regulation of freshwater nitrogen fixation can seldom be explained by one environmental variable. Most likely, a complicated interplay of physical, chemical and biological conditions is responsible for initiating and sustaining nitrogen-fixing genera. In short, our understanding of environmental factors controlling and enhancing nitrogen fixation is primitive in a great majority of lakes and streams.

In the open ocean and estuarine systems, nitrogen fixation appears localized. In aphotic oceanic waters little evidence of fixation exists and on theoretical grounds, little is expected. Eukaryotic–prokaryotic assemblages provide potential sites for fixation. With the exception of some *Oscillatoria* blooms, the *Nodularia spumigena* and the algal flats of coral reefs, most investigators have concluded that nitrogen fixation in the sea can account for, at most, only a few per cent of the fixed nitrogen required by plants. There is a need to evaluate the role of recycling in nitrogen budgets before any firm conclusions can be made, however. There is also an urgent need to evaluate the factors that may control nitrogen-fixation rates in nature and to learn more about the controls they exert on in-situ fixation rates. Finally, we still cannot answer the question that has been posed by many marine scientists: since fixed nitrogen appears to limit primary production in much of the sea, why isn't there much more nitrogen fixation taking place?

References

ALLISON, F. E., HOOVER, S. R. and MORRIS, H. J. (1937). *Bot. Gaz.* **98**, 433–63.
BAAS-BECKING, L. G. M. (1951). *Akad. Wetenschap. Proc. C* **54**, 213–25.
BARBER, R. T. and RYTHER, J. H. (1969). *J. exp. Mar. Biol. Ecol.* **3**, 191–9.
BATTERTON, J. C. and BAALEN, C. VAN (1971). *Arch. Mikrobiol.* **76**, 151–65.
BECKING, J. H. (1976). In *Proc. 1st Int. Symp. Nitrogen Fixation* (ed. W. E. Newton) pp. 551–67. Washington State University Press, Pullman.
BEIJERINCK, M. W. and DELDEN, A. VAN (1902). *Zentbl. Bakt. parasitKde Abt. II* **9**, 3–43.
BILLAUD, V. A. (1967). In *Environmental requirements of blue–green algae*, pp. 35–53. US Dept. Interior Fed. Control Admin., Northwest Region, Corvallis, Oregon.
BOHLOOL, B. B. and WIEBE, W. J. (1979). *Can. J. Microbiol.* in press.
BORTELS, H. (1940). *Arch. Mikrobiol.* **11**, 155–86.
BREZONIK, P. L. and HARPER, C. L. (1969). *Science, N.Y.* **164**, 1277–9.
BROUGHTON, W. J. and PARKER, C. A. (1978). In *Global impacts of applied microbiology* (ed. W. R. Stanton and E. Da Silva) Vol. V, p. 113–22 and Suppl. UNEP/UNESCO/ ICRO, Kuala Lumpur.
BULEN, W. A. and LECOMTE, J. R. (1966). *Proc. natn. Acad. Sci., U.S.A.* **56**, 979–86.
BUNT, J. S., COOKSEY, K. E., HEEB, M. A., LEE, C. C., and TAYLOR, B. F. (1969). In *Scientists in the sea* (ed. J. W. Miller, J. G. Van Derwalker, and R. A. Wallers) US Department of Interior, Washington, D.C.
—— —— —— —— —— (1970). *Nature, Lond.* **227**, 1163–4.
BURRIS, R. H. (1976). *Aust. J. Pl. Physiol.* **3**, 41–51.
CALVIN, M. and LYNCH, V. (1952). *Nature, Lond.* **169**, 455–6.
CAMERON, R. E. and FULLER, W. M. (1960). *Soil Sci. Soc. Am. Proc.* **24**, 353–6.
CAPONE, D. G. (1978). *Proc. Int. Coral Reef Symp.* **1**, 337–42.
—— and TAYLOR, B. F. (1977). *Mar. Biol.* **40**, 19–28.
—— TAYLOR, D. L. and TAYLOR, B. F. (1977). *Mar. Biol.* **40**, 29–32.
CARPENTER, E. J. (1973). *Deep Sea Res.* **20**, 285–8.
—— and CULLINEY, J. L. (1975). *Science, N.Y.* **187**, 551–2.
—— and MCCARTHY, J. J. (1975). *Limnol. Oceanogr.* **20**, 389–401.
—— and PRICE IV, C. C. (1976). *Science, N.Y.* **191**, 1278–80.
—— —— (1977). *Limnol. Oceanogr.* **22**, 60–72.
CARR, N. G. and WHITTON, B. A. (1973). In *The biology of blue–green algae* (ed. N. G. Carr and B. A. Whitton) Blackwell, Oxford.
CHALMERS, A. (1977). Salt marsh N pools. Ph.D. Thesis, University of Georgia.
CHRISTIAN, R. R., BANKROFT, K., and WIEBE, W. J. (1975). *Soil Sci.* **119**, 89–97.
COX, R. M. and FAY, P. (1969). *Proc. R. Soc. B* **172**, 357–66.
CROSSLAND, C. J. and BARNES, D. J. (1976). *Limnol. Oceanogr.* **21**, 153–5.
DALTON, H. (1974). *C.R.C. Critical Reviews* **3**, 183–251.
DE LA CRUZ, A. A. and GABRIEL, B. C. (1974). *Ecology* **55**, 882–6.
DIXON, R. O. D. (1972). *Arch. Mikrobiol.* **85**, 193–201.

Drouet, F. (1968). *Monogr. Acad. nat. Sci. Philad.* 15.
Dugdale, R. C., Goering, J. J., and Ryther, J. H. (1964). *Limnol. Oceanogr.* **9**, 507–10.
—— Menzel, D. W., and Ryther, J. H. (1961). *Deep Sea Res.* **7**, 297–300.
Dugdale, V. A. and Dugdale, R. C. (1962). *Limnol. Oceanogr.* **7**, 170–7.
Dunstan, W. M. and Hosford, J. (1978). *Bull. mar. Sci.* **27**, 824–9.
Fay, P. (1965). *J. gen. Microbiol.* **39**, 11–20.
—— and Kulasooriya, S. A. (1972). *Arch. Mikrobiol.* **87**, 341–52.
—— and Walsby, A. E. (1966). *Nature, Lond.* **209**, 94–5.
—— Stewart, W. D. P., Walsby, A. E., and Fogg, G. E. (1968). *Nature, Lond.* **220**, 810–12.
Flett, R. J. (1976). Nitrogen fixation in some Canadian Shield Lakes. Ph.D. Thesis, University of Manitoba, Winnipeg.
Fogg, G. E. (1942). *J. exp. Biol.* **19**, 78–87.
—— (1956). *A. Rev. Pl. Physiol.* **7**, 51–70.
—— (1968). *Photosynthesis*. English Universities Press, London.
—— (1974). In *Algal physiology and biochemistry* (ed. W. D. P. Stewart) pp. 560–82. University of California Press, Berkeley.
—— Stewart, W. D. P., Fay, P., and Walsby, A. E. (1973). In *The blue–green algae*. Academic Press, London.
Frank, B. (1889). *Ber. dt. bot. Ges.* **7**, 34–42.
Fujita, Y., Ohama, M., and Hattori, A. (1964). *Pl. Cell Physiol., Tokyo* **5**, 305–14.
Gallagher, J. L. (1971). Algal productivity and some aspects of the ecological physiology of the edaphic communities of Canary Creek tidal marsh. Ph.D. Thesis, University of Delaware.
—— (1975). *Am. J. Bot.* **62**, 644–8.
—— Linthurst, R. A., and Pfeiffer, W. J. (1980). *Ecology* in press.
—— Reimold, R. J., and Thompson, D. E. (1972). *Proc. 38th Ann. Meeting, Am. Soc. Photogrammetry*, Falls Church, Virginia, pp. 338–48.
Gallon, J. R., Kurz, W. G. W., and LaRue, T. A. (1975). In *Nitrogen fixation by free-living microorganisms* (ed. W. D. P. Stewart) p. 159. Cambridge University Press.
Geiselman, J. A. (1977). *Bull. Mar. Sci.* **27**, 821–4.
Goering, J. J. and Neess, J. C. (1964). *Limnol. Oceanogr.* **9**, 530–9.
—— and Parker, P. L. (1972). *Limnol. Oceanogr.* **17**, 320–3.
—— Dugdale, R. C., and Menzel, D. W. (1966). *Limnol. Oceanogr.* **11**, 614–20.
Gogotov, J. N. and Schlegel, H. G. (1974). *Archs Microbiol.* **97**, 359–62.
Goldman, C. R. (1961). *Ecology* **42**, 282–8.
—— (1964). *Verh. int. Verein. theor. angew. Limnol.* **15**, 365–74.
Goldner, L. L. (1978). Nitrogen fixation in shallow water marine environments of the West Indies. Unpublished MS.
Gotto, J. W. and Taylor, B. F. (1976). *Appl. Environ. Microbiol.* **31**, 781–3.
Granhall, U. and Lundgren, A. (1971). *Limnol. Oceanogr.* **16**, 711–19.
Guerinot, M. L., Fong, W., and Patriquin, D. G. (1977). *J. Fish. Res. Bd Can.* **34**, 416–20.
Haines, E., Chalmers, A., Hanson, R., and Sherr, B. (1977). In *Estuarine processes*, Vol. II, pp. 241–54. Academic Press, London.
Hanson, R. B. (1977*a*). *Appl. Environ. Microbiol.* **33**, 596–602.
—— (1977*b*). *Appl. Environ. Microbiol.* **33**, 846–52.
—— and Gundersen, K. R. (1976*a*). *Pacif. Sci.* **30**, 385–93.
—— —— (1976*b*). *Appl. Environ. Microbiol.* **31**, 942–8.
—— —— (1977). *Estuar. Coastal Mar. Sci.* **5**, 437–44.
Hartwig, E. O. and Stanley, S. O. (1978). *Deep Sea Res.* **25**, 411–18.
Head, W. D. and Carpenter, E. J. (1975). *Limnol. Oceanogr.* **20**, 815–23.

HERBERT, R. A. (1975). *J. exp. Mar. Biol. Ecol.* **18**, 215–26.
HERRISET, A. (1952). *Bull. Soc. Chim. biol.* **34**, 532–7.
HITCH, C. J. B. and STEWART, W. D. P. (1977). *New Phytol.* **72**, 509–24.
HOLM-HANSEN, O., GERLOFF, G. C., and SKOOG, F. (1954). *Physiologia Pl.* **7**, 665–75.
HORNE, A. J. and FOGG, G. E. (1970). *Proc. R. Soc. B* **175**, 351–66.
—— and GOLDMAN, C. R. (1972). *Limnol. Oceanogr.* **17**, 678–92.
—— and VINER, A. B. (1971). *Nature, Lond.* **232**, 417–18.
—— DILLARD, J. E., FUJITA, D. K., and GOLDMAN, C. R. (1972). *Limnol. Oceanogr.* **17**, 693–703.
HOWARD, D. L., FREA, J. I., PFISTER, R. M., and DUGAN, P. R. (1970). *Science, N.Y.* **169**, 61–92.
HUTCHINSON, G. E. (1957). In *A treatise on limnology*, Vol. 2; *Introduction to lake biology and the limnoplankton.* John Wiley, New York.
JAKOB, H. (1954). *C. r. hebd. Séanc. Acad. Sci., Paris* **238**, 928–30.
JANNASCH, H. W. and PRITCHARD, P. H. (1972). In *Detritus and its role in aquatic ecosystems. Proc. IBP-UNESCO Symp.* Pallanza, Italy (ed. U. Melchiarri-Santollini and J. W. Hopton). *Mem. 1st Ital. Idrobiol.* **29**, Suppl.
JONES, K. (1974). *J. Ecol.* **62**, 553–65.
—— and STEWART, W. D. P. (1969). *J. mar. biol. Ass. U.K.* **49**, 475–88.
KAMEN, M. D. and GEST, H. W. (1949). *Science, N.Y.* **109**, 560–1.
KATOR, H. and HERWIG, R. (1976). In *Proc. 1976 Conf. Prevention and Control of Oil Pollution.* American Petroleum Institute, Washington, DC.
KAWAI, A. and SUGAHARA, I. (1971*a*). *Bull. Jap. Soc. scient. Fish.* **37**, 747–54.
—— —— (1971*b*). *Bull. Jap. Soc. scient. Fish.* **37**, 981–5.
—— —— (1972). *Bull. Jap. Soc. scient. Fish.* **38**, 291–7.
KEEFE, C. W. and BOYNTON, W. R. (1973). *Chesapeake Sci.* **14**, 117–23.
KEIRN, M. A. and BREZONIK, P. L. (1971). *Limnol. Oceanogr.* **16**, 720–31.
KELLAR, P. E. (1978). M.Sc. Thesis, University of California, Davis.
KENYON, C. N., RIPPKA, R., and STANIER, R. Y. (1972). *Arch. Mikrobiol.* **83**, 216–36.
KNOWLES, R. and WISHART, C. (1977). *Environ. Pollut.* **13**, 133–49.
KOWALSKI, M. (1978). M.Sc. Thesis, Virginia Institute of Marine Science, Gloucester Point, Virginia.
KUHL, A. (1968). In *Algae, man and environment* (ed. D. F. Jackson) pp. 37–52. Syracuse University Press.
KUZNETZOV, S. I. (1959). In *Die Rolle der Mikroorganismen im Stoffkreislauf der Seen*, Deutscher Verlag der Wissenschaften, Berlin.
—— (1968). *Limnol. Oceanogr.* **13**, 211–24.
LANG, N. J. and FAY, P. (1971). *Proc. R. Soc. B* **178**, 193–203.
LANGE, W. (1971). *Can. J. Microbiol.* **17**, 304–14.
LINDSTROM, E. S., BURRIS, R. H., and WILSON, P. W. (1949). *J. Bact.* **58**, 313–16.
—— LEWIS, S. M., and PINSKY, M. J. (1951). *J. Bact.* **61**, 481–7.
MACRAE, I. C. (1977). *Aust. J. biol. Sci.* **30**, 593–6.
MCROY, P., GOERING, C., and CHANEY, J. J. (1973). *Limnol. Oceanogr.* **18**, 998–1002.
MAGUE, T. H. and HOLM-HANSEN, O. (1975). *Phycologia* **14**, 87–92.
—— MAGUE, F. C., and HOLM-HANSEN, O. (1977). *Mar. Biol.* **41**, 213–27.
—— WEARE, N. W., and HOLM-HANSEN, O. (1974). *Mar. Biol.* **41**, 109–19.
MARSHO, T. V., BURCHARD, R. P., and FLEMING, R. (1975). *Can. J. Microbiol.* **21**, 1348–56.
MARUYAMA, Y., SUZUKI, T., and OTOBE, K. (1974). In *Effect of the ocean environment on microbial activities. Proc. 2nd United States–Japan Conf. Marine Microbiology*, Baltimore (ed. R. R. Colwell and R. Y. Morita) pp. 341–53.
—— TAGA, N. and MATSUDA, O. (1970). *J. oceanogr. Soc. Japan* **26**, 360–6.

MENDELSSOHN, I. A. and MARCELLUS, K. (1976). *Chesapeake Sci.* **17**, 15–23.
MULDER, E. G. (1975). In *Nitrogen fixation by free-living microorganisms* (ed. W. D. P. Stewart) p. 3. Cambridge University Press.
MURPHY, T. P., LEAN, D. R. S., and NALEWJKO, C. (1976). *Science, N.Y.* **192**, 900–2.
NALEWJKO, C., DUNSTALL, T. G., and SHEAR, H. (1976). *J. Phycol.* **12**, 1–5.
NAUWERCK, A. (1963). *Symb. bot. Upsal.* **17**, 1–163.
NIXON, S. W. and OVIATT, C. A. (1973). *Ecol. Monogr.* **43**, 463–98.
ODUM, E. P. and DE LA CRUZ, A. A. (1967). In *Estuaries* (ed. G. H. Lauff) pp. 383–8. American Association for the Advancement of Science.
OSTROM, B. (1976). *Remote Sensing Environ.* **4**, 305–10.
PAERL, H. W. (1976). *J. Phycol.* **12**, 431–5.
—— (1978). *Oecologia* **32**, 135–9.
—— (1979). *Oecologia* **38**, 275–90.
—— and KELLAR, P. E. (1978). *J. Phycol.* **14**, 254–60.
PATRIQUIN, D. G. (1972). *Mar. Biol.* **15**, 35–46.
—— and KNOWLES, R. (1972). *Mar. Biol.* **16**, 49–58.
—— —— (1975). *Mar. Biol.* **32**, 49–62.
PATTNAIK, H. (1966). *Ann. Bot.* **30**, 231–8.
PEARSALL, W. H. (1932). *J. Ecol.* **20**, 241–62.
PETERS, G. A. and MAYNE, B. C. (1974). *Plant Physiol., Lancaster* **53**, 820–4.
PETERSON, R. B. and BURRIS, R. H. (1976). *Analyt. Biochem.* **73**, 404–10.
—— FRIBERG, E. E., and BURRIS, R. H. (1977). *Plant Physiol., Lancaster* **59**, 74–80.
PFENNIG, N. (1967). *A. Rev. Microbiol.* **21**, 285–324.
POSTGATE, J. R. (1971*a*). In *The chemistry and biochemistry of nitrogen fixation* (ed. J. R. Postgate) pp. 161–90. Plenum Press, London.
—— (1971*b*). *Symp. Soc. gen. Microbiol.* **21**, 287–307.
POTTS, M. and WHITTON, B. A. (1977). *Oecologia* **27**, 275–84.
PRICE, C. A. (1968). *A. Rev. Pl. Physiol.* **19**, 239–48.
QASIM, S. Z. (1970). *Deep Sea Res.* **17**, 655–60.
RALPH, R. D. (1975). Blue–green algae of the shores and marshes of southern Delaware. Coll. Mar. Stud. Publ. No. Del. SG 20-75. University of Delaware, Newark.
RAMAMURTHY, V. D. (1974). *Sci. Cult.* **40**, 500–1.
REYNOLDS, C. S. (1971). *Fld Stud.* **3**, 409–32.
RIPPKA, R., NEILSON, A., KUNISAWA, R., and COHEN-BAZIRE, G. (1971). *Arch. Mikrobiol.* **76**, 341–8.
SCHINDLER, D. W. (1974). *Science, N.Y.* **184**, 897–8.
SCHMIDT-LORENZ, N. and RIPPEL-BALDES, A. (1957). *Mikrobiologiya* **28**, 45–68.
SCHOPF, J. W. (1970). *J. Paleontol.* **44**, 1–6.
SHAH, V. K., DAVIS, L. C., and BRILL, W. J. (1975). *Biochim. biophys. Acta* **384**, 353–9.
SHAPIRO, J. (1973). *Science, N.Y.* **179**, 382–4.
SHUBERT, K. R. and EVANS, H. J. (1976). *Proc. natn. Acad. Sci. U.S.A.* **73**, 1207–11.
SIMPSON, F. B. and NEILANDS, J. B. (1976). *J. Phycol.* **12**, 44–8.
SINGH, R. N. (1955). *Verh. int. Verein. theor. angew. Limnol.* **12**, 831–6.
SIRENKO, L. A., STENTSENKO, N. M., ARENDARCHUK, V. V., and KUZMENKO, M. I. (1968). *Mikrobiologia* **37**, 199–202.
SQUIERS, E. R. and GOOD, R. E. (1974). *Chesapeake Sci.* **15**, 63–71.
STACEY, G., VAN BAALEN, C., and TABITA, F. R. (1977). *Archs. Microbiol.* **114**, 197–201.
STANIER, R. Y. and COHEN-BAZIRE, G. (1977). *A. Rev. Microbiol.* **31**, 225–74.
STEWART, W. D. P. (1962). *Ann. Bot.* **26**, 439–47.
—— (1964). *J. gen. Microbiol.* **36**, 415–22.
—— (1965). *Ann. Bot.* **29**, 229–39.
—— (1967*a*). *Ann. Bot.* **31**, 385–407.

—— (1967*b*). *Nature, Lond.* **214**, 603–4.
—— (1968). In *Algae, man and environment* (ed. D. F. Jackson) pp. 53–74. Syracuse University Press.
—— (1969). *Proc. R. Soc. B* **172**, 367–88.
—— (1970*a*). *Phycologia* **91**, 261–8.
—— (1970*b*). *Pl. Soil* **32**, 555–88.
—— (1971). In *Fertility of the sea* (ed. J. D. Coslow) pp. 537–64. Gordon and Breach, London.
—— (1973). *A. Rev. Microbiol.* **27**, 283–316.
—— (1977*a*). *Ambio* **6**, 166–73.
—— (1977*b*). *Br. Phycol. J.* **12**, 89–115.
—— and Alexander, G. (1971). *Freshwater Biol.* **1**, 389–404.
—— and Pearson, H. W. (1970). *Proc. R. Soc. B* **175**, 293–311.
—— and Pugh, G. C. F. (1963). *J. mar. Biol. Ass. U.K.* **43**, 309–17.
—— Fitzgerald, G. P., and Burris, R. H. (1967). *Proc. natn. Acad. Sci. U.S.A.* **58**, 2071–8.
—— —— —— (1970). *Proc. natn. Acad. Sci. U.S.A.* **66**, 1104–11.
Sugahara, I., Sawada, T., and Kawai, A. (1971). *Bull. Jap. Soc. scient. Fish.* **37**, 1093–9.
Sullivan, M. J. and Daiber, F. C. (1974). *Chesepeake Sci.* **15**, 121–3.
Taylor, B. F., Lee, C. C., and Bunt, J. S. (1973). *Archs. Microbiol.* **88**, 205–12.
Tel-Or, E. and Stewart, W. D. P. (1976). *Biochim. biophys. Acta* **423**, 189–95.
—— Luick, L. W., and Packer, L. (1977). *FEBS Lett.* **78**, 49–52.
Thomas, J. (1970). *Nature, Lond.* **228**, 181–3.
Thomson, A. D. (1977). M.Sc. Thesis, Virginia Institute of Marine Science, Gloucester Point, Virginia.
Tjepkema, J. D. and Evans, H. J. (1976). *Soil Biol. Biochem.* **8**, 505–9.
Tsuda, R. T. (1977). *Bull. Jap. Soc. Phycol.* **25**, 155–8.
Valiela, I. and Teal, J. M. (1974). In *Ecology of halophytes* (ed. R. J. Reinmold and W. H. Queen) pp. 547–63. Academic Press, New York.
—— —— and Persson, N. Y. (1976). *Limnol. Oceanogr.* **21**, 245–52.
—— —— and Sass, W. (1973). *Est. Coastal Mar. Sci.* **1**, 261–9.
—— —— —— (1975). *J. appl. Ecol.* **12**, 973–81.
—— Vince, S., and Teal, J. M. (1976). In *Estuarine Research* (ed. M. Wiley) Vol. 1, pp. 234–53. Academic Press, New York.
Vance, B. D. (1966). *J. Phycol.* **5**, 1–3.
Van Gemerden, H. (1974). *Microbial Ecol.* **1**, 109–14.
Vankataraman, G. S. (1961). *Indian J. agric. Sci.* **31**, 213–15.
Van Raalte, C., Valiela, I., and Teal, J. M. (1976). *Limnol. Oceanogr.* **21**, 862–72.
—— —— Carpenter, E. J., and Teal, J. M. (1974). *Est. Coastal Mar. Sci.* **2**, 301–5.
Volk, S. L. and Phinney, H. K. (1968). *Can. J. Bot.* **46**, 619–30.
Vollenweider, R. A. (1968). In Scientific fundamentals of the eutrophication of lakes and flowing waters with particular reference to nitrogen and phosphorus as factors in eutrophication. OECD Directorate for Scientific Affairs, Paris.
Waksman, S. A., Hotchkiss, M., and Carey, C. L. (1933). *Biol. Bull.* **65**, 137–67.
Walsby, A. E. (1968). *New Scient.* **40**, 436–7.
—— (1974). *Br. Phycol. J.* **9**, 371–81.
Watanabe, A. (1951). *Archs. Biochem. Biophys.* **34**, 50–5.
—— Shirota, M., Endo, H., and Yamamoto, Y. (1971). In *Proc. 3rd Int. Conf. Global Impacts of Applied Microbiology*, pp. 53–64. University of Bombay.
Webb, K. L., DuPaul, W. D., Wiebe, W. J., Sottile, W. and Johannes, R. E. (1975). *Limnol. Oceanogr.* **20**, 198–210.
Webber, E. (1967). Blue–green algae from a Massachusetts salt marsh. *Bull. Torrey bot.*

Club **94**, 99–106.
WHITNEY, E. E., WOODWELL, G. M., and HOWARTH, R. W. (1975). *Limnol. Oceanogr.* **20**, 640–3.
WHITTON, B. A. (1973). In *The biology of blue–green algae* (ed. N. G. Carr and B. A. Whitton). Blackwell, Oxford.
WIEBE, W. J., JOHANNES, R. E., and WEBB, K. L. (1975). *Science, N.Y.* **188**, 257–9.
WINKENBACK, F. and WOLK, C. P. (1973). *Pl. Physiol., Lancaster* **52**, 480–3.
WOLK, C. P. (1973). *Bact. Rev.* **37**, 32–101.
—— and WOJCIUCH, E. (1971). *J. Phycol.* **7**, 339–44.
—— THOMAS, J., SHAFFER, P. W., AUSTIN, S. M., and GABONSKI, A. (1976). *J. biol. Chem.* **251**, 5027–34.
WOOD, E. F. J. (1963). In *Marine microbiology* (ed. C. H. Oppenheimer) pp. 28–39. C. C. Thomas, Springfield.
—— (1965). In *Marine microbiol ecology*. Chapman and Hall, London.
WYATT, J. T. and SILVEY, J. K. G. (1969). *Science, N.Y.* **165**, 908–9.
WYNNE-WILLIAMS, D. D. and RHODES, M. E. (1974*a*). *J. appl. Bact.* **37**, 217–24.
—— —— (1974*b*). *J. appl. Bact.* **37**, 203–16.
ZUBERER, D. A. and SILVER, W. S. (1978). *Appl. Environ. Microbiol.* **35**, 567–75.

7 Paddy fields

I. WATANABE AND S. BROTONEGORO

7.1 Flooded rice soils as habitats of nitrogen-fixing organisms

Paddy fields are managed in a special way for the wet cultivation of rice. During most of the growth period the fields are submerged. Before the rice is planted, levelling the land and puddling is performed, which greatly affects the biochemical and biological reactions during the flooding. Before or after harvest, the water is generally drained, but in an ill-drained field, the flooding sometimes continues longer than the growth period of rice. In well-drained and irrigated rice fields, the depth of flood water is maintained at 5–10 cm, but the depth of water can reach 2 to 5 m as found in deep-water rice areas. On the other hand, in rain-fed rice areas, the rice is subjected to frequent moisture stress during its growth. Principal characteristics of the paddy rice fields therefore originate from the flooding of the soil and the presence of rice plant.

The most important change caused by flood is aeration of the soil. As oxygen moves ten thousand times slower through a water phase than through a gaseous phase, the capacity of a soil to exchange gases with the atmosphere will decrease as the soil becomes water saturated. Waterlogging of a soil will, therefore, lead quickly to anaerobic conditions. Anaerobic soil conditions develop a few millimetres beneath the surface soil, as the oxygen supply in this layer is already insufficient for respiring micro-organisms in the soil, but the surface soil is in an oxidized state. Differentiation of the two layers was first recognized by the limnologists (Pearsall and Mortimer 1939) and the implications for rice culture were recognized by the Japanese scientists Shioiri (1941) and Mitsui (1954). The overlying flood water is also an important site for the growth of organisms. Rice plants themselves provide another niche for microbial activity, including root and the rhizosphere.

In short, the environments in paddy fields are divided into four sites: (i) flood water, (ii) surface oxidized soil, (iii) reduced soil, and (iv) rice root and the rhizosphere. The reader is referred to reviews on the chemistry of the submerged soil (Ponnamperuma 1972) and the microbiology of the flooded soil (Yoshida 1975; Brotonegoro 1977).

Flood water

The flood water is inhabited with aquatic communities including prokaryotic and eukaryotic algae, aquatic weeds, and heterotrophic microflora and fauna. There are intense interactions among members of an aquatic community (Kikuchi, Furusaka, and Kurihara), but little is known about their nature. During the day, photosynthetic activities surpass heterotrophic activities and as a result, oxygen concentration and pH increase and free carbon dioxide concentration decreases. At night the reverse reactions occur. Diurnal changes in pH and carbonic acid components have been studied by Mikkelsen, De Datta, and Obcemea (1978). They showed that pH increases up to 9.5 during daytime in neutral Maahas soil (soil pH 6.7) and 7.5 in acidic Luisiana soil (pH 5.0) in the Philippines. Oxygen evolution and consumption were studied in neutral Maahas soil by Saito and Watanabe (1978). Daytime oxygen concentration rose to 20 p.p.m. and fell to 3 p.p.m. at night. Using diurnal curves of oxygen content in the flood water, Saito and Watanabe calculated net production in an aquatic community of flood water. As the rice plant grows, diurnal changes become less marked due to the shading by plant canopy.

Surface soil

The surface soil is maintained at a redox potential of higher than + 300 mV. The depth of this layer is dependent upon reducing capacities or the oxygen consuming capacity of soil owing to microbial respiration and Fe^{2+} (Howeler and Bouldin 1971). This ranges from 2 to 20 mm. In this layer, NO_3^-, Fe^{3+}, SO_4^{2-}, and carbon dioxide are stable (Pearsall and Mortimer 1939). Aerobic bacteria predominate (Ishizawa and Toyota 1960; Hayashi, Asatsuma, Nagatsuka, and Furusaka 1978). It has been reported that oxygen consumption by nitrifying bacteria accounted for about 40 per cent of oxygen uptake at the surface in the submerged soil after the intense oxygen uptake which occurs immediately after flooding (Patrick and Reddy 1976).

Anaerobic soil

In anaerobic soil, reduction process predominate. E_h is usually less than 300 mV and falls down to − 200 mV. Takai and Kamura (1966) divided the reducing process of paddy soil into two stages: one before and the other after the reduction of iron is completed. In the first stage absorption of oxygen, nitrate reduction, manganese reduction, and iron reduction proceed in this order as the reduction goes on. Ammonia and carbon dioxide are liberated. In the second stage, organic acids, methane, hydrogen, and hydrogen sulphide are produced and the population of anaerobic bacteria increases. Formic, acetic, propionic, and butyric acids are mainly detected in reduced paddy soils (Takijima 1963; Chandrasekaran and Yoshida 1973). Methane formation is accompanied by the decrease of organic acids and carbon dioxide. The pH values of acid soils increase on submergence, while those of calcareous soils decrease. In most soils, the fairly stable pH attained after several weeks of flooding is between 6.5 to 7.0 (Ponnamperuma 1972). On draining the soil, the pH returns to its original value. The anaerobic bacteria in the anaerobic paddy soil were studied by Furusaka and his associates (Furusaka 1977; Hayashi *et al.* 1978). The population of anaerobic bacteria before submergence was 2×10^5 and increased up to 1×10^6 after submergence, then decreased to the level of $3–4 \times 10^5$ per g dry soil. Almost 70 per cent of the isolates are facultative anaerobes belonging to the genera *Aeromonas*, *Bacillus*, *Erwinia*, *Escherichia*, and *Streptococcus*. The population of strict anaerobes was nearly constant throughout the season at about $7 \times 10^4–3 \times 10^5$ per g dry soil. Almost 85 per cent of the isolates were *Clostridium*. To explain the flora composition of anaerobic soil, Nagatsuka and Furusaka (1976) studied the composition of aerobic, facultative anaerobic, and strict anaerobic bacteria in soil suspensions which were exposed continuously to 159 mm Hg, 1–2 mm Hg, and 0 mm Hg oxygen tension. At higher oxygen tensions, strict anaerobes did not grow, while in the absence of oxygen, strict anaerobes predominated. Composition of three groups at 1–2 mm Hg oxygen tensions resembled that found in anaerobic layers of paddy fields. Perhaps the microbial environment in anaerobic layers in the paddy fields is not strictly anaerobic. The reasons are not well understood, but the presence of rice roots excreting oxygen and the interchange of aerobic and anaerobic soil by the movement of soil fauna (Kikuchi, Furusaka, and Kurihara 1977), might be factors in regulating soil oxygen tensions. Furusaka's group (Furusaka 1977) have found higher percentages of strict anaerobes by increasing the incubation period for counting anaerobic bacteria. Longer incubations revealed the presence of *Propionibacterium*-type anaerobes at higher numbers than *Clostridium*. Still a half of anaerobes were of the facultative type. Strictly anaerobic bacteria as the predominant organism in paddy soils is unlikely.

Rice root and the rhizosphere

The old concept of the rhizosphere suggests that the plant exudes organic substances on which soil micro-organisms grow and, consequently, the root in the soil is surrounded by the micro-organisms. It now seems, however, that soil adjacent to the root (rhizosphere, in the strict sense), mucilagenous layers developed on the surface of the epidermis, intercellular spaces among epidermis layers, and the inner tissues of the epidermis and cortex are also inhabited by microbial colonies and these spaces provide a more or less continuous media for their activities (Foster and Rovira 1976; Old and Nicolson 1975). Secretion of carbon compounds by roots not only provides energy sources for microbial growth but is greatly accelerated by the presence of micro-organisms (Martin 1977; Barber and Lynch 1977). Invasion of bacteria and other fungi and protozoa in wetland rice roots was quite frequently observed at later stages of rice growth under the light microscope (Miyashita, Wada, and Takai 1977).

MacRae and Castro (1966) reported that rice seedlings excrete various amino acids and carbohydrates through the roots. As in other marsh plants, roots of rice plants receive oxygen from aerial parts of the plant. This essential element is usually transported so efficiently that some oxygen is excreted into the soil. The brownish colour of rice roots indicates the oxidation of ferrous iron to ferric and its precipitation along the root surface (Van Raalte 1940; Armstrong 1969; Arashi and Nitta 1955). By the same mechanism, nitrogen gas (Yoshida and Broadbent 1975) and acetylene (Lee and Watanabe 1977) are transported to the rooting medium from the atmosphere. The only aerobic bacteria found in the rice rhizosphere were sulphur-oxidizing ones (Joshi and Hollis 1977). Rice roots grown in submerged soil have fewer and shorter root hairs, which are straighter than those found in dry soil (Kawata, Ishihara, and Shioya 1964). Rhizosphere to soil ratios (R/S) of various groups of bacteria in the rice rhizosphere have been studied by Kimura, Wada, and Takai (1977). They were generally lower in wetland rice than in upland rice. There must be a difference in the amount, diffusion, and utilization of the root exudate between rice grown in submerged soil and in dryland soil, but little is known about these factors.

7.2 Importance of nitrogen fixation in wetland rice

Relative yields of non-nitrogen-fertilized plots to those receiving nitrogen, phosphorus, and potassium fertilizer are higher in wetland rice than in dryland rice (Mitsui 1954). Rice yield can be maintained at considerable levels without fertilizer nitrogen. This fact indicates that in the flooded condition nitrogen fertility is kept higher. Knowing this De (1936) studied nitrogen-fixing blue–green algae in paddy soils and attributed nitrogen fertility to these organisms in the flood water.

To correctly estimate the nitrogen gains in submerged rice soils, nitrogen balance studies are essential. Despite many studies of nitrogen balance in dryland soils (Allison 1965), there is a dearth of data for wetland rice soils. Two nitrogen balance sheets with an almost complete set of nitrogen credit and debit data are available in Japan, and these data show nitrogen gains of 15–38 kg ha^{-1} a^{-1} (Koyama and App 1979). In both cases, rice was followed by fallow (Aomori) or rye (Kagawa). Low nitrogen gains in both cases might have been due to the dryland condition after rice. Okuda (1948) conducted lysimeter experiments with dry fallow and with continuous flooding. Rice took up more nitrogen from continuously flooded plots than from dry fallow plots, but soil nitrogen contents after five years were higher in the continuously flooded plots. In the flooded condition, nitrogen accumulates in the soil, as has been found at IRRI. In Table 7.1 some data of nitrogen balance in the flooded soils are shown. As the data of De and Sulaiman (1950*a*) in Table 7.1 shows quite a large amount of nitrogen was fixed if algal growth is enhanced. Similar results were obtained in IRRI pot experiments. Consequently, it seems that flooded soil culture has an advantage of keeping and increasing soil nitrogen fertility by utilizing non-symbiotic nitrogen fixation.

7.3 Heterotrophic nitrogen-fixing micro-organisms

Distribution

The distribution of *Azotobacter* has been studied extensively. Yamagata (1924) reported higher frequencies in neutral soils in Japan and slightly higher frequencies in paddy soils than in dryland soils. Ishizawa, and Toyota (1960) also reported higher frequencies of *Azotobacter* in paddy soils than in dryland soils. This difference may be explained by the higher pH of paddy soils; yet *Azotobacter* populations seldom exceed 10^4 per gram. Uppal, Patel, and Daji (1939), Rangaswami and Subbaraja (1962), and Rouquerol (1962) studied the seasonal changes of *Azotobacter* populations which showed great fluctuations from 10 to 10^6 per gram or more. Flooding the soil decreased the population in the work of Rangaswami *et al.*, but the reverse was true in Rouquerol's case.

Ishizawa, Suzuki, and Araragi (1975) found few differences in the population between the surface (0–1 cm) and in the lower soil (1–4 cm). In Egypt and Iraq where alkaline soils predominate, *Azotobacter* is abundant. In Iraqi soils, the average population estimated by most probable number (MPN) methods is 10^6 per gram soil (Hamdi, Yousef, Al-Azawi, Al-Tai, and Al-Baquasi 1978). Lowest average populations were found in rice soils, in non-saline or moderately saline soils, and in slightly alkaline soils. In Egypt, the population was more than 10^5 per gram (Mahmoud, El-Sawy, Ishac, and El-Safty 1978) and in rice soils the population decreased from 10^6 to 10^5 during flooding (Abd-el-Malek 1971).

The higher counts in Arabic countries may be partly due to the

TABLE 7.1

Nitrogen balance in wetland rice-soil system

Period	Field or pot	Condition and soil	Analysis†			Balance		Assumption	Reference
			Soil before	Soil after	plant	mg kg soil^{-1} for whole period	kg ha^{-1} (period)		
21 years (21 crop)	Field	Aomori, Japan							
		No nitrogen	A	A	A		14.9 (yr)		Koyama and App (1979)
		Chemical N					−35.4 (yr)		
21 years (42 crops)	Field	Kagawa, Japan							
		No nitrogen	A	A	A		+ 37.7 (yr)		Koyama and App (1979)
		Chemical N					− 28.4 (yr)		
22 years (22 crops)	Field	Ishizawa, Japan							
		Non-fertilizer	A	A	E		+ 19 (yr)	(1)	Konishi and Seino (1961)
		PK plot					+ 34 (yr)		
		PKCa plot					+ 39 (yr)		
		NPKCa plot					+ 0.4 (yr)		
12 years (24 crops)	Field	Maahas, Philippines							
		Non-fertilizer	A	A	E		+ 50 (crop)	(2)	Koyama and App (1979)
8.5 years (17 crops)	Field	Maligaya, Philippines							
		Non-fertilizer	A	A	E		+ 35 (crop)		

5 years (5 crops)	Pot	Furidpur, Bangladesh				(only in soil)	(including plant uptake)		De and Sulaiman (1950*a*)
		Nutrient + light	A	A	E	+ 180	119 (yr)	(3)	
		Nutrient + dark				− 45	30 (yr)		
		Tippera, Bangladesh							
		Nutrient + light	A	A	E	+ 220	250 (yr)		
						− 80	72 (yr)		
120 days	Pot	Crowly silt loam, USA							
		PK + plant	A	A	A	+ 36.5	72 (120 d)	(4)	Willis and Green (1948)
		PK − plant				+ 27.7	54 (120 d)		
		Lake Charles clay loam, USA							
		PK + plant				+ 28.1	56 (120 d)		
		PK − plant				+ 9.6	−19 (120 d)		
6 crops (480 days)	Pot	Maligaya soil, Phil.							
		Planted, light				+ 56.0 ± 7.4	18.1 (crop)	(5)	IRRI (unpublished)
		Planted, dark	A	A	A	+ 16.5 ± 5.2	5.3 (crop)		
		Planted, fallow				+ 17.0 ± 8.5	5.6 (crop)		

Assumption: (1) One tonne grains absorb 25 kg N; (2) N% in grain is 1.1% and in straw 0.6%, grain and straw ratio 1:0.67; (3) To calculate plant N-uptake = N% in grain (1.2%), straw N (0.7%), soil in 1 ha = 2×10^9 g; (4) 1 ha = 2.0×10^9 g; (5) 1 ha = 2.5×10^5 pots.
† A: analysed; E: Estimated.

alkalinity of the soil and to the MPN method that workers in these countries adopted. MPN methods give higher counts than plate methods (Abd-el-Malek 1968).

Acid-tolerant *Beijerinckia* was first isolated by Starkey and De (1939) from rice soils in Bangladesh. Derx (1950), Becking (1961), and others (see Becking 1978) surveyed the occurrence of *Beijerinckia* and pointed out that these bacteria are mainly tropical. Derx (1950) showed a sharp drop in numbers of *Beijerinckia* below the first 5 cm of top soil in irrigated rice soils. However, Becking reported much lower populations of *Beijerinckia* in 1971–2 in the lateritic paddy soils, where Derx found quite higher numbers in 1948–9. This has been ascribed to the heavy applications of pesticides since then (Becking 1978).

Ishizawa *et al.* (1975) reported quite high frequencies of *Beijerinckia* in Japanese wetland and dryland soils. Suto (1954) also reported the presence of *Beijerinckia* in phosphate-deficient volcanic ash soils. The actual significance of these heterotrophic aerobic nitrogen-fixing organisms in nitrogen fixation of paddy soils therefore remains to be elucidated.

Dilution or plate counts sometimes give counts of aerobic nitrogen-fixing organisms of paddy soils as high as 10^5 or more (Ishizawa *et al.* 1975; Watanabe, Lee, and Alimagno 1978). Most were neither *Azotobacter* nor *Beijerinckia.* However, little is known about predominant aerobic (probably micro-aerobic) nitrogen-fixing bacteria in paddy soils.

Yamagata (1924) reported that nitrogen-fixing *Clostridium* occurred in 95 per cent of Japanese paddy soils and their population was higher than that of *Azotobacter.* Ishizawa *et al.* (1975) and Matsuguchi, Tangcham and Patiyuth (1975) observed higher numbers of nitrogen-fixing *Clostridium* than *Azotobacter.*

In laboratory scale flooded soils incubated with straw, the increase of nitrogen fixation in anaerobic conditions is accompanied by the multiplication of *Clostridium* type nitrogen-fixing bacteria. *Clostridium* spp were the only nitrogen-fixing organisms among the isolates (Rice and Paul 1972).

Sulphate-reducing bacteria were abundant in reduced soils (10^4 per gram or more, Furusaka, Sato, Yamaguchi, Hattori, Nioh, Nioh, and Nishio 1969) and active in sulphide formation (Furusaka and Hattori 1956). *Desulfovibrio* and *Desulfotomaculum* are known to have nitrogen-fixing activity (Postgate 1970; Reider-Henderson and Wilson 1970).

Durbin and Watanabe (1980) found that nitrogen fixation with straw in the flooded soil was increased by the addition of sulphate. This was accompanied by sulphide formation and the multiplication of sulphur-reducing bacteria. One gram of sulphate reduction increased nitrogen fixation by 1 to 2 mg. Obviously sulphate-reducing bacteria play a role in nitrogen fixation in submerged rice soils, but it is of little significance.

Methane and hydrogen oxidation take place at the surface of paddy soils (Harrison and Aiyer 1916). Methane-oxidizing bacteria are known to fix nitrogen (Davies, Coty, and Stanley 1964; de Bont and Mulder 1976). De Bont, Lee, and Bouldin (1978) studied the distribution of methane-oxidizing bacteria in tropical paddy soils and counted $2\text{–}18 \times 10^6$ bacteria in the oxidized layer and $1\text{–}2 \times 10^7$ in the rice rhizosphere soils; there were fewer methane-oxidizing organisms in the reduced layer. Because acetylene blocks methane oxidation, the increase of methane evolution by acetylene addition is ascribed to methane oxidation. Using this technique they found that methane oxidation continued at a low rate both in the surface soil and in the rhizosphere and presumed that nitrogen fixation due to methane oxidation might not be appreciable.

Biochemical reactions such as sulphur oxidation and methane formation are, however, quite important in flooded rice soil. Because *Thiobacillus* (Mackintosh 1971) and methane-forming bacteria (Pine and Barker 1954) have been reported to fix nitrogen, the roles of these bacteria need to be examined.

Anaerobiosis and oxygen supply

As nitrogenase is oxygen labile and the nitrogenases of most nitrogen-fixing bacteria function in semi-aerobic or strictly anaerobic conditions, it is reasonable to assume that rates of nitrogen fixation in soils are higher in anaerobic or semi-anaerobic conditions. In straw-amended soil incubated in the laboratory, flooding increases nitrogen fixation (Brouzes, Lasik, Knowles 1969; Rice, Paul, and Wetler 1967). Rice and Paul (1972) suggested that *Clostridium* is responsible for nitrogen fixation in straw-amended flood soils. One reason why nitrogen fixation increases in flooded conditions is anaerobiosis. O'Toole and Knowles (1973) compared the efficiency of nitrogen fixation of soil amended with glucose or mannitol under different oxygen concentrations. The efficiency of nitrogen fixation estimated by acetylene-reduction rates was always higher at $p_{O_2} = 0.0$ and higher at low substrate levels. The data of Chang and Knowles (1965) also showed $^{15}N_2$ fixation was higher in anaerobic incubations. Yoneyama, Lee, and Yoshida (1977) reported lower E_h values (less than $-$ 200 mV) were favourable for nitrogen fixation (acetylene reduction) in straw-amended flooded soils, which suggested the importance of strongly anaerobic bacteria. In isolated bacteria, nitrogen-fixing efficiency is always higher with aerobic bacteria (Hill 1978). In mixed cultures as in soils, less intense competition for available organic substances between ammonia-assimilating and nitrogen-fixing flora (Broadbent and Nakashima 1970) may be one of the factors explaining higher nitrogen-fixing efficiency in anaerobic soils.

In mixed culture, intermediates of fermentation such as acetate and butyrate may be further used for nitrogen fixation. Research by Rice and Paul (1972) and Magdoff and Bouldin (1970) indicated that in flooded soil, the aerobic–anaerobic interface is an active site of nitrogen fixation. Magdoff and Bouldin also suggested that substances formed from the cellulose in the anaerobic zone diffuse up to the aerobic zone and support nitrogen fixation by aerobes. On the other hand, Rice and Paul suggested that substances formed from straw in the aerobic zone support nitrogen fixation by anaerobes. Both mechanisms may function, according to the type of organic substances and predominant bacteria.

Organic matter

Addition of organic matter increases dark nitrogen fixation. Table 7.2 shows enhancement of nitrogen fixation by the addition of organic matter in anaerobic or flooded soils. All of these data are based on laboratory incubations.

TABLE 7.2
Efficiency of nitrogen fixation in anaerobic soil

Substrate	Concentration	p_{O_2}	Efficiency	Author
Glucose	1.5 (% w/w)		12.3†	O'Toole and Knowles (1973)
	2.0 (% w/w)	0	14.2	
	3.0 (% w/w)		10.6	
Manitol	1.5 (% w/w)		10.8	
	2.0 (% w/w)	0	10.3	
	3.0 (% w/w)		7.7	
Straw	0.4 g straw† 0.1 g soil 2 g sand–clay mixture	Waterlogged	12.8† (2.1)‡	Rice and Paul (1972)
Straw	2 g soil + 0.4 g straw	Waterlogged	16.1† (2.2)‡	
Straw	1 (% w/w)	Waterlogged	7.07‡	Rao (1976)
	2 (% w/w)	Waterlogged	1.72	
Cellulose	1 (% w/w)	Waterlogged	4.2–20†	Rao, Kalinskaya, and Miller (1973)

† mg N/g substrate used.
† mg N/g added substrate.

Mineral nitrogen

In isolated bacteria, ammonia is an inhibitor of nitrogen fixation, but in the soil system, the inhibitory effects of ammoniacal nitrogen are alleviated by factors such as ammonia adsorption to soil clays, absorption by non-nitrogen-fixing microflora and by the presence of other bacteria. Under field conditions the enhancement of growth of the rice plant and aquatic community which increases organic matter supply may be other complicating factors. Knowles and Denike (1974) studied the effect of mineral nitrogen on nitrogen fixation in anaerobic soils and found that the inhibitory effect of mineral nitrogen on nitrogen fixation is decreased by amendment with organic substances. Matsuguchi (1979) reported that a top dressing of ammonia in paddy fields slightly depressed heterotrophic nitrogen fixation in the 3–10 cm soil layer for ten days and then subsequently increased nitrogen-fixing activity. Kalinskaya, Rao, Volkova, and Ippolitov (1973) also reported the enhancement of nitrogen fixation by 50 kg N ha^{-1} in plots amended with straw.

Soil pH

After flooding, the soil pH is restricted to about pH 6.5–7.0. Therefore, nitrogen-fixing bacteria requiring neutral or alkaline conditions (e.g. *Azotobacter*) are prevalent in paddy fields in areas of acidic soils (Yamagata 1924). The reverse is true in areas of alkaline soil (Hamdi *et al.* 1978; Mahmoud, El-Sawy, Ishac, and El-Safty 1978). Matsuguchi *et al.* (1975) found a high correlation between soil pH and the population of *Azotobacter* in Thailand soils.

Rhizosphere micro-organisms

Heterotrophic nitrogen-fixing bacteria develop in association with the roots of higher plants. This association, termed *associative symbiosis* by Burns and Hardy (1975) is thought to be a possible source of nitrogen. Associative symbiosis in grasses were reviewed by Neyra and Döbereiner (1977). Wetland rice is also known to have this association (Dommergues and Rinaudo 1979). The presence of the rice plant may also exert effects on nitrogen fixation and soil organic matter. Uppal *et al.* (1939) suggested that the living rice root accelerates nitrogen fixation of *Azotobacter* on glucose-mineral medium. This aspect, however, has received little attention. Yoshida and Ancajas (1973) applied the acetylene-reduction assay to wetland rice soil, dryland rice soil, and unplanted soil and found that the rate of acetylene reduction in soil was much higher in the presence of the rice plant than in its absence. Their data must not, however, be interpreted as the evidence of associative nitrogen fixation.

Until recently, little was known about the composition of the nitrogen-fixing micropopulations in association with rice. Wetland rice roots are colonized by *Spirillum* (*Azospirillum*) (Kumari, Kavimandan, and Subba Rao 1976), *Enterobacter* sp, and other unidentified micro-aerophilic bacteria (Watanabe, Lee, Alimagno, Sato, Del Rosario, and De Guzman 1977). *Azotobacter* and *Beijerinckia* occur only sporadically

(Diem, Rougier, Hamad-Fares, Balandreau, and Dommergues 1978). Populations of nitrogen-fixing *Clostridium* are often smaller than that of aerobic nitrogen-fixing bacteria (Trolldenier 1977).

To study nitrogen-fixing populations, the most probably number (MPN) method can be used to determine the population of nitrogen-fixing bacteria associated with wetland rice grown in the field. Different dilutions were inoculated simultaneously on to glucose and malate semi-solid media (Day and Döbereiner 1976) supplemented with yeast extract. Positive tubes were detected by acetylene-reduction rates under aerobic conditions. Root segments were vigorously shaken with glass beads to remove rhizosphere bacteria on the root surface. The roots were then macerated to release bacteria in the root tissues (histosphere).

There were more nitrogen-fixing rhizoplane bacteria in the glucose yeast-extract medium (from 2 to 6 $\times$ 10^7 g fr. wt^{-1}) than in the malate yeast-extract medium (from 2 to 15 $\times$ 10^5 g fr. wt^{-1}). There were more nitrogen-fixing histosphere bacteria than rhizoplane bacteria, and more in the glucose medium (1 to 18 $\times$ 10^7 g fr. wt^{-1}) than in the malate medium (3 to 72 $\times$ 10^5 g fr. wt^{-1}). Most of the bacteria in the malate medium resembled *Spirillum*, but the bacteria in the glucose media seemed to differ from any of the reported nitrogen-fixing bacteria. The population of nitrogen-fixing bacteria associated with the roots was not related to acetylene reduction associated with plants. It increased with the growth of rice (Watanabe, Barraquio, De Guzman, and Cabrera 1979).

Because the MPN of nitrogen-fixing organisms is close to the plate counts of aerobic heterotrophs, the colonies of aerobic heterotrophs on 0.1 per cent tryptic soy plates inoculated with a dilution from histosphere samples of IR26 were picked out and tested for nitrogenase on the yeast-extract-amended, semi-solid glucose medium. About 80 per cent of the isolates were positive in nitrogen fixation, but none of them could be grown on nitrogen-free medium. Amino acids (0.1 g l^{-1}) could substitute for yeast extract in derepressing nitrogenase in the isolates. They resembled *Achromobacter* (Watanabe and Barraquio 1979). These findings suggest the need to re-examine the occurrence of nitrogen-fixing bacteria through use of nitrogen-containing culture media for the isolation of nitrogenase-positive bacteria.

7.4 Factors affecting nitrogen fixation

Measurement of acetylene-reduction activities is used to study factors affecting associative nitrogen fixation. As a quantitative tool, the acetylene-reduction technique has limitations, but it is useful for semi-quantitative or relative measurements of the nitrogen-fixing rate of natural systems. In submerged soils, slow gas transfer in the liquid phase is one of the problems in acetylene-reduction technique. The problems of acetylene-reduction assays are discussed by Watanabe and Cholitkul (1979). The elimination of the contribution of phototrophic nitrogen-fixing organisms is another problem in assessing nitrogen fixation associated with rice. Watanabe, Lee, and De Guzman (1978) used the water-replacement technique to remove the bulk of phototrophic acetylene reduction. Stem portions were covered with aluminium foil to avoid phototrophic acetylene reduction associated with stem portions of wetland rice. Herbicides were also used to depress algal activity (Habte and Alexander, personal communication). Covering of the field or pot surface with black cloth may have an indirect effect on plant growth and does not depress algal activity completely.

The use of the excised roots to detect nitrogenase associated with roots as used by Bülow and Döbereiner (1975) has little relevance to phenomena in intact plants. Watanabe and Cabrera (1979) developed a technique to study nitrogen fixation associated with rice in water culture. Rice plants grown in the field were transferred to water culture and grown for a while. Factors such as mineral nitrogen, gas composition of rooting medium, antiobiotics and so on are controlled easily with this technique.

Plant age

Lee, de Castro, and Yoshida (1977) found the peak of nitrogen-fixing activity (per plant) at the heading stage. Watanabe, Lee, and De Guzman (1978) also observed similar trends. Differences in rates of acetylene reduction associated with the two rice varieties was also dependent on the growth stage.

Rice varieties and species

By screening collections of rice germplasm, one can select the most desirable genotypes according to their ability to associate themselves with nitrogen-fixing bacteria. Dommergues and Rinaudo (1979) reported intra- and intervarietal variability in acetylene reduction assayed for rice seedlings. As the rate of acetylene reduction varies greatly according to the growth stage, differences at the seedling stage are of little significance. At the International Rice Research Institute over 70 varieties and several wild *Oryza* sp have been tested for acetylene-reduction activity, but none of these was consistently high. Further improvement of the technique is needed to study varietal differences.

Mineral nitrogen

In water culture (Watanabe and Cabrera 1979), 0.33 mM ammonium depresses acetylene reduction associated with wetland rice, but in the field, the application of ammonium fertilizer not only did not depress, but increased the rate of acetylene reduction probably due to the increase in plant size (Matsuguchi 1979).

Climatic factors

Balandreau, Ducerf, Hamad-Fares, Weinhard, Rinaudo, Miller and Dommergues (1978) discussed the climatic factors and showed high light intensity (40 klx) and low-air water deficit stimulated acetylene reduction associated with rice. Dommergues and Rinaudo (1979) also showed great differences in rhizospheric acetylene reduction among various soils. The reasons for this are not yet clear.

Site of fixation

By removing the roots from the hill, Watanabe, Lee, and De Guzman (1978) showed that some acetylene reduction was associated with stem or crown portions of rice. The expression of activity on the basis of root weight, is, therefore, misleading. The relative contribution among various sites around and in the root tissue are unknown. There must be a close interaction of pectinolytic bacteria and nitrogen-fixing bacteria in the rhizosphere, because cross-inoculation with nitrogen-fixing bacteria from rice root and pectinolytic non-nitrogen-fixing bacteria on sterile rice plants greatly increased acetylene reduction as compared to single inoculation (Dommergues and Rinaudo 1979). The colonization of nitrogen-fixing bacteria on gnotobiotic rice culture showed that the bacteria are embedded in mucigels around the root surface (Diem *et al.* 1978). The surface sterilization of new roots killed nitrogen-fixing bacteria, but that of basal and old roots did not kill nitrogen-fixing bacteria (Diem *et al.* 1978), indicating deep colonization of nitrogen-fixing bacteria in old root tissues as shown for the general microflora by Miyashita *et al.* (1977).

7.5 Autotrophic nitrogen fixation in paddy fields

Blue–green algae (Cyanobacteria)

De (1936) first recognized the importance of nitrogen-fixing blue–green algae in paddy fields as the factor responsible for maintaining nitrogen fertility. Since then, an enormous volume of material on this subject has been published. Extensive reviews have been written by Singh (1961), Venkataraman (1975), Stewart, Rowell, Ladha, and Sampaio (1979), and Roger and Reynaud (1979).

Among four groups of blue–green algae—*Chroococacean*, *Pleurocapsalean*, non-heterocystous filamentous forms, and heterocystous filamentous forms—heterocystous filamentous forms and some *Chroococacean* spp can fix nitrogen under aerobic conditions, but other forms can fix nitrogen only under anaerobic or semi-aerobic conditions (Stewart *et al.* 1979). As discussed before, the oxygen concentration of the flood water exceeds the saturation level during daytime when algal activity is high, and so aerobic nitrogen-fixing blue–green algae may be of great ecological significance.

Distribution

Floristic descriptions of blue–green algae in paddy fields are abundant in India (Singh 1961; Pandey 1965; Gupta 1966; Venkataraman 1975), in Japan (Okuda and Yamaguchi 1956), in Indonesia (Jutono 1973), in Australia (Bont 1961), in Senegal (Reynaud and Roger 1978), in South-East Asia (Watanabe and Yamamoto 1971), in Iraq (Al-Kaisi 1976), in the Philippines (Pantastico and Suayan 1973), in Egypt (el Nawawy and Hamdi 1975), in Morocco (Renault, Sasson, Pearson and Stewart 1975) and so on. Watanabe and Yamamoto (1971) stated that out of 612 paddy soil samples from South-East and East Asia nitrogen-fixing blue–green algae were observed at 2 per cent frequency from soils north of 30 °N latitude and at 12 per cent frequency from south of 30 °N latitude. The frequency of nitrogen-fixing blue–green algae was higher in tropical or subtropical regions and some of them (*Tolypothrix tenuis*) were quite active in nitrogen fixation. In India, the Indian Agricultural Research Institute conducted an extensive survey on the occurrence of blue–green algae in paddy soils and an average of 33 per cent detection was obtained out of 2213 samples (Venkataraman 1975). The abundance of nitrogen-fixing blue–green algae differed among states, and states with acid soils like Kerala, and Tamil Nadu showed the lowest incidence. Among these surveys in wide areas of the world, heterocystous filamentous blue–green algae such as *Anabaena* and *Nostoc* seem quite ubiquitous. In the floristic descriptions of Singh (1961) for the Uttar Pradesh and Bihar regions, some of the non-heterocystous filamentous or unicellular blue–green algae were abundant inhabitants of paddy soils (*Aphanothece pallida*, *Schizothrix*, *Porphyrosiphon*, and *Microcoleus*). The nitrogen-fixing activities of all these organisms except *Aphanothece* has not been confirmed.

Seasonal variation

In most rainfed rice fields, the soils are quite dry during the non-crop season. After the monsoon rains come, the paddy fields are flooded. Sometimes puddling is carried out during the flooding. After the rice harvest, the soil is again dried. Within this cycle, seasonal variations of blue–green algae and other algae occur. Singh (1961) described these changes. Algae lies dormant as spores in the dry season. After flooding, the spores germinate, but puddling temporarily retards the algal growth.

A month after transplanting, the algae forms a thick yellowish-blue–green mass on the surface of the water. This mat is mainly composed of *Aulosira fertilissima*. This mass dominates the field for about three months and after the rice harvest, the algal mats remain. Lately, Reynaud and Roger (1978) described biomass changes in the rice cycle at Senegal. After flooding, the succession of predominant flora proceeds from Diatom and green algae (tillering stage) → filamentous green algae and non-heterocystous blue–green algae (panicle initiation) → filamentous green algae and blue–green algae (heading to maturity at weak plant cover) or blue–green algae (heading to maturity at dense plant cover). Probably light intensity controls the succession type of algae and blue–green algae. Many factors determine the algal succession (see Roger and Reynaud 1979). In rice culture, chemical changes after soil reduction (release of NH_4^+, phosphorus, and iron) and the presence of the rice plant canopy may be of importance for algal succession.

Assessment of nitrogen-fixing blue–green algae

Most of the descriptive studies of blue–green algae in paddy fields have been qualitative. Estimation of algal biomass and nitrogen-fixing rates in the field were conducted by Prasad (1949) at South Bihar taking algal crustations from the field and determining algal weight and nitrogen content. Nitrogen content was compared to that beneath algal crustations. Prasad estimated the nitrogen input at about 14 kg ha^{-1} (see Singh 1961). Recent use of the acetylene-reduction assay has enabled field evaluation of algal nitrogen fixation (Watanabe and Cholitkul 1979; Watanabe 1978*a*, *b*), although some caution is needed in analysing the data. In long-term fertility plots without fertilizers at the International Rice Research Institute, Watanabe, Lee, and Alimagno (1978) estimated nitrogen fixation in the field using the acetylene-reduction assay. In the dry season, algal nitrogen-fixing activity was higher than in the wet season. Algal nitrogen-fixing activity showed two peaks: one at the early stage of rice growth and another at the end or after harvest. The later peak was higher and at that time the field was fully covered by slimy colonies of *Gloeotrichia*—fresh mass 24 ton ha^{-1}. The nitrogen-fixing (C_2H_2, 3:1) activity was about 0.5 kg N ha^{-1} day^{-1}.

Roger and Reynaud (1976) developed a technique of biomass evaluation of non-nitrogen-fixing algae, nitrogen-fixing non-filamentous algae, and nitrogen-fixing filamentous algae, principally based on selective culture media and cell number and volume of algal germs in soil dilutions. In ten out of 17 soils tested, the nitrogen-fixing algal mass was within 100–500 kg wet weight ha^{-1} and their nitrogen-fixing activity corresponded to 0.8–40 kg N ha^{-1} $season^{-1}$. The highest value was 2.2 ton ha^{-1} fresh algal mass and the corresponding nitrogen-fixing rate was 18 kg N ha^{-1} $season^{-1}$.

Factors affecting the distribution and activity of nitrogen-fixing blue–green algae

pH

Most blue–green algae do not tolerate low pH levels. The optimum pH is usually neutral or alkaline with activity falling off markedly below a pH of about 6.6 (Fogg, Stewart, Fay, and Walsby 1973; Okuda and Yamaguchi 1956). The optimum pH of blue–green algae is sometimes higher than the optimum pH of rice growth in paddy soils (pH 6.5). De (1936), Shioiri, Hatsumi, and Nishigaki (1943), and Okuda and Yamaguchi (1955) showed the enhancement of algal nitrogen fixation by liming.

Phosphorus

Stewart, Fitzgerald, and Burris (1970) observed rapid reaction of algal nitrogen (acetylene) reduction by the addition of 3–200 μM orthophosphate in eutrophic lakes. De and Sulaiman (1950*b*) increased nitrogen fixation by the addition of superphosphate to the extent of 0.5 kg N kg $P_2O_5^{-1}$ applied. Shioiri *et al.* (1943) also observed similar values. Phosphorus application may eliminate, at least partially, the need for fertilizer nitrogen.

Molybdenum

Bortels (1940) and Wolfe (1954) demonstrated the need for molybdenum for optimum growth (0.2 p.p.m.) of blue–green algae. Under waterlogged conditions, molybdenum may be solubilized to the optimum level. Subramanyan, Relwani, and Manna (1965) suggested the addition of sodium molybdate (0.28 kg ha^{-1}) to soil to improve nitrogen-fixing algal growth.

Light

Blue–green algae are sensitive to high light intensity (Reynaud and Roger 1978) and may be regarded as low light requiring species (Brown and Richardson 1968). In South-East Asia, light intensity in the dry season is as high as 100 klx at noon. In Senegal, under light of similar intensities partial shading of the unplanted paddy field (up to 93 per cent shading) favoured growth of blue–green algae over green algae.

The predominance of nitrogen-fixing blue–green algae in later stages of dense plant cover may be explained by the decrease of light intensity. The active growth of blue–green algae near the dry season harvest in the Philippines (Watanabe, Lee, and Alimagno 1978) may partly be due to weak light under the plant canopy.

Ammonia

In long-term fertility plots at the International Rice Research Institute, algal nitrogen fixation was more pronounced in non-fertilized plots than in nitrogen–phosphorus–potassium-fertilized plots (Watanabe, Lee, and Alimagno 1978). Probably the surface application of ammoniacal fertilizer depressed growth of blue–green algae and favoured the growth of green algae. In contrast, heterotrophic nitrogen fixation in these plots was not depressed by fertilization (Watanabe, unpublished results).

Plant

De and Sulaiman (1950*b*) conducted nitrogen fixation of blue–green algae in culture media suspended with soil and showed that the rice plant stimulated algal nitrogen fixation. They attributed this action to carbon dioxide release by the plant.

Field assays at the International Rice Research Institute showed that in unplanted plots algal nitrogen fixation was much less than in planted plots (Watanabe, Lee, and De Guzman 1978). The stimulative effects of the plant may be attributed to shading, carbon dioxide supply, and physical support for algal growth. The nodal roots that grow over the soil and the surface of stems that stand in the flood water are inhabited with blue–green algae, but the intimate relations of blue–green algae and rice plants have not yet been examined.

7.6 Photosynthetic bacteria

In paddy fields, high populations of non-sulphur purple bacteria (up to 10^6 per gram of soil) and sulphur bacteria (mostly 10^2–10^3 per gram of soil) were found in Japanese soils (Okuda, Yamaguchi, and Kamata 1957), and tropical soils (Kobayashi, Takahashi, and Kawaguchi 1967). Kobayashi and Haque (1971) reported the population of both groups approached their peak near heading in the rhizosphere of wetland rice.

Photosynthetic bacteria require anaerobic conditions and organic acids or sulphide which are formed by anaerobic metabolism for their light dependent growth. It is unlikely that the surface soil or flood water meets these requirements. However, associations of photosynthetic bacteria and aerobic heterotrophic bacteria as demonstrated by Okuda, Yamaguchi and Kobayashi (1960) and Kobayashi, Katayama, and Okuda (1965*a*, *b*) do not require anaerobic conditions. If we assume the presence of this association under natural conditions, a role for photosynthetic bacteria is feasible. Nothing is known, however, about the roles of these bacteria in nitrogen fixation of the paddy fields.

Algae in symbiosis with Azolla

Anabaena azollae lives in the cavities of the dorsal lobes of *Azolla*—a water fern—fixes nitrogen symbiotically and supports growth of the host without combined nitrogen (Moore 1969). *Azolla* is widely distributed in the aquatic habitats of the world. *A. pinnata* is found in tropical and subtropical conditions. Daily nitrogen-fixing rates of the *Azolla–Anabaena* complex are as high as 3 kg N ha^{-1} a^{-1}. Due to this high potential for nitrogen fixation, *Azolla* is extensively used as green manure in rice culture in Vietnam and China (Dao and Tran 1979; Liu 1979). *A. filiculoides* and *A. mexicana* have been tried in North America (Talley, Talley, and Rains 1977). Reviews of recent development are given by Becking (1979), Brotonegoro and Abdulkadir (1976, 1978), and Watanabe (1978*a*, *b*).

7.7 Concluding remarks

Rice as a cereal crop is as important as wheat, supporting the life of a billion people in the world. Rice can support high population densities because of its relatively high and stable yield with no or little fertilizers. This unique property stems from the submerged growth of rice. The submerged soil provides suitable sites for biological nitrogen fixation; first in the flood water where phototrophic diazotrophs predominate, and second in the anaerobic soil where anerobic nitrogen fixation predominates.

Most of the rice growers in less developed countries can not afford expensive chemical fertilizers. Measures that can mitigate the fertilizer shortage or help reduce the consumption of chemical fertilizers for rice production include (i) increasing efficiency of use of the existing fertilizers and (ii) expanding the use of native sources of nitrogen (soil and organic nitrogen).

If we are to exploit currently recognized sources of biological nitrogen in rice fields, the phototrophic nitrogen-fixing organisms—blue–green algae and *Azolla* must be considered first. Despite extensive trials with these organisms in some countries (Venkataraman 1975; El Nawawy and Hamdi 1975), our basic knowledge on the ecology of the introduced organisms is minimal.

Questions such as the amount of nitrogen fixation in the field, the relative importance of various nitrogen-fixing agents and the fate of biologically fixed nitrogen remain to be answered.

Exploitation of associative nitrogen fixation in rice has recently been investigated, but we need to accumulate basic information about the extent and mechanism of this association before it can be managed agrotechnically.

References

ABD-EL-MALEK, Y. (1968). *J. appl. Bact.* **31**, 267.
—— (1971). *Pl. Soil* Special Volume, 423.
AL-KAISI, K. A. (1976). *Nova Hedwigia* **27**, 813.
ALLISON, F. E. (1965). In *Soil nitrogen* (ed. W. V. Bartholomew and F. E. Clark) p. 573. Agronomy Society of America, Madison, Wisconsin.
ARASHI, K. and NITTA, N. (1955). *Proc. Crop Sci. Soc. Japan* **24**, 78. [In Japanese, English summary.]
ARMSTRONG, W. (1969). *Physiologia Pl.* **22**, 296.
BALANDREAU, J., DUCERF, P., HAMAD-FARES, I., WEINHARD, P., RINAUDO, G., MILLER, C., and DOMMERGUES, Y. (1978). In *Limitations and potentials for biological nitrogen fixation in the tropics* (ed. J. Döbereiner, R. H. Burns, and A. Hollander) p. 275. Plenum Press, New York.
BARBER, D. A. and LYNCH, J. M. (1977). *Soil Biol. Biochem.* **9**, 305.
BECKING, J. H. (1961). *Pl. Soil* **14**, 49.
—— (1978). *Ecol. Bull., Stockh.* **26**, 116.
—— (1979). In *Nitrogen and rice*, p. 345. IRRI, Manila, The Philippines.
DE BONT, J. A. M. and MULDER, E. G. (1976). *Appl. Environ. Microbiol.* **31**, 640.
—— LEE, K. K., and BOULDIN, D. F. (1978). *Ecol. Bull., Stockh.* **26**, 91.
BORTELS, H. (1940). *Archs Mikrobiol.* **11**, 155.
BROADBENT, F. E. and NAKASHIMA, T. (1970). *Soil Sci. Soc. Am. Proc.* **34**, 218.
BROTONEGORO, S. (1977). *BioIndon.* **4**, 1–13.
—— and ABDULKADIR, S. (1976). *Ann. Bogor.* **6**, 69.
—— —— (1978). *Ann. Bogor.* **6**, 169.
BROUZES, R., LASIK, J., and KNOWLES, R. (1969). *Can. J. Microbiol.* **15**, 899.
BROWN, T. E. and RICHARDSON, F. L. (1968). *J. Phycol.* **4**, 38.
BÜLOW, J. F. W. and DÖBEREINER, J. (1975). *Proc. natn. Acad. Sci. U.S.A.* **72**, 2389.
BUNT, J. S. (1961). *Nature, Lond.* **192**, 479.
BURNS, R. C. and HARDY, R. W. F. (1975). *Nitrogen fixation in the bacteria and higher plants.* Springer-Verlag, Berlin.
CHADRASEKARAN, S. and YOSHIDA, T. (1973). *Soil Sci. Pl. Nutr.* **19**, 39.
CHANG, P. C. and KNOWLES, R. (1965). *Can. J. Microbiol.* **11**, 29.
DAO, T. T. and TRAN, Q. T. (1979). In *Nitrogen and rice*, p. 395. IRRI, Manila, The Philippines.
DAVIES, J. B., COTY, V., and STANLEY, J. P. (1964). *J. Bacteriol.* **88**, 468.
DAY, J. M. and DÖBEREINER, J. (1976). *Soil Biol. Biochem.* **8**, 45.
DE, P. K. (1936). *Indian J. agric. Sci.* **6**, 1237.
—— and SULAIMAN, L. N. (1950*a*). *Indian J. agric. Sci.* **20**, 327.
—— —— (1950*b*). *Soil Sci.* **70**, 137.
DERX, H. G. (1950). *Ann. bogor.* **1**, 1.
DIEM, G., ROUGIER, M., HAMAD-FARES, I., BALANDREAU, J. P., and DOMMERGUES, Y. (1978). *Ecol. Bull., Stockh.* **26**, 305.
DOMMERGUES, Y. and RINAUDO, G. (1979). In *Nitrogen and rice*, p. 241. IRRI, Manila, The Philippines.
DURBIN, K. and WATANABE, I. (1980). *Soil Biol. Biochem.* **12**, 11.
EL-NAWAWY, A. S. and HAMDI, Y. A. (1975). In *Nitrogen fixation by free living micro-organisms* (ed. W. D. P. Stewart) p. 219. Cambridge University Press.
FOGG, G. E., STEWART, W. D. P., FAY, P., and WALSBY, A. E. (1973). *The blue–green algae.* Academic Press, London.
FOSTER, R. C. and ROVIRA, A. D. (1976). *New Phytol.* **76**, 343.
FURUSAKA, C. (1977). *Proc. Int. Symp. Soil Environment and Fertilizer Management in Intensive Agriculture*, Tokyo. Japanese Society for Soil Manure, Tokyo.

—— and HATTORI, T. (1956). *Bull. Inst. agric. Res. Tôhoku Univ.* **8**, 35. [In Japanese, English summary.]

—— —— SATO, K., YAMAGUCHI, H., HATTORI, R., NIOH, I., NIOH, T., and NISHIO, M. (1969). *Rep. Inst. agric. Res. Tôhoku Univ.* **20**, 87.

GUPTA, A. B. (1966). *Hydrobiologia* **28**, 213.

HAMDI, Y. A., YOUSEF, A. N., AL-AZAWI, S., AL-TAI, A., and AL-BAQUARI, M. S. (1978). *Ecol. Bull., Stockh.* **26**, 110.

HARRISON, W. H. and AIYER, P. A. S. (1916). *Mem. Dep. Agric. India chem. Ser.* **4**, 135.

HAYASHI, S., ASATSUMA, K., NAGATSUKA, T., and FURUSAKA, C. (1978). *Rep. Inst. agric. Res. Tôhoku Univ.* **29**, 19.

HILL, S. (1978). *Ecol. Bull., Stockh.* **26**, 130.

HOWELER, R. H. and BOULDIN, D. R. (1971). *Soil Sci. Soc. Am. Proc.* **35**, 202.

INTERNATIONAL RICE RESEARCH INSTITUTE (1977). Annual Report for 1976. Manila, The Philippines.

ISHIZAWA, S. and TOYOTA, H. (1960). *Bull. natn. Inst. agric. Sci., Tokyo* **14B**, 204. [In Japanese, English summary.]

—— SUZUKI, T., and ARARAGI, M. (1975). *JIBP Synth.* **12**, 41.

JOSHI, M. M. and HOLLIS, J. (1977). *Science, N.Y.* **195**, 179.

JUTONO (1973). *Soil Biol. Biochem.* **5**, 91.

KALINSKAYA, T. A., RAO, V. R., VOLKOVA, T. N., and IPPLITOV, L. T. (1973). *Mikrobiologiya* **42**, 481. [In Russian, English summary.]

KAWATA, S., ISHIHARA, K., and SHIOYA, T. (1964). *Proc. Jap. Crop Sci. Soc.* **32**, 250. [In Japanese, English summary.]

KIKUCHI, E., FURUSAKA, C., and KURIHARA, Y. (1975). *Rep. Inst. agric. Res. Tôhoku Univ.* **26**, 25.

—— —— —— (1977). *Jap. J. Ecol.* **27**, 113. [In Japanese, English summary.]

KIMURA, M., WADA, H., and TAKAI, Y. (1977). *J. Sci. Soil Manure, Tokyo* **48**, 91. [In Japanese.]

KNOWLES, R. and DENIKE, D. (1974). *Soil Biol. Biochem.* **6**, 353.

KOBAYASHI, M. and HAQUE, M. Z. (1971). *Pl. Soil* Special Volume, 443.

—— KATAYAMA, T., and OKUDA, A. (1965*a*). *Soil Sci. Pl. Nutr.* **11**, 74.

—— —— —— (1965*b*). *Soil Sci. Pl. Nutr.* **11**, 200.

—— TAKAHASHI, H., and KAWAGUCHI, K. (1967). *Soil Sci.* **104**, 113.

KOCH, B. L. and OYA, J. (1975). *Soil Biol. Biochem.* **6**, 363.

KONISHI, C. and SEINO, K. (1961). *Bull. Hokuriku agric. Exp. Stn* **2**, 41. [In Japanese, English summary.]

KOYAMA, T. and APP, A. (1979). In *Nitrogen and rice,* p. 95. IRRI, Manila, The Philippines.

KUMARI, L. M., KAVIMANDAN, S. K., and SUBBA RAO, N. S. (1976). *Indian J. exp. Biol.* **14**, 638.

LEE, K. K. and WATANABE, I. (1977). *Appl. Environ. Microbiol.* **34**, 654.

—— DE CASTRO, T., and YOSHIDA, T. (1977). *Pl. Soil* **43**, 613.

LIU, C. C. (1979). In *Nitrogen and rice*, p. 375. IRRI, Manila, The Philippines.

MACKINTOSH, M. E. (1971). *J. gen. Microbiol.* **66**, i.

MACRAE, I. C. and DE CASTRO, T. F. (1966). *Phyton* **23**, 95.

MAGDOFF, F. R. and BOULDIN, D. R. (1970). *Pl. Soil* **33**, 49.

MAHMOUD, S. A. Z., EL-SAWY, M., ISHAC, Y. Z., and EL-SAFTY, M. M. (1978). *Ecol. Bull., Stockh.* **26**, 99.

MARTIN, J. K. (1977). *Soil Biol. Biochem.* **9**, 1.

MATSUGUCHI, T. (1979). In *Nitrogen and rice*, p. 207. IRRI, Manila, The Philippines.

—— TANGCHAM, B., and PATIYUTH, S. (1975). *Japan agric. Res. Q.* **8**, 253.

MIKKELSEN, D. S., DE DATTA, S. K., and OBCEMEA, W. N. (1978). *Soil Sci. Soc. Am. Proc.* **42**, 725.

MITSUI, S. (1954). *Inorganic nutrition, fertilization and soil amelioration for lowland rice.* Yokendo, Tokyo.
MIYASHITA, K., WADA, H., and TAKAI, Y. (1977). *J. Sci. Soil Manure, Tokyo* **44**, 558. [In Japanese.]
MOORE, A. W. (1969). *Bot. Rev.* **35**, 17.
NAGATSUKA, T. and FURUSAKA, C. (1976). *Soil Sci. Plant Nutr.* **22**, 287.
NEYRA, C. A. and DÖBEREINER, J. (1977). *Adv. Agron.* **29**, 1.
OKUDA, A. (1948). *Nogaku* **2**, 306. [In Japanese.]
—— and YAMAGUCHI, M. (1955). *Soil Sci. Pl. Nutr.* **1**, 102.
—— —— (1956). *Soil Sci. Pl. Nutr.* **2**, 4.
—— —— and KAMATA, S. (1957). *Soil Sci. Pl. Nutr.* **2**, 131.
—— —— and KOBAYASHI, M. (1960). *Soil Sci. Pl. Nutr.* **6**, 35.
OLD, K. M. and NICOLSON, T. H. (1975). *New Phytol.* **74**, 51.
O'TOOLE, P. and KNOWLES, R. (1973). *Soil Biol. Biochem.* **5**, 789.
PANDEY, D. C. (1965). *Nova Hedwigia* **9**, 299.
PANTASTICO, J. B. and SUAYAN, Z. A. (1973). *Philipp. Agric.* **57**, 313.
PATRICK, W. H. and REDDY, K. R. (1976). *J. Environ. Microbiol.* **5**, 469.
PEARSALL, W. H. and MORTIMER, C. H. (1939). *J. Ecol.* **27**, 485.
PINE, M. J. and BARKER, H. A. (1954). *J. Bact.* **68**, 589.
PONNAMPERUMA, F. N. (1972). *Adv. Agron.* **24**, 29.
POSTGATE, J. R. (1970). *J. gen. Microbiol.* **63**, 137.
PRASAD, S. (1949). *J. Proc. Inst. Chem.* **21**, 135.
VAN RAALTE, M. H. (1940). *Ann. bot. Gdn Buitenz.* **50**, 99.
RANGASWAMI, B. and SUBBARJA, K. T. (1962). *Ind. Acad. Sci. Proc.* **56B**, 174.
RAO, V. R. (1976). *Soil Biol. Biochem.* **8**, 445.
—— KALINSKAYA, T. A., and MILLER, U. M. (1973). *Mikrobiologiya* **42**, 729. [In Russian.]
REIDER-HENDERSON, M. A. and WILSON, P. W. (1970). *J. gen. Microbiol.* **61**, 27.
RENAUT, J., SASSON, A., PEARSON, H. W., and STEWART, W. D. P. (1975). In *Nitrogen fixation by free-living microorganisms* (ed. W. D. P. Stewart) p. 229. Cambridge University Press.
REYNAUD, P. A. and ROGER, P. A. (1978). *Ecol. Bull., Stockh.* **26**, 148.
RICE, W. A. and PAUL, E. A. (1972). *Can. J. Microbiol.* **18**, 715.
—— —— and WETTER, L. R. (1967). *Can. J. Microbiol.* **13**, 829.
ROGER, P. A. and REYNAUD, P. A. (1976). *Rev. Ecol. Biol. Sol.* **13**, 545.
—— —— (1979). In *Nitrogen and rice*, p. 287. IRRI, Manila, The Philippines.
ROUQUEROL, T. (1962). *Ann. Agron.* **13**, 325.
SAITO, M. and WATANABE, I. (1978). *Soil Sci. Pl. Nutr.* **24**, 427.
SHIOIRI, M. (1941). *Kagaku* **11**, 24. [In Japanese.]
—— HATSUMI, T., and NISHIGAKI, S. (1943). *J. Sci. Soil Manure, Tokyo* **17**, 287. [In Japanese.]
SINGH, R. N. (1961). *Role of blue–green algae in nitrogen economy of Indian agriculture.* Indian Council for Agricultural Research, New Delhi.
STARKEY, R. L. and DE, P. K. (1939). *Soil Sci.* **47**, 329.
STEWART, W. D. P., FITZGERALD, G. P., and BURRIS, R. H. (1970). *Proc. natn. Acad. Sci. U.S.A.* **66**, 1104.
—— ROWELL, P., LAHDA, J. K. and SAMPAIO, M. J. A. M. (1979). In *Nitrogen and rice*, p. 263. IRRI, Manila, The Philippines.
SUBRAMANYAN, R., RELWANI, L. L., and MANNA, G. B. (1965). *Proc. natn. Acad. Sci. Ind.* **35**, 382.
SUTO, T. (1954). *Sci. Rep. Res. Insts Tôhoku Univ. D* **6**, 25.
TAKAI, Y. and KAMURA, T. (1966). *Folia Microbiol., Praha* **11**, 304.

TAKIJIMA, Y. (1963). *Bull. natn. Inst. agric. Sci., Tokyo* **31B**, 117. [In Japanese, English summary.]
TALLEY, S. N., TALLEY, B. J., and RAINS, D. W. (1977). In *Genetic engineering for nitrogen fixation* (ed. A. Hollaender) p. 259. Plenum Press, New York.
TROLLDENIER, G. (1977). *Pl. Soil* **47**, 203.
UPPAL, B. N., PATEL, M. K., and DAJI, J. A. (1939). *Indian J. agric. Sci.* **9**, 689.
VENKATARAMAN, G. S. (1975). In *Nitrogen fixation by free living microorganisms* (ed. W. D. P. Stewart) p. 207. Cambridge University Press.
WATANABE, A. and YAMAMOTO, Y. (1971). *Pl. Soil* Special Volume, 403.
WATANABE, I. (1978*a*). In *Soils and rice*, p. 465. IRRI, Manila, The Philippines.
—— (1978*b*). *Tsuchi to Biseibutsu*, Japan **20**, 1.
—— and BARRAQUIO, W. L. (1979). *Nature, Lond.* **277**, 565.
—— and CABRERA, D. A. (1979). *Appl. Environ. Microbiol.* **37**, 373.
—— and CHOLITKUL, W. (1979). In *Nitrogen and rice*, p. 223. IRRI, Manila, The Philippines.
—— LEE, K. K., and ALIMAGNO, B. V. (1978). *Soil Sci. Pl. Nutr.* **24**, 1.
—— —— and DE GUZMAN, M. R. (1978). *Soil Sci. Pl. Nutr.* **24**, 550.
—— BARRAQUIO, W. L., DE GUZMAN, M. R., and CABRERA, D. A. (1979). *Appl. Environ. Microbiol.* **37**, 813.
—— LEE, K. K., ALIMAGNO, B. V., SATO, M., DEL ROSARIO, D. C., and DE GUZMAN, M. R. (1977). IRRI Res. Paper Ser. 3. Manila, The Philippines.
WILLIS, W. H. and GREEN, V. R. (1948). *Proc. Soil Sci. Soc. Am.* **13**, 229.
WOLFE, M. (1954). *Ann. Bot.* **18**, 299.
YAMAGATA, U. (1924). *J. agric. Chem. Soc. Japan* **1**, 85. [In Japanese, English summary.]
YONEYAMA, K., LEE, K. K., and YOSHIDA, T. (1977). *Soil Sci. Plant Nutr.* **23**, 287.
YOSHIDA, T. (1975). In *Soil biochemistry* (ed. E. A. Paul and A. D. MacLaren) Vol. 3, p. 83. Marcel Dekker, New York.
—— and ANCAJAS, R. R. (1973). *Soil Sci. Soc. Am. Proc.* **37**, 42.
—— and BROADBENT, F. E. (1975). *Soil Sci.* **120**, 288.

8 Forage legumes

E. F. HENZELL

8.1 Introduction

Nitrogen fixation is an essential function of forage legumes but not the only important one. Their feeding value, using that term broadly to include absence of deleterious substances as well as content of essential nutrients, is of equal significance. A number of legume species are efficient fixers of nitrogen but are toxic, e.g. *Indigofera spicata* grows well in some subtropical environments, but contains the hepatotoxic amino acid indospicine (Hutton 1970). The nitrogen fixed by forage legumes is important not only for their own growth but also as a source of combined nitrogen for grasses and non-legume crops.

It is convenient, particularly when discussing the measurement of nitrogen fixation, to separate pasture legumes, i.e. those that are used for grazing, from fodder legumes, i.e. those that are harvested mechanically or manually for hay, silage, or green feed. Fodder legumes, such as *Medicago sativa* (lucerne) are often grown in monoculture. In some cases the same legume species may be used either for pasture or for fodder.

The first part of this chapter examines existing knowledge of the main ecological factors that influence the growth and nitrogen fixation of pasture and fodder legumes. Water, temperature, light, mineral nutrients, consumption by herbivores, and diseases have an important influence on legume persistence and yield in both mixed plant communities and in monocultures. Growth and nitrogen fixation by pasture legumes are strongly influenced by competition from grasses and forbs (the term is used here for all herbs other than grasses and nodulated legumes) for light, water, combined nitrogen, and other nutrients. Fodder legumes may suffer competition from weeds. In pastures only inedible or toxic plants can truly be considered as weeds.

Man has played a major role in the ecology of forage legumes, through the propagation of naturally occurring or improved genotypes, and through the manipulation of factors such as water, mineral nutrients, grazing or cutting, plant competition, and nodulation. The formation of an effective *Rhizobium* symbiosis is an essential feature of successful forage legumes. It is assumed in this chapter that all legumes are effectively nodulated and capable of efficient symbiotic nitrogen fixation. Factors affecting the establishment of an efficient symbiosis are discussed in Chapter 4.

The second part examines the methods that are available for measuring nitrogen fixation in the field and reviews some published estimates of the amounts fixed.

8.2 Ecological factors

Most of the knowledge of environmental factors and their influence on legume nitrogen fixation has come from two sources: first, from studies carried out in nitrogen-free media (until the introduction of the acetylene-reduction assay, the only simple method for measuring nitrogen fixation) or in small chambers (acetylene reduction, or sometimes $^{15}N_2$ fixation); secondly, by deduction from a large body of empirical evidence on the ecology of legume growth in the field as nitrogen fixation being strongly linked with growth.

In neither case is the evidence particularly satisfactory. Plants grown in nitrogen-free media may behave differently from those grown with access to combined nitrogen in soil and in competition with non-legume plants. The 'nitrogen-free' systems are rarely completely free of combined nitrogen because of gases such as ammonia in the atmosphere. Ammonia levels in uncontaminated laboratory air are usually $<10\ \mu g\ m^{-3}$ (Bremner 1965), but ammonia is absorbed by plants at such concentrations (Denmead, Freney, and Simpson 1976). Nevertheless, the generally poor growth of uninoculated legumes in nitrogen-free systems indicates that uptake of combined nitrogen from the atmosphere is usually very small.

In small chambers it is very difficult to duplicate the conditions of soil and climate found in the field. In contrast, physical conditions are realistic in the field but it is impracticable to measure nitrogen fixation directly.

There are two problems in the use of measurements of legume growth, say herbage dry-matter yield, as estimates of fixation. First, legume herbage yield, or even the total quantity of nitrogen in the legume herbage, are not an exact measure of the total amount of nitrogen assimilated. In fact, it is extremely difficult in practice to measure the total flux of nitrogen through legume growth in the field (p. 280). Dry matter and nitrogen yields are often closely correlated (Norris and Date 1976) although the concentration of nitrogen in legume plants decreases with increasing maturity and can be reduced also by environmental stresses that affect symbiotic fixation directly, such as molybdenum deficiency or extreme temperatures.

Secondly, the proportion of nitrogen fixation in that total nitrogen flux is variable and usually unknown. Efficiently nodulated legumes will take up available combined nitrogen from soil and this uptake is not easily measured (p. 281). With mixed pastures, and with weedy stands of fodder legumes, forbs and grasses compete strongly with the legumes for soil nitrogen and the legumes have to rely mainly on fixation. Thus Vallis, Henzell, and Evans (1977) found that the proportion of fixed nitrogen in legume herbage from mixed pastures averaged 94 per cent at one field site in Queensland and 92 per cent at another, and was independent of legume yield. Where legumes are grown in monoculture, the proportion of legume nitrogen derived from fixation can be expected to vary widely depending on the availability of combined nitrogen in the soil (p. 270).

While there are many gaps in the present knowledge of factors influencing nitrogen fixation by forage legumes in the field, the major ecological factors and their importance in commercial agriculture are known with reasonable certainty.

Mineral elements

Deficiencies or excesses of mineral elements may affect nitrogen fixation directly, through effects on nodule function, or indirectly, through effects on assimilate supply from the host plant. Two methods are used to distinguish direct effects of environmental factors from their indirect effects (Robson 1978): first, determination of the interaction between the factor and the level of combined nitrogen (addition of combined nitrogen overcomes direct effects on nitrogen fixation); secondly, measurement of the factor's influence on plant nitrogen concentration (direct effects on nitrogen fixation influence the concentration; indirect effects do not).

Every one of the twelve essential plant nutrients (other than nitrogen) must be classed as of potential significance to nitrogen fixation through its indirect effects on growth of the host plant (Pate 1977). At least two of these elements influence fixation directly. Molybdenum is a constituent of the nitrogenase enzyme and the primary effect of molybdenum deficiency is on nitrogen fixation. The host plant has a smaller molybdenum requirement than the symbiosis. Cobalt deficiency also has a direct effect on nitrogen fixation, through its role in the vitamin B_{12} coenzymes (Bergersen 1977). Iron is a constituent of nitrogenase but there is no evidence that nodulated legumes have a special requirement for this element (Bergersen 1977). There is evidence that pH, calcium, phosphorus, sulphur, copper, and zinc may have some direct effect on nitrogen fixation (Munns 1977; Robson 1978), though their main effect appears to be on the host plant.

In agriculture, natural soil fertility and fertilizer practices have a major influence on legume growth and nitrogen fixation. Soils are most commonly deficient in phosphorus, potassium, and sulphur (Munns 1977). Increased legume growth and nitrogen content have been reported also from molybdenum fertilization in many parts of the world, but only rarely from cobalt and apparently never from iron. In some parts of the world atmospheric accessions of sulphur from industrial sources are sufficient to satisfy plant requirements. On acid tropical soils, the principal nutrient deficiencies affecting pasture legumes are of phosphorus, potassium, calcium, and sulphur (plus molybdenum); deficiencies of magnesium, copper, zinc, and boron may be locally important (Kerridge 1978).

Fertilization with mineral nutrients is a common means of altering the environment to promote growth and nitrogen fixation by pasture legumes. Table 8.1 shows a situation where 20 years of fertilization of a pasture with superphosphate resulted in legume dominance. In most cases, however, the period of legume dominance is short and followed by increased growth of other pasture species, or there is alternation of legume dominance and grass dominance (Rossiter 1966).

TABLE 8.1
The effect of single superphosphate on botanical composition of a pasture (from Crocker and Tiver 1948)

Species	Superphosphate (kg $ha^{-1} a^{-1}$)		
Component	nil	100	200
Grasses	80	25	14
Clovers	3	45	63
Other forbs	17	30	23

Arising from this evidence there is a common misconception that legumes are nutritionally more demanding than other plants (Munns 1977). Nodulation (Chapter 4) and nitrogen fixation (p. 266) have some special requirements, but 'there is no good evidence that legumes grown non-symbiotically differ consistently in their nutritional needs from other similarly large diverse groups of plants' (Munns 1977, p. 356). Nevertheless, there is wide variation amongst legumes and amongst grasses in nutrient absorption at low levels of supply and in the concentrations of nutrients required at functional sites in the plants (Robson and Loneragan 1978).

How is this generalization about relative nutritional needs to be reconciled with the field evidence that pasture legumes are favoured by mineral fertilizers? It appears that: (i) grasses have an advantage in inter-specific competition and resistance to grazing and so are dominant on unfertilized soils; (ii) nitrogen deficiency prevents the grasses from responding to an increased supply of other nutrients in many situations; in contrast, growth of effectively nodulated legumes is not limited by nitrogen supply; (iii) studies of nutrient requirements of grasses and legumes in monoculture may not give an accurate picture of competitive relations, where speed of uptake may be the critical attribute; (iv) further research may show legumes to be less efficient competitors for one or more limiting nutrients; and (v) some legumes do have a higher requirement for mineral nutrients than some grasses.

Support for the hypothesis that nitrogen deficiency often prevents grasses in mixed pastures from exercising their competitive advantage comes from the observation that additions of combined nitrogen as fertilizer or by symbiotic fixation usually swing the balance towards the grasses (and non-legume weeds). Also some soils provide sufficient combined nitrogen at least temporarily, for grasses to respond to mineral fertilizers. Fig. 8.1 illustrates the response of a *Stylosanthes guianensis* (stylo)–*Panicum maximum* (guinea grass) pasture on an ultisol newly cleared from jungle in peninsular Malaysia. Application of superphosphate caused strong *P. maximum* dominance in the first year but the mown sward reverted to a balance between grass and legume after 20 months (presumably because the supply of combined nitrogen had decreased). Without superphosphate the grass made little growth. Similar responses by grasses to superphosphate have occurred elsewhere (Henzell 1977).

Insufficient information has been published on responses to soil acidity (pH, excess aluminium or manganese) and salinity to decide whether or not there is a consistent difference between grasses and legumes. Whereas sodium chloride had a greater effect on the growth of nodulated *Glycine wightii* plants than on that of ammonium nitrate-fed controls (Wilson 1970), indicating a direct effect on nitrogen fixation, it depressed equally the yields of nodulated and of nitrate-fed lucerne plants (Bernstein and Ogata 1966). Excess manganese tends to limit nitrogen fixation and legume growth to about the same degree (Munns 1977).

Combined nitrogen

Addition of ammonium- and nitrate-nitrogen to nodulated legume plants has a deleterious effect on nodule structure and nitrogen fixation (Dart 1977). Pate (1977) has suggested that the added nitrogen reduces the supply of carbohydrates to underground parts of the plant and that,

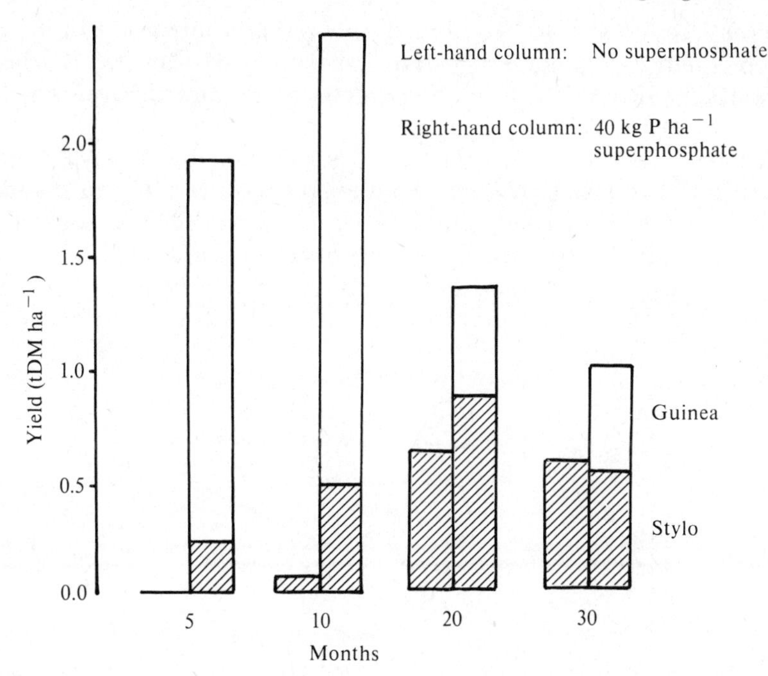

FIG. 8.1. The response to an initial application of triple superphosphate (40 kg P ha^{-1}) by a *Stylosanthes guianensis* cv. Schofield and *Panicum maximum* sward (Kerridge and Tham, unpublished results).

under such conditions, the nodules are affected more severely than the roots. Ammonium and nitrate suppress symbiotic nitrogen fixation in *M. sativa* only at concentrations much higher than those needed for suppression of nitrogenase activity in free-living nitrogen-fixing bacteria. On the other hand, low concentrations of nitrite-nitrogen quickly suppress nitrogenase activity (Kamberger 1977). Chen and Phillips (1977) have concluded that products of nitrate reduction, rather than a shortage of photosynthate, are responsible for the depression of nitrogen fixation by additions of combined nitrogen. There is no evidence that amino acid concentrations in the nodule regulate nitrogenase activity directly (Kamberger 1977; Pate 1977).

Research with nodulated legumes in monoculture has shown that yield and total nitrogen content are increased by an initial low level of combined nitrogen supply (Gibson 1977). This response is usually confined to the period of nodule formation, though there is also some evidence of a response in later phases of growth by crop legumes such as *Glycine max* (soya beans). Fig. 8.2 summarizes the effect of combined nitrogen on nitrogen fixation during the first season's growth of a forage legume. Above level 'A' combined nitrogen and nitrogen fixation are complementary. There is a good deal of evidence (Allos and Bartholomew 1959; Gibson 1977) that nodulated legumes use nitrogen which is available in the soil and fixation occurs only when the supply is inadequate. However, this conclusion is disputed by the evidence of Hardy and Havelka (1976) who found that exposure of *G. max* shoots to high carbon dioxide concentrations increased symbiotic nitrogen fixation and reduced the uptake of soil nitrogen (Table 8.2). This also occurs with *Trifolium repens* (white clover); it may be a specific effect of carbon dioxide acting via transpiration (Masterson and Sherwood 1978).

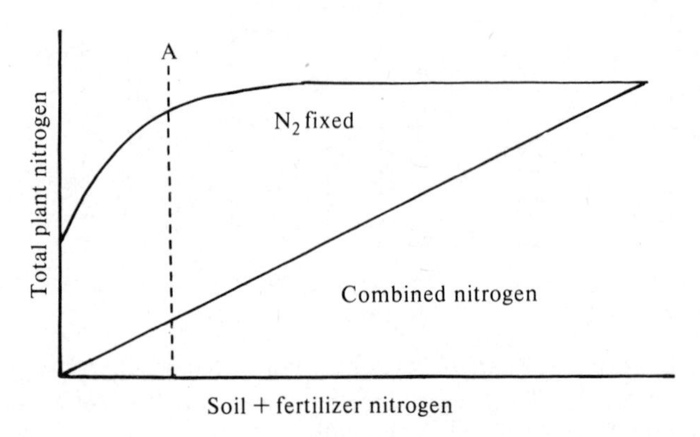

FIG. 8.2. The effects of combined nitrogen on the nitrogen assimilated by legumes over a growing season (from Gibson 1977).

TABLE 8.2

Effect of carbon dioxide enrichment on nitrogen assimilation by soya bean plants (from Hardy and Havelka 1976)

	Soil nitrogen (kg N ha^{-1})	Nitrogen fixed (kg N ha^{-1})	Proportion fixed (%)
Air-grown	219	76	26
CO_2-enriched	84	427	84

The supply of combined nitrogen is one of the most important ecological factors regulating growth and nitrogen fixation by legumes in mixed pastures (Donald 1963). Even in pastures containing a significant proportion of legume, grasses frequently have low nitrogen concentrations and respond to an increased supply of combined nitrogen (Ball, Molloy, and Ross 1978; Hoglund and Brock 1978) indicating that the non-legume component of such pastures is nitrogen deficient.

Furthermore, herbaceous pasture legumes usually do not persist in significant proportions in association with grasses that have been given an optimum supply of combined nitrogen (Whitehead 1970).

The superior competitive ability of grasses for other environmental factors appears to be the main reason why grasses have to be restricted by nitrogen deficiency to maintain legume growth and nitrogen fixation in mixed pastures. There is evidence also that, when nitrogen supply is the major factor limiting grass growth in pasture mixtures, grasses obtain the major share of the combined soil nitrogen (Fig. 8.3).

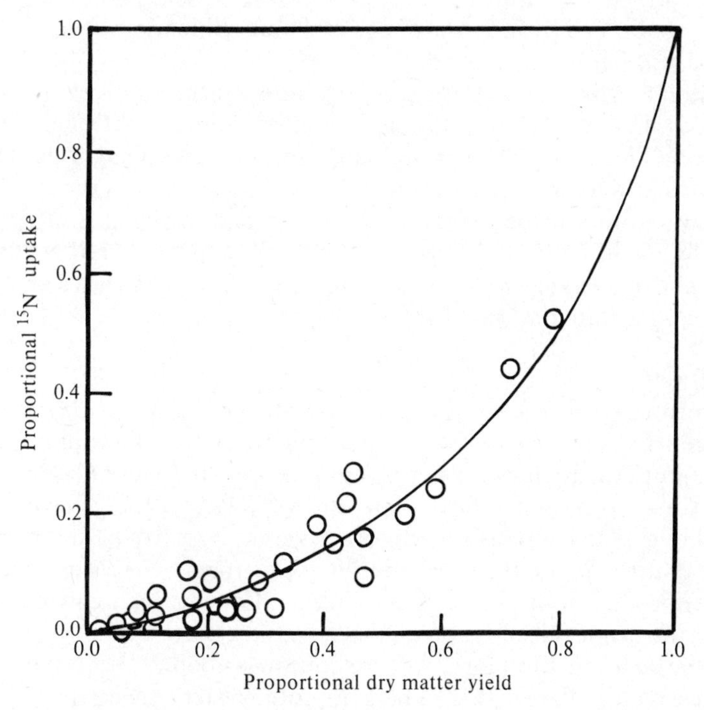

FIG. 8.3. The relation between ^{15}N uptake and yield of *Trifolium repens* cv. Ladino expressed as proportions of total sward uptake or yield (from Vallis 1978).

Attempts have been made to improve legume growth and nitrogen fixation by nitrogen fertilization, particularly at establishment (starter nitrogen). While this objective is theoretically sound (Fig. 8.2), the main practical effect of fertilizing mixed pastures or weedy stands of fodder legumes seems to be to intensify competition against the legume. Most of the fertilizer nitrogen is likely to be taken up by the grasses and weeds (Fig. 8.3). Because nitrogen levels in many soils are adequate to overcome nitrogen starvation during nodule formation, there is often no response to nitrogen fertilization even by legume monocultures (Gibson 1977).

The arguments in this section imply that the processes of nitrogen fixation and transfer of combined nitrogen to associated grasses in mixed pastures should eventually destroy the legume. This is not so in practice, probably because both the main pathways of nitrogen transfer (through excreta of grazing livestock, and through dead legume residues) are subject to large losses to air and water. Also, nitrogen from legumes may be rapidly immobilized into slowly-available organic forms in grassland soils. It appears that, provided nitrogen fixation is approximately balanced by nitrogen losses and immobilization, mixed pastures can be stable for long periods.

When fodder legumes are grown in monoculture, nitrogen fixation is reduced linearly as uptake of combined mineral nitrogen replaces fixation (Fig. 8.2). In theory the same applies to weedy stands of fodder legumes and to pastures, but the field evidence suggests that the reduction in legume nitrogen fixation by combined nitrogen is attributable mainly to competitive depression of host-plant growth and persistence. Over short periods, however, combined nitrogen substitutes for fixation (Moustafa, Ball, and Field 1969).

Temperature

Laboratory studies show that the chief effect of temperature on legume nitrogen fixation occurs via the host plant. Hence, the optimum constant temperature for nitrogen fixation usually lies close to or coincides with the optimum for host plant growth (Pate 1977). Nevertheless, nodulated plants are affected more severely by extreme temperatures than plants dependent on combined nitrogen (Mulder, Lie, and Houwers 1977) and affected plants have low nitrogen concentrations suggesting a specific effect on nitrogen fixation (Pate 1977). In temperate species the lower limit for nitrogen fixation is about 2 °C and in tropical species 10 °C (Pate 1977). The symbiotic system seems to be able to compensate to some extent for the adverse effect of moderately low temperature by increased nodule weight or activity (Gibson 1977).

Most of the above information pertains to root or whole-plant temperatures. Relatively little attention has been given to the effect of shoot temperature *per se* but there is evidence that it does have an effect (Gibson 1977), presumably mediated via photosynthesis. Short-term acetylene-reduction assays have demonstrated a very wide temperature tolerance by nitrogenase, the optimum spanning a range of 15 or 20 °C (Pate 1977).

Temperature is a prominent ecological factor determining the distribution, persistence, and yield of forage legumes in different parts of the world. For example, *T. repens* is best adapted to temperate latitudes and fails in the tropics. The converse is true of tropical pasture legumes. These differences in adaptation correlate with differences in optimum constant temperature for growth (20–25 °C for temperate legume; 25–33 °C for tropicals) and particularly with their tolerance of extreme low or high temperatures (McWilliam 1978). Moreover, with the exception of the equatorial tropics where monthly mean temperatures may vary by no more than 1 °C over the year, seasonal variations of temperature (as much as of rainfall or of light) determine annual patterns of legume growth and nitrogen fixation. The annual pattern recorded for nitrogen (acetylene) fixation in a grazed *T. repens* sward in Eire is shown in Fig. 8.4.

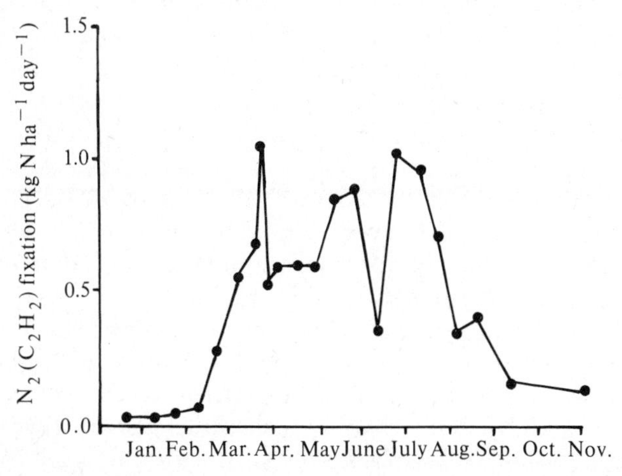

FIG. 8.4. The seasonal pattern of nitrogen (acetylene) fixation by a grazed *T. repens*–grass sward (from Masterson and Murphy 1976).

While there is no true grass–legume competition for temperature or heat energy, the different temperature responses of grasses and legumes can have important indirect competitive effects on legume productivity. Brougham, Ball, and Williams (1978) pointed out the complementary nature of the temperature responses of *T. repens* and temperate grasses in New Zealand, resulting in summer clover dominance following grass dominance in late autumn, winter, and spring. Blaser and Brady (1950) suggested that the ability of grasses to grow at lower temperatures than clovers in spring gives the grasses an advantage in competition for potassium.

Light

The effects of light on nitrogen fixation seem to be mainly due to variations in host-plant photosynthesis. Symbiotic nitrogen fixation exhibits a marked dependence on current photosynthesis (Pate 1977).

Within five hours of transferring *Trifolium subterraneum* plants from an illuminance of 620 $\mu Em^{-2} s^{-1}$ to one of 165 $\mu Em^{-2} s^{-1}$ the rate of nitrogen (acetylene) fixation fell by 40 per cent (Gibson 1976). Fixation does not cease immediately when photosynthesis stops. Detached nodules and nodulated roots continue to reduce acetylene for a time and nodules of some legumes reduce acetylene during the night (Bergersen 1977).

Light is usually a limiting factor for growth of pasture and fodder legumes and hence for their nitrogen fixation. Although photosynthesis by legume leaves shows saturation at 30–50 per cent of maximum sunlight on a clear day (Ludlow 1978) they commonly receive irradiances lower than that because of (i) atmospheric interception by cloud or haze; (ii) diurnal or seasonal variation of solar elevation; and (iii) self-shading. Except for the early stages of pasture establishment or seasonal re-establishment, or for regrowth following severe defoliation or seasonal dormancy, legumes are light-limited for most of the time (Ludlow 1978).

In the canopies of mixed pastures, shading by grasses and forbs further reduces the irradiance received by legume leaves (Donald 1963). Height is the main factor determining preferential access to light (Ludlow 1978). Light intensity falls sharply as it penetrates into pasture canopies and a short legume such as *T. subterraneum* is at a considerable disadvantage compared with a taller grass such as *Lolium rigidum* (Wimmera rye grass) unless grass growth is restricted by nitrogen deficiency (Fig. 8.5). Taller legume shrubs, such as *Leucaena leucocephala* (leucaena) or *Stylosanthes scabra* (shrubby stylo) and legumes that climb over other plants, e.g. *Macroptilium atropurpureum* (siratro) are affected less and may even dominate the grasses (Tothill and Jones 1977). The profiles of light intensity and leaf area for the sprawling tropical legume *Desmodium intortum* and the tall grass *Setaria anceps* in Fig. 8.6 contrast with those in Fig. 8.5. The photosynthetic rate of the *D. intortum* leaves was closely correlated with light levels in the canopy.

The widespread practical finding that short legumes such as the clovers and *Stylosanthes humilis* (Townsville stylo) are favoured by frequent low cutting or grazing of the sward (Donald 1963; Gillard and Fisher 1978) is usually interpreted as evidence of strong competition for light. However, the possibility of below-ground competition for water and nutrients has probably been greatly underestimated, because it is invisible. The problem is to delineate these different forms of competition (Rhodes and Stern 1978).

Nitrogen fixation is usually greater under longer than shorter days, but it is difficult to say if this is a direct effect of photoperiod or an indirect effect via flowering or yield (Ludlow 1978). A reduction of nitrogen (acetylene) fixation in June (Fig. 8.4) coincided with flowering

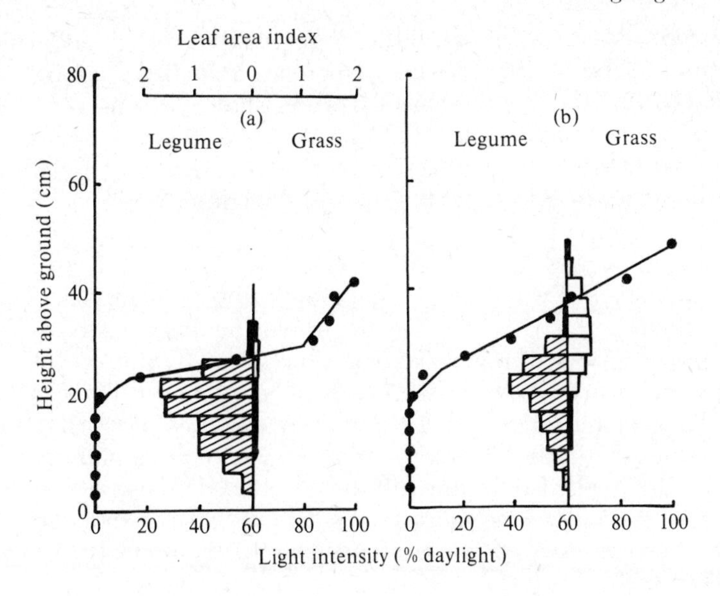

FIG. 8.5. The vertical distribution of leaf area and of light intensity in swards of *Trifolium subterraneum* cv. Tallarook and *Lolium rigidum* cv. Wimmera with a low (a) or a higher (b) supply of combined nitrogen (from Donald 1963).

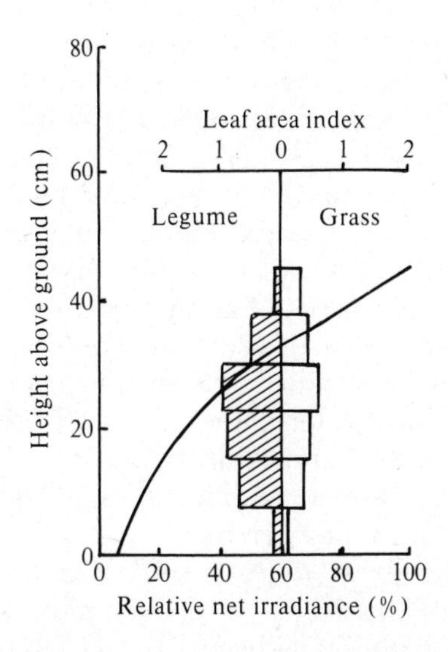

FIG. 8.6. The vertical distribution of leaf area and irradiance in a sward of *Desmodium intortum* cv. greenleaf and *Setaria anceps* cv. Nandi (Wilson and Ludlow, unpublished results).

of *T. repens*, but studies with other legumes showed no such effect of flowering (Gibson 1977), while Hoglund and Brock (1978) have suggested that the depression of fixation shown in Fig. 8.4 may be attributable to uptake of mineral nitrogen from the soil. Fruiting and senescence bring an end to seasonal growth and nitrogen fixation by determinate types of legumes but not by indeterminate types.

Water

Water deficits and waterlogging have both to be considered. Surprisingly little research has been done to unravel the mechanisms by which water stress affects nitrogen fixation (Sprent 1976; Gibson 1977). In *G. max* the primary effect is on host-plant photosynthesis (Huang, Boyer, and Vanderhoef 1975), and it is assumed here that forage legumes behave in the same way. Water stress reduces photosynthesis first by reducing leaf area (due to reduced expansion), then by stomatal closure, and finally, with more severe deficits, by leaf and stem death (Hsiao, Fereres, Acevedo, and Henderson 1976; Johns 1978; Turner and Begg 1978).

Water deficits restrict the yield and nitrogen fixation of forage legumes in many parts of the world. Legumes show a number of adaptations that enable them to survive the water deficits caused by lack of rainfall and intensified by competition. They may avoid the deficit by drawing on water at depths inaccessible to other plants. *M. sativa* uses water to depths of 6 metres or more, where soil and climate permit (Bolton, Goplen, and Baenziger 1972). They may reduce the impact of soil water deficits by restricting water loss from the plant. *M. atropurpureum* does this by shedding leaves and closing its stomata (Peake, Stirk, and Henzell 1975). Conversely, the inferior herbage yields of *T. repens* under dry field conditions in northern New South Wales (Johns and Lazenby 1973) have been attributed to the marked inability of its stomata to control leaf hydration (Johns 1978). Finally, they may be able to tolerate desiccation. The seeds of annual pasture legumes are usually highly resistant to seasonal desiccation. Some tropical legumes are quite tolerant of arid environments; for instance, *Stylosanthes scabra* and *S. viscosa* show pre-dawn water potentials as low as −40 bars and still retain green leaf (McCown and Wall 1977).

Grasses and weeds compete with legumes for soil water and thereby reduce legume yield and nitrogen fixation. The key factor in competition appears to be the ability to reach and use water first (Donald 1963; Gillard and Fisher 1978). New root growth plays a large part in this, even with perennial plants (Caldwell 1976; Hsiao *et al.* 1976). Pasture plants may differ also in their ability to reduce the water potential of the root zone; they certainly differ in plant water potentials recorded during drought. On one occasion, dawn leaf water potentials of *Cenchrus ciliaris* (buffel grass) *M. sativa*, and *M. atropurpureum* associated in the same pasture were −25, −44, and −6 bars respectively (Peake *et al.* 1975).

Cultivated forage legumes are generally not well adapted to waterlogging and herbaceous legumes are not commonly found growing in very wet habitats in nature (Pate 1976), but there are exceptions. *Neptunia oleracea* is an aquatic plant with nodules that grow freely in water (Schaede 1940). The same paper refers to nodulation of *Aeschynomene paniculata* plants on very wet sites. Waterlogging depresses nitrogen fixation by cultivated species largely as a result of oxygen deficiency (Sprent 1976). Although effects of waterlogging on host plant performance cannot be discounted, plants supplied with nitrate grow almost normally when waterlogged (Pate 1976).

Soil gases

A sufficient supply of nitrogen gas to the nodules is an obvious requirement for nitrogen fixation, but there is virtually nothing in the literature to indicate that soil N_2 concentrations limit fixation in practice. Symbiotic fixation is apparently independent of the pN_2 above 0.10 to 0.15 atmospheres (Wilson 1940). In contrast, oxygen supply does seem to be an important factor (Sprent 1976). Nitrogen fixation by legumes in water culture is maximum at pO_2 of 0.20–0.21 and it is suppressed by both sub- and supra-optimal oxygen levels (Pate 1977) with differences between species (Bergersen 1977). In the field, soil oxygen levels may be lowered by poor permeability, waterlogging, or rapid respiration. Responses of nitrogen fixation to variations in carbon dioxide concentration surrounding legume roots have been inconsistent (Gibson 1977). High carbon dioxide concentrations are common in soils.

Grazing and cutting

Forage crops are grazed by a variety of animals ranging from vertebrates (domesticated and feral) through insects to nematodes. The short-term effects of grazing and mechanical defoliation of legumes are explicable chiefly in terms of reduced host-plant photosynthesis and supply of assimilates to root nodules (Pate 1977). There are a few examples of insects or nematodes attacking nodules preferentially (Gibson 1977), but this selectivity is believed to be uncommon.

The immediate effect of removing photosynthetic tissue from legume plants by grazing, treading, or cutting is a reduction of nitrogen fixation in the nodules. Severe defoliation causes nodules to die and decay (Gibson 1977). The initiation of new nodules and regrowth of nodule tissue are usually rapid. Nodulated *M. sativa* and *T. subterraneum* plants regrew as rapidly as those supplied with combined nitrogen (Rogers 1969; Gibson 1977), though Sasaki and Kudo (1966) found that *T. repens* made better regrowth with combined nitrogen.

The intensity of grazing and cutting has marked effects on legume productivity, and the frequency of grazing may also be important. Thus, legumes such as *L. leucocephala* and *M. sativa* produce less under continuous than under rotational grazing (Jones 1973; Leach 1978). The basic philosophy of grazing management is to harvest sufficient plant material for efficient livestock production and to maintain the stability and productivity of the pasture. Cutting schedules for fodder crops such as *M. sativa* have the same aims. In addition to the direct effects of grazing and cutting on the plants and soil there are indirect effects on the physical environment in a sward, e.g. on the temperature and humidity.

There are also large indirect effects on plant competition. The lower-growing legumes such as the clovers and *S. humilis* are favoured by heavy close grazing or cutting (within limits—Rossiter 1978) and the taller ones are not (Brougham *et al.* 1978; Gillard and Fisher 1978; Jones and Jones 1978). All animals graze selectively, given the opportunity (Watkin and Clements 1978) and this must affect nitrogen fixation by pasture legumes through its influence on competition for environmental factors. A strong preference by cattle or sheep for the legume component of mixed pastures makes it more difficult to maintain legume productivity. Conversely, selective grazing of grasses and forbs favours the legumes. Some of the most successful pasture legumes, such as *T. subterraneum*, *M. atropurpureum*, and *S. humilis*, seems to be slightly unpalatable during active vegetative growth.

Insects also graze preferentially above and below ground, some species preferring legumes and others grasses (Wallace 1970; Brougham *et al.* 1978). Insects consume only a small proportion of plant production, except when their numbers reach plague proportions. The trouble with some insects is that they damage far more tissue than they consume. This is particularly true of those that feed on stems and roots, for instance the stemborer (*Zaratha* sp) that attacks woody-stemmed *Stylosanthes* spp in Colombia (CIAT 1976), and of those that transmit diseases.

Diseases

Forage legumes are attacked by a variety of pathogenic fungi, bacteria, viruses, and mycoplasma. So far as is known, they have no direct effect on symbiotic nitrogen fixation (Gibson 1977), but they can have a critical indirect influence on fixation through their effects on growth and persistence of the host plants, as illustrated by the following examples. Several fungal and viral diseases damage *T. repens* populations severely (Brougham *et al.* 1978). It is likely that up to 0.5×10^6 ha of *T. subterraneum* pastures are affected by the disease clover scorch (*Kabatiella caulivora*) in southern Australia (Rossiter 1978). In the USA, diseases cause estimated annual losses of 24 per cent of the forage and 9 per cent of the seed yield in *M. sativa* (Graham, Kreitlow, and Faulkner 1972). Foliar blight caused by *Rhizoctonia solani* is thought to be a major reason why *M. atroporpureum* fails in the wet tropics (Jones and Jones 1978). The susceptibility to anthracnose disease (*Colletotrichum gloeosporioides* and *C. dematium*) of the Australian cultivars of *S. guianensis* (CIAT 1972) has prevented their commercial use in South America.

Through their selective effects on legumes, grasses, and forbs, diseases can have large indirect effects on the ability of legumes to compete for light, water, and nutrients. Control of diseases affecting pasture and forage legumes is based almost entirely on quarantine to prevent their spread and on selection of resistant plant genotypes.

Man

Man's activities have had profound effects on the ecology of forage legumes. In fact, forage legumes are rarely prominent except in commercial agriculture. While there are a few reports of unimproved pastures containing a high proportion of legume in the herbage (*Desmodium barbatum* comprised 40 per cent of the herbage in Honduras and *Zornia glochidiata* 38 and 77 per cent of the herbage in two natural pastures in Senegal—Döbereiner and Campelo 1977), most natural grazing lands or grazing lands induced by removal of trees and shrubs contain only a small proportion of herbaceous legumes. This common scarcity of pasture legumes can not be attributed entirely to the effects of soil or climate, because legumes often make up a substantial proportion of the woody vegetation in the same environment. This is true of the *Acacia*-dominant woodlands of inland Australia, and legumes constitute a high proportion of the trees in the African savannas and woodlands (Baker 1978) and in the edaphic (Cerrado) savannas of Brazil (Döbereiner and Campelo 1977).

Man has manipulated the environment to favour the growth and nitrogen fixation of forage legumes by inoculation of seed (Chapter 4); mineral fertilization; irrigation; management of grazing and cutting; control of pests, diseases, and weeds; and by the breeding and selection of superior genotypes. Perhaps most important of all man's activities has been the harvesting and spreading of seed. Thus *T. pratense* (red clover) seed was taken from the continent of Europe to England, where it had a major impact on agriculture (Cooper 1962), and to North America, where it became the cornerstone of cropping in the eastern half of the continent (Fergus and Hollowell 1960).

Similarly, *T. repens* was taken to New Zealand and it is now the major legume in nearly 9×10^6 ha of sown pastures that have made a major contribution to livestock production (Daly 1973). Fertilized pastures of introduced subterranean clover and annual medics (*Medicago* spp) have been sown on about 16×10^6 ha in southern Australia (Carter 1966), of which more than 12 × 10^6 ha carry *T. subterraneum* (Rossiter 1978). Seed of *M. sativa* has been taken from its centre of origin in the Near East and Central Asia to all parts of the world. It is now the world's most important fodder crop occupying an area estimated at 33 × 10^6 ha (Bolton *et al.* 1972).

The use of legumes for improvement of tropical pastures is a more recent development (Skerman 1977). *S. humilis* is endemic in Central and South America (Gillard and Fisher 1978). Seed was brought to northern Australia at the beginning of this century and *S. humilis* pastures now occupy an area of about 0.6 × 10^6 ha. Purposeful introduction of seed of tropical legumes has resulted in the domestication of other species in Australia. About 0.1 × 10^6 ha of a bred variety of *M. atropurpureum* have been sown (Shaw and Whiteman 1977).

For pasture legumes to be profitable they must have the ability, with satisfactory management, to maintain the stand of plants indefinitely by nature re-seeding or vegetative reproduction. Seed is then sown only to introduce new genotypes, to rejuvenate mismanaged pastures, or to re-establish pasture after crops. Fodder legumes have to be re-sown at regular intervals. *M. sativa* is potentially a long-lived perennial, and a few stands have persisted for 25 years in southern Australia, but the average life for intensively-used fodder stands in Europe and North America is about 4 years (Leach 1978). Amongst the pasture and fodder legumes, the longest lived individual plants occur in the browse shrubs such as *L. leucocephala*. Exactly how long, is not yet known.

8.3 Measuring fixation in the field

One of the major weaknesses in current knowledge of the ecology of nitrogen fixation by pasture and forage legumes is in the understanding of fixation in the field. Fig. 8.7 represents the main pathways for flow of nitrogen in grazed pastures, and serves as a framework for discussion of the three groups of methods used to measure fixation in the field, which are based on total nitrogen contents, acetylene reduction, and the use of $^{15}N_2$ as a tracer.

Total nitrogen contents

Nearly all the published fixation rates have been obtained from analyses of plants and soils for total nitrogen by Kjeldahl digestion. The two main problems lie in measuring accurately the total quantity of nitrogen used in legume growth and in measuring the uptake of combined nitrogen from soil (and atmosphere?).

Measurement of the total quantity of nitrogen used in growth of a legume between two sampling times in the field is not a simple matter. It is easiest for an annual sowing of a fodder legume that is harvested,

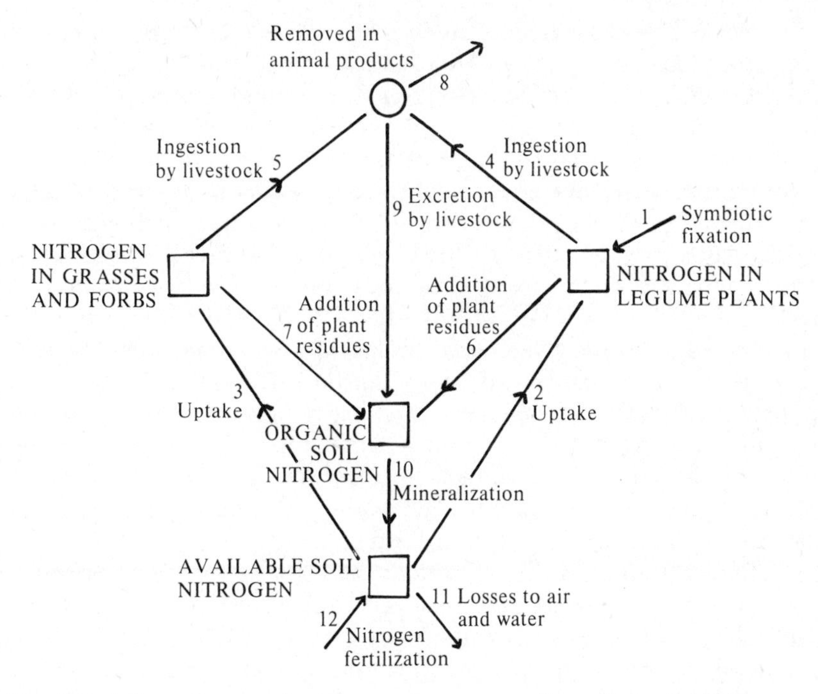

FIG. 8.7. A flow-diagram for nitrogen in pastures (from Henzell 1970).

and sampled, just once at the peak of vegetative growth. Subtraction of the nitrogen content of the seed from the total nitrogen in the harvested herbage, stubble, and roots gives the amount of nitrogen assimilated by the crop. It is most difficult for grazed perennial pastures. There, the nitrogen in the plants at a particular time comprises combined nitrogen that was in the plants at the previous time of sampling plus fixation and uptake during the intervening period (pathways p1 and p2, Fig. 8.7), less the nitrogen removed in plant material that is grazed or shed (p4 and p6) and less any nitrogen lost from the remaining tissues. Because of the operational difficulty of measuring intake by grazing animals, measurements usually are made in the temporary absence of grazing, and because no one has yet discovered how to measure accurately the amount of nitrogen used in growth of roots of perennial pasture plants, usually only the shoots are sampled.

Attempts have been made to measure uptake of combined nitrogen by including parallel treatments sown to the same legume without nodulation, or to a grass. The weakness of this procedure is the assumption that the parallel treatments take up combined nitrogen with the same efficiency as the nodulated legumes. It is virtually certain that they do not. For example, even with a small volume of soil in pots, effectively nodulated temperate pasture legume monocultures typically take up only 50–80 per cent as much labelled fertilizer nitrogen as grass monocultures (Vallis 1978). There is published evidence also of differences between an ineffectively nodulated legume and a grass (Nutman 1976).

Another method that has been used on pastures is to determine the accumulation of soil nitrogen. Allowance may also be made for nitrogen removed by livestock, nitrogen added in rainfall, and any change in herbage nitrogen. The defect of this method is that no account is taken of nitrogen losses (Simpson, Bromfield, and Jones 1974). There is growing evidence of important losses of ammonia and dinitrogen oxide from grazed pastures (Denmead, Simpson, and Freney 1974; Catchpoole 1976; Freney, Denmead, and Simpson 1978) and by inference, of important losses of nitrogen gas. Leaching of nitrate occurs where soil and climate permit. Hence the increase in total soil nitrogen usually underestimates the rate of symbiotic nitrogen fixation. Indeed, there may be little change in total soil nitrogen as the soil organic matter content approaches equilibrium (Jackman 1964).

Drainage lysimeters were formerly thought to provide the most accurate estimates of legume nitrogen fixation (Henzell and Norris 1962), but this is true only if gaseous losses are negligibly small. Lysimeter experiments are practicable only with mechanically or manually harvested fodder legumes.

Acetylene reduction

The acetylene-reduction method for measuring nitrogenase activity to estimate nitrogen fixation by legumes has been used widely since it was first described in 1966 (Koch and Evans 1966). The analysis of acetylene, and of the reduction product ethylene, is routinely performed by gas chromatography. The whole technique is simple, inexpensive, and sensitive (Burris 1974; Hardy and Holsten 1977). In practice, a series of short-term rates are measured and integrated over time to estimate the amount of nitrogen fixed, usually designated as nitrogen (acetylene) fixed to identify the origin of the results.

The acetylene-reduction method has two important limitations. First, the roots of the legumes have to be enclosed for assay, so it is impossible to use it on undisturbed swards in the field. The technique preferred at present encloses the plants and a core of topsoil in a small chamber with the environment as close to ambient field conditions as possible. Secondly, it is an indirect method that has to be calibrated against a direct measurement of nitrogen fixation. The conversion factor often deviates quite widely from the theoretical ratio of 1 mole of nitrogen to 3 moles of acetylene (Hardy, Burns, and Holsten 1973). It is recommended that the factor applicable to each system be determined experimentally (Hardy and Holsten 1977). Masterson and Murphy

(1976) used a ratio of 1:3 to calculate a nitrogen fixation rate of 183 kg N ha^{-1} from the results given in Fig. 8.4.

$^{15}N_2$ as a tracer

The first way of using ^{15}N is as a tracer nitrogen gas (p1, Fig. 8.7). This technique has the same limitation as acetylene reduction, in that the soil:plant system has to be enclosed in a gas-tight chamber for exposure to $^{15}N_2$. The sensitivity of measurement of $^{15}N_2$ reduction is less than that of acetylene reduction, though far greater than the sensitivity of measurement of changes in total plant nitrogen. A disadvantage of the $^{15}N_2$ technique is the high cost of the stable isotope and the mass spectrometer required for its accurate analysis. Nevertheless, measurement of $^{15}N_2$ fixation remains the absolute standard to which other methods, including acetylene reduction, are referred (Burris 1974).

The second way of using ^{15}N is as a tracer for combined mineral nitrogen (p2, Fig. 8.7). For many years ^{15}N-labelled fertilizer has been used in pot experiments with nodulated legumes to distinguish fixation from uptake of combined nitrogen (McAuliffe, Chamblee, Uribe-Arango, and Woodhouse 1958). Later the technique has been extended by using small additions of $^{15}NH_4^+$ (e.g. 0.3 kg N ha^{-1} $month^{-1}$, 30 per cent atom excess) or $^{15}NO_3^-$ to label combined soil nitrogen taken up by grass–legume mixtures in the field (Haystead and Lowe 1977; Vallis *et al.* 1977; Goh, Edmeades, and Robinson 1978). The same technique has been developed from the concept of A values (Fried and Broeshart 1975; Fried and Middelboe 1977) and used to measure nitrogen fixation by *T. subterraneum* in field plots (Phillips and Bennett 1978) and lysimeters (Williams, Jones, and Delwiche 1977).

The calculation of the proportion of fixed nitrogen in the total legume nitrogen depends on three assumptions: that the legume and non-legume components absorb $^{15}N_2$ and combined nitrogen in the same ratio; that during the test none of the legume nitrogen is transferred to the associated non-legumes (McAuliffe *et al.* 1958), and that there is no nitrogen fixation associated with the non-legumes. The first assumption may be invalidated if the foliage absorbs combined nitrogen from the atmosphere. There is published evidence of an ammonia cycle within pasture canopies (Denmead, Freney, and Simpson 1976), but the significance of advective ammonia is largely unknown. Nevertheless, the technique is attractive for use with pasture legumes in the field because the only disturbance is the exclusion of grazing during the test. However, there is still the problem, mentioned above, of measuring accurately the total flux of nitrogen through the legume plants.

Amount of fixation

Published estimates tend to be biased because most of them have been obtained from mown or infrequently grazed swards dominated by the legumes. Under commercial agricultural conditions both the legume yields and the nitrogen fixation rates are likely to be lower (Vallis 1972).

Fixation rates in excess of 500 kg N ha^{-1} a^{-1} for *T. repens* in New Zealand were estimated from the total nitrogen contents of herbage (Melville and Sears 1953) and of herbage and a nitrogen-deficient soil (Sears, Goodall, Jackman, and Robinson 1965). Legume dominance is usually a transitory phase quickly replaced under normal grazing by grass dominance and lower rates of legume growth and nitrogen fixation (Brougham *et al.* 1978). While the estimates of $>$ 500 kg N ha^{-1} a^{-1} are too high, the figure of 120 kg N ha^{-1} a^{-1} from soil nitrogen changes in New Zealand (Walker, Thapa, and Adams 1959) is probably too low. Hoglund and Brock (1978) estimated nitrogen (acetylene) fixation rates of 180 kg N ha^{-1} a^{-1} in a dry warm summer and 230 kg N ha^{-1} a^{-1} in a moist cool summer. For a *T. repens* sward in Northern Ireland the annual nitrogen (acetylene) fixation rate was calculated to be 270 kg N ha^{-1} (Halliday and Pate 1976).

Fixation rates by *T. subterraneum* in southern Australia probably fall between the 24–80 kg N ha^{-1} a^{-1} estimated from soil nitrogen changes (Henzell 1970; Nutman 1976) and the 160–340 kg N ha^{-1} a^{-1} estimated for total plant nitrogen flux (Lapins and Watson 1970; Simpson *et al.* 1974; Nutman 1976). Comparable figures for tropical pastures in Queensland were a net decrease (during drought) to 80 kg ha^{-1} a^{-1} soil nitrogen change (Vallis 1972) and 40–336 kg ha^{-1} a^{-1} total plant nitrogen flux (Henzell 1968).

Mowing and removal is a realistic form of management for estimating nitrogen fixation by fodder legumes. Table 8.3 shows the total nitrogen contents of selected legumes in Georgia, USA, at the end of 176 days of 'excellent growing conditions'. As no estimate was made of the uptake of combined nitrogen by these monocultures, the figures represent the upper limit of potential nitrogen fixation. Nitrogen-fixation

TABLE 8.3

Nitrogen in tops and roots of legume plants at Tifton, Georgia (from Burton 1976)

Legume	Tops (kg ha^{-1})	Roots (kg ha^{-1})	Total (kg ha^{-1})
Lupinus angustifolius (cv. Frost)	383	47	430
Trifolium incarnatum (cv. Chief)	211	15	226
T. pratense (cv. Ken star)	175	111	286
T. repens (cv. SI)	209	46	255
T. subterraneum (cv. Mt. Barker)	199	48	247
T. vesiculosum (cv. Meechee)	253	38	291
Vicia hirsuta	326	69	395

rates of 90–430 kg N ha^{-1} a^{-1} by *M. sativa* monocultures in Britain were estimated from herbage nitrogen contents with allowance for soil nitrogen uptake (Nutman 1976). Nutman quotes similar values for lucerne in the United States.

There are no statistics for nitrogen fixation by pasture and fodder legumes in agriculture, corresponding, say, to statistics for grain yield. The existing data on nitrogen fixation have been obtained from so few sites to be almost anecdotal. Unless there is a technical breakthrough in measurement of fixation in the field, it may be best to approach the problem through computer simulation, using as inputs available meterological and fertilizer data, and more easily measurable crop attributes such as yield of dry matter and plant populations. There have been preliminary attempts to relate nitrogen-fixation rates to legume yields (Henzell 1968; Hoglund and Brock 1978).

8.4 Concluding comments

Because of the scarcity of specific information about the effects of environmental factors on nitrogen fixation by forage legumes in the field, this chapter has drawn heavily on the literature dealing with the ecology of legume growth. While legume growth must set an upper limit to fixation, and the two will usually be closely correlated, it is desirable to find out much more about actual rates of fixation and the factors influencing them, particularly the substitution of uptake of combined nitrogen for fixation. Another surprising area of ignorance is the mechanisms of competition in mixed pastures, where the balance between legumes, grasses, and forbs has a dominant influence on legume growth and nitrogen fixation. It should be easier to maximize nitrogen fixation if more were known about competition for light, water, combined nitrogen, and other nutrients in the field. It may be easier to maintain a stable pasture with a legume that is a stronger competitor for combined nitrogen as Vallis (1978) has suggested. Would it fix more nitrogen or less?

The importance of forage legumes has declined during recent years in Europe and parts of North America, though not elsewhere. Their future role depends very much on the price of synthetic nitrogen fertilizers, in turn dependent on the cost of hydrogen from fossil fuels. The present indications are that synthetic nitrogen will remain expensive. There have been major developments recently in the search for tropical pasture legumes (Hutton 1970) and it is in the developing countries of the tropics that future advances may occur. There is tremendous scope in Africa, Central America, South America, and even in the intensively-farmed lands of Asia, especially if legume nitrogen fixation can be exploited to raise the yield of food crops.

References

Allos, H. F. and Bartholomew, W. V. (1959). *Soil Sci.* **87**, 61.

Baker, H. G. (1978). In *Plant relations in pastures* (ed. J. R. Wilson) p. 368. CSIRO, Melbourne.

Ball, R., Molloy, L. F., and Ross, D. J. (1978). *N.Z. Jl agric. res.* **21**, 47.

Bergersen, F. J. (1977). In *A treatise on dinitrogen fixation.* Section III *Biology* (ed. R. W. F. Hardy and W. S. Silver) p. 519. John Wiley, New York.

Bernstein, L. and Ogata, G. (1966). *Agron. J.* **58**, 201.

Blaser, R. E. and Brady, N. C. (1950). *Agron. J.* **42**, 128.

Bolton, J. L., Goplen, B. P., and Baenziger, H. (1972). In *Alfalfa, science and technology.* Agronomy Monograph No. 15 (ed. C. H. Hanson) p. 1. American Society of Agronomy, Madison.

Bremner, J. M. (1965). In *Methods of soil analysis.* Part 2. *Chemical and microbiological properties.* Agronomy Monograph No. 9 (ed. C. A. Black) p. 1179. American Society of Agronomy, Madison.

Brougham, R. W., Ball, P. R., and Williams, W. M. (1978). In *Plant relations in pastures* (ed. J. R. Wilson) p. 309. CSIRO, Melbourne.

Burris, R. H. (1974). In *The biology of nitrogen fixation* (ed. A. Quispel) *Frontiers of Biology,* Vol. 33, p. 9. North-Holland, Amsterdam.

Burton, G. W. (1976). In *Biological N fixation in forage–livestock systems.* ASA Special Publication No. 28 (ed. C. S. Hoveland) p. 55. American Society of Agronomy, Madison.

Caldwell, M. M. (1976). In *Water and plant life.* Problems and Modern Approaches. Ecological Studies 19 (ed. O. L. Lange, L. Kappen, and E. D. Schulze) p. 63. Springer Verlag, Berlin.

Carter, E. D. (1966). *Proc. 9th Int. Grassland Congr.*, São Paulo, 1965, Vol. 2, p. 1027.

Catchpoole, V. R. (1976). Aust. CSIRO Div. Trop. Crops and Past. Div. Rep. 1975–6, p. 55.

Chen, P. C. and Phillips, D. A. (1977). *Pl. Physiol., Lancaster* **59**, 440.

CIAT (1972). Colombia, Centro Internacional de Agricultura Tropical Ann. Rep. 1972, p. 14.

—— (1976). Colombia, Centro Internacional de Agricultura Tropical Ann. Rep. 1976, p. C7.

Cooper, M. McG. (1962). *Grass farming.* Dairy Farmers Books, Ipswich.

Crocker, R. L. and Tiver, N. S. (1948). *J. Br. Grassld Soc.* **3**, 1.

Daly, G. T. (1973). In *Pastures and pasture plants* (ed. R. H. M. Langer) p. 1. A. H. & A. W. Reed, Wellington.

Dart, P. J. (1977). In *A treatise on dinitrogen fixation.* Section III *Biology* (ed. R. W. F. Hardy and W. S. Silver) p. 367. John Wiley, New York.

Denmead, O. T., Freney, J. R., and Simpson, J. R. (1976). *Soil Biol. Biochem.* **8**, 161.

—— Simpson, J. R., and Freney, J. R. (1974). *Science, N.Y.* **185**, 609.

Döbereiner, J. and Campelo, A. B. (1977). In *A treatise on dinitrogen fixation.* Section IV *Agronomy and ecology* (ed. R. W. F. Hardy and A. H. Gibson) p. 191. John Wiley, New York.

Donald, C. M. (1963). *Adv. Agron.* **15**, 1.

Fergus, E. N. and Hollowell, E. A. (1960). *Adv. Agron.* **12**, 365.

Freney, J. R., Denmead, O. T., and Simpson, J. R. (1978). *Nature, Lond.* **273**, 530.

Fried, M. and Broeshart, H. (1975). *Pl. Soil* **43**, 707.

—— and Middelboe, V. (1977). *Pl. Soil* **47**, 713.

Gibson, A. H. (1976). In *Symbiotic nitrogen fixation in plants* (ed. P. S. Nutman) p. 385. Cambridge University Press.

—— (1977). In *A treatise on dinitrogen fixation*. Section IV *Agronomy and ecology* (ed. R. W. F. Hardy and A. H. Gibson) p. 393. John Wiley, New York.
GILLARD, P. and FISHER, M. J. (1978). In *Plant relations in pastures* (ed. J. R. Wilson) p. 340. CSIRO, Melbourne.
GOH, K. M., EDMEADES, D. C., and ROBINSON, B. W. (1978). *Soil Biol. Biochem.* **10**, 13.
GRAHAM, J. H., KREITLOW, K. W., and FAULKNER, L. R. (1972). In *Alfalfa science and technology*. Agronomy Monograph No. 15 (ed. C. H. Hanson) p. 497. American Society of Agronomy, Madison.
HALLIDAY, J. and PATE, J. S. (1976). *J. Br. Grassld Soc.* **31**, 29.
HARDY, R. W. F. and HAVELKA, U. D. (1976). In *Symbiotic nitrogen fixation in plants* (ed. P. S. Nutman) p. 421. Cambridge University Press.
—— and HOLSTEN, R. D. (1977). In *A treatise on dinitrogen fixation*. Section IV *Agronomy and ecology* (ed. R. W. F. Hardy and A. H. Gibson) p. 451. John Wiley, New York.
—— BURNS, R. C., and HOLSTEN, R. D. (1973). *Soil Biol. Biochem.* **5**, 47.
HAYSTEAD, A. and LOWE, A. G. (1977). *J. Br. Grassld Soc.* **32**, 57.
HENZELL, E. F. (1968). *Trop. Grasslds* **2**, 1.
—— (1970). *Proc. 11th Int. Grassld Congr.*, Surfers Paradise, 1970, p. A112.
—— (1977). In *Tropical forage legumes* (ed. P. J. Skerman) p. 86. FAO, Rome.
—— and NORRIS, D. O. (1962). In *A review of nitrogen in the tropics with particular reference to pastures*. Bull. 46. Commonw. Bur. Past. Fld Crops, p. 1.
HOGLUND, J. H. and BROCK, J. L. (1978). *N. Z. Jl agric. Res.* **21**, 73.
HSIAO, T. C., FERERES, E., ACEVEDO, E., and HENDERSON, D. W. (1976). In *Water and plant life. Problems and modern approaches*. Ecological Studies 19 (ed. O. L. Lange, L. Kappen, and E. D. Schulze) p. 281. Springer Verlag, Berlin.
HUANG, C. Y., BOYER, J. S., and VANDERHOEF, L. N. (1975). *Pl. Physiol., Lancaster* **56**, 228.
HUTTON, E. M. (1970). *Adv. Agron.* **22**, 1.
JACKMAN, R. H. (1964). *N.Z. Jl agric. Res.* **7**, 445.
JOHNS, G. G. (1978). *Aust. J. Pl. Physiol.* **5**, 113.
—— and LAZENBY, A. (1973). *Aust. J. agric. Res.* **24**, 797.
JONES, R. J. (1973). Aust. CSIRO Div. Trop. Agron. Ann. Rep. 1972–3, p. 15.
—— and JONES, R. M. (1978). In *Plant relations in pastures* (ed. J. R. Wilson) p. 353. CSIRO, Melbourne.
KAMBERGER, W. (1977). *Arch. Mikrobiol.* **115**, 103.
KERRIDGE, P. C. (1978). In *Mineral nutrition of legumes in tropical and sub-tropical soils* (ed. C. S. Andrew and E. J. Kamprath) p. 395. CSIRO, Melbourne.
KOCH, B. and EVANS, H. J. (1966). *Pl. Physiol., Lancaster* **41**, 1748.
LAPINS, P. and WATSON, E. R. (1970). *Aust. J. exp. Agric. Anim. Husb.* **10**, 599.
LEACH, G. J. (1978). In *Plant relations in pastures* (ed. J. R. Wilson) p. 290. CSIRO, Melbourne.
LUDLOW, M. M. (1978). In *Plant relations in pastures* (ed. J. R. Wilson) p . 35. CSIRO, Melbourne.
McAULIFFE, C., CHAMBLEE, D. S., URIBE-ARANGO, H., and WOODHOUSE, W. W. Jr (1958). *Agron. J.* **50**, 334.
McCOWN, R. L. and WALL, B. H. (1977). Aust. CSIRO Div. Trop. Crops and Past. Div. Rep. 1976–7, p. 100.
McWILLIAM, J. R. (1978). In *Plant relations in pastures* (ed. J. R. Wilson) p. 17. CSIRO, Melbourne.
MASTERSON, C. L. and MURPHY, P. M. (1976). In *Symbiotic nitrogen fixation in plants* (ed. P. S. Nutman) p. 299. Cambridge University Press.
—— and SHERWOOD, M. T. (1978). *Pl. Soil* **49**, 421.
MELVILLE, J. and SEARS, P. D. (1953). *N.Z. Jl Sci. Technol.* **35A**, Suppl. 1, 30.

Moustafa, E., Ball, R., and Field, T. R. O. (1969). *N.Z. Jl agric. Res.* **12**, 691.

Mulder, E. G., Lie, T. A., and Houwers, A. (1977). In *A treatise on dinitrogen fixation.* Section IV *Agronomy and ecology* (ed. R. W. F. Hardy and A. H. Gibson) p. 221. John Wiley, New York.

Munns, D. N. (1977). In *A treatise on dinitrogen fixation.* Section IV *Agronomy and ecology* (ed. R. W. F. Hardy and A. H. Gibson) p. 353. John Wiley, New York.

Norris, D. O. and Date, R. A. (1976). In *Tropical pasture research—principles and methods* (ed. N. H. Shaw and W. W. Bryan). Bull. 51, Commonw. Bur. Past. Fld Crops, p. 134.

Nutman, P. S. (1976). In *Symbiotic nitrogen fixation in plants* (ed. P. S. Nutman) p. 211. Cambridge University Press.

Pate, J. S. (1976). In *Symbiotic nitrogen fixation in plants* (ed. P. S. Nutman) p. 335. Cambridge University Press.

—— (1977). In *A treatise on dinitrogen fixation.* Section III *Biology* (ed. R. W. F. Hardy and W. S. Silver) p. 473. John Wiley, New York.

Peake, D. C. I., Stirk, G. B., and Henzell, E. F. (1975). *Aust. J. exp. Agric. Anim. Husb.* **15**, 645.

Phillips, D. A. and Bennett, J. P. (1978). *Agron. J.* **70**, 671.

Rhodes, I. and Stern, W. R. (1978). In *Plant relations in pastures* (ed. J. R. Wilson) p. 175. CSIRO, Melbourne.

Robson, A. D. (1978). In *Mineral nutrition of legumes in tropical and sub-tropical soils* (ed. C. S. Andrew and E. J. Kamprath) p. 277. CSIRO, Melbourne.

—— and Loneragan, J. F. (1978). In *Plant relations in pastures* (ed. J. R. Wilson) p. 128. CSIRO, Melbourne.

Rogers, V. E. (1969). Aust. Commonw. Sci. Ind. Res. Org., Div. Plant Ind., Field Stat. Rec., **8**, 37.

Rossiter, R. C. (1966). *Adv. Agron.* **18**, 1.

—— (1978). In *Plant relations in pastures* (ed. J. R. Wilson) p. 325. CSIRO, Melbourne.

Sasaki, N. and Kudo, K. (1966). *Bull. Fac. Agric. Hirosaki Univ.* **12**, 1. [Taken from Gibson 1977.]

Schaede, R. (1940). *Planta* **31**, 1.

Sears, P. D., Goodall, V. C., Jackman, R. H., and Robinson, G. S. (1965). *N.Z. Jl agric. Res.* **8**, 270.

Shaw, N. H. and Whiteman, P. C. (1977). *Trop. Grasslds* **11**, 7.

Simpson, J. R., Bromfield, S. M., and Jones, O. L. (1974). *Aust. J. exp. Agric. Anim. Husb.* **14**, 487.

Skerman, P. J. (1977). *Tropical forage legumes.* FAO, Rome.

Sprent, J. I. (1976). In *Symbiotic nitrogen fixation in plants* (ed. P. S. Nutman) p. 405. Cambridge University Press.

Tothill, J. C. and Jones, R. M. (1977). *Trop. Grasslds* **11**, 55.

Turner, N. C. and Begg, J. E. (1978). In *Plant relations in pastures* (ed. J. R. Wilson) p. 50. CSIRO, Melbourne.

Vallis, I. (1972). *Aust. J. exp. Agric. Anim. Husb.* **12**, 495.

—— (1978). In *Plant relations in pastures* (ed. J. R. Wilson) p. 190. CSIRO, Melbourne.

—— Henzell, E. F., and Evans, T. R. (1977). *Aust. J. agric. Res.* **28**, 413.

Walker, T. W., Thapa, B. K., and Adams, A. F. R. (1959). *Soil Sci.* **87**, 135.

Wallace, M. M. H. (1970). In *Australian grasslands* (ed. R. M. Moore) p. 361. Australian National University Press, Canberra.

Watkin, B. R. and Clements, R. J. (1978). In *Plant relations in pastures* (ed. J. R. Wilson) p. 273. CSIRO, Melbourne.

Whitehead, D. C. (1970). *The role of nitrogen in grassland productivity.* Bull. 48 Commonw. Bur. Past. Fld Crops.

WILLIAMS, W. A., JONES, M. B., and DELWICHE, C. C. (1977). *Agron. J.* **69**, 1023.
WILSON, J. R. (1970). *Aust. J. agric. Res.* **21**, 571.
WILSON, P. W. (1940). *The biochemistry of symbiotic nitrogen fixation*. The University of Wisconsin Press, Madison.

Index